# NANOTECHNOLOGY HORIZONS IN FOOD PROCESS ENGINEERING

*Volume 1*

*Food Preservation, Food Packaging, and Sustainable Agriculture*

**Nanotechnology Horizons in Food Process Engineering, Set of 3 Vols**
ISBN: 978-1-77491-064-1 (hbk)
ISBN: 978-1-77491-065-8 (pbk)
ISBN: 978-1-00330-545-3 (ebk)

**Nanotechnology Horizons in Food Process Engineering, Vol 1: Food Preservation, Food Packaging, and Sustainable Agriculture**
ISBN: 978-1-77491-060-3 (hbk)
ISBN: 978-1-77491-061-0 (pbk)
ISBN: 978-1-00330-537-8 (ebk)

**Nanotechnology Horizons in Food Process Engineering, Vol 2: Scope, Biomaterials, and Human Health**
ISBN: 978-1-77491-062-7 (hbk)
ISBN: 978-1-77491-063-4 (pbk)
ISBN: 978-1-00330-538-5 (ebk)

**Nanotechnology Horizons in Food Process Engineering, Vol 3: Trends, Nanomaterials, and Food Delivery**
ISBN: 978-1-77491-098-6 (hbk)
ISBN: 978-1-77491-099-3 (pbk)
ISBN: 978-1-00330-540-8 (ebk)

*Innovations in Agricultural and Biological Engineering*

# NANOTECHNOLOGY HORIZONS IN FOOD PROCESS ENGINEERING

*Volume 1*

*Food Preservation, Food Packaging, and Sustainable Agriculture*

*Edited by*

**Megh R. Goyal, PhD, PE**

**Junaid Ahmad Malik, PhD**

**Satish Kumar, PhD**

**Ritesh B. Watharkar, PhD**

AAP | APPLE ACADEMIC PRESS

First edition published 2023

**Apple Academic Press Inc.**
1265 Goldenrod Circle, NE,
Palm Bay, FL 32905 USA

760 Laurentian Drive, Unit 19,
Burlington, ON L7N 0A4, CANADA

**CRC Press**
6000 Broken Sound Parkway NW,
Suite 300, Boca Raton, FL 33487-2742 USA

4 Park Square, Milton Park,
Abingdon, Oxon, OX14 4RN UK



**Library and Archives Canada Cataloguing in Publication**

Title: Nanotechnology horizons in food process engineering / edited by Megh R. Goyal, PhD, PE, Junaid Ahmad Malik, PhD, Satish Kumar, PhD, Ritesh B. Watharkar, PhD.
Names: Goyal, Megh R., editor. | Malik, Junaid Ahmad, 1987- editor. | Kumar, Satish (Lecturer in food science and technology), editor. | Watharkar, Ritesh B., editor. | Mishra, Santosh K., editor.
Series: Innovations in agricultural and biological engineering.
Description: First edition. | Series statement: Innovations in agricultural and biological engineering | Includes bibliographical references and indexes. | Contents: Volume 1. Food preservation, food packaging, and sustainable agriculture.
Identifiers: Canadiana (print) 20220408009 | Canadiana (ebook) 20220408033 | ISBN 9781774910641 (set) | ISBN 9781774910603 (v. 1 ; hardcover) | ISBN 9781774910610 (v. 1 ; softcover) | ISBN 9781003305378 (v. 1 ; ebook)
Subjects: LCSH: Food—Biotechnology. | LCSH: Food industry and trade—Technological innovations. | LCSH: Nanotechnology.
Classification: LCC TP248.65.F66 N36 2023 | DDC 664—dc23

**Library of Congress Cataloging-in-Publication Data**

CIP data on file with US Library of Congress

ISBN: 978-1-77491-060-3 (hbk)
ISBN: 978-1-77491-061-0 (pbk)
ISBN: 978-1-00330-537-8 (ebk)

# ABOUT THE BOOK SERIES: INNOVATIONS IN AGRICULTURAL AND BIOLOGICAL ENGINEERING

Under this book series, Apple Academic Press Inc. is publishing book volumes over a span of 8–10 years in the specialty areas defined by the American Society of Agricultural and Biological Engineers (<asabe.org>). Apple Academic Press Inc. aims to be a principal source of books in agricultural and biological engineering. We welcome book proposals from readers in areas of their expertise.

The mission of this series is to provide knowledge and techniques for agricultural and biological engineers (ABEs). The book series offers high-quality reference and academic content on agricultural and biological engineering (ABE) that is accessible to academicians, researchers, scientists, university faculty and university-level students, and professionals around the world.

Agricultural and biological engineers ensure that the world has the necessities of life, including safe and plentiful food, clean air and water, renewable fuel and energy, safe working conditions, and a healthy environment by employing knowledge and expertise of the sciences, both pure and applied, and engineering principles. Biological engineering applies engineering practices to problems and opportunities presented by living things and the natural environment in agriculture.

ABE embraces a variety of the following specialty areas (<asabe.org>): aquaculture engineering, biological engineering, energy, farm machinery and power engineering, food, and process engineering, forest engineering, information, and electrical technologies, soil, and water conservation engineering, natural resources engineering, nursery, and greenhouse engineering, safety, and health, and structures and environment.

For this book series, we welcome chapters on the following specialty areas (but not limited to):

- Academia to industry to end-user loop in agricultural engineering;
- Agricultural mechanization;
- Aquaculture engineering;
- Biological engineering in agriculture;

- Biotechnology applications in agricultural engineering;
- Energy source engineering;
- Farm to fork technologies in agriculture;
- Food and bioprocess engineering;
- Forest engineering;
- GPS and remote sensing potential in agricultural engineering;
- Hill land agriculture;
- Human factors in engineering;
- Impact of global warming and climatic change on agriculture economy;
- Information and electrical technologies;
- Irrigation and drainage engineering;
- Nanotechnology applications in agricultural engineering;
- Natural resources engineering;
- Nursery and greenhouse engineering;
- Potential of phytochemicals from agricultural and wild plants for human health;
- Power systems and machinery design;
- Robot engineering and drones in agriculture;
- Rural electrification;
- Sanitary engineering;
- Simulation and computer modeling;
- Smart engineering applications in agriculture;
- Soil and water engineering;
- Micro-irrigation engineering;
- Structures and environment engineering;
- Waste management and recycling;
- Any other focus areas.

For more information on this series, readers may contact:
***Megh R. Goyal, PhD, PE***
*Book Series Senior Editor-in-Chief: Innovations in Agricultural and Biological Engineering*
*E-mail: goyalmegh@gmail.com*

# OTHER BOOKS ON AGRICULTURAL AND BIOLOGICAL ENGINEERING BY APPLE ACADEMIC PRESS, INC.

**Management of Drip/Trickle or Micro Irrigation**
*Megh R. Goyal, PhD, PE, Senior Editor-in-Chief*
**Evapotranspiration: Principles and Applications for Water Management**
*Megh R. Goyal, PhD, PE and Eric W. Harmsen, PhD Editors*

**Book Series: Research Advances in Sustainable Micro Irrigation**
***Senior Editor-in-Chief: Megh R. Goyal, PhD, PE***

- **Volume 1:** Sustainable Micro Irrigation: Principles and Practices
- **Volume 2:** Sustainable Practices in Surface and Subsurface Micro Irrigation
- **Volume 3:** Sustainable Micro Irrigation Management for Trees and Vines
- **Volume 4:** Management, Performance, and Applications of Micro Irrigation Systems
- **Volume 5:** Applications of Furrow and Micro Irrigation in Arid and Semi-Arid Regions
- **Volume 6:** Best Management Practices for Drip Irrigated Crops
- **Volume 7:** Closed Circuit Micro Irrigation Design: Theory and Applications
- **Volume 8:** Wastewater Management for Irrigation: Principles and Practices
- **Volume 9:** Water and Fertigation Management in Micro Irrigation
- **Volume 10:** Innovation in Micro Irrigation Technology

**Book Series: Innovations and Challenges in Micro Irrigation**
***Senior Editor-in-Chief: Megh R. Goyal, PhD, PE***

- Engineering Interventions in Sustainable Trickle Irrigation: Water Requirements, Uniformity, Fertigation, and Crop Performance
- Management Strategies for Water Use Efficiency and Micro Irrigated Crops: Principles, Practices, and Performance

- Micro Irrigation Engineering for Horticultural Crops: Policy Options, Scheduling, and Design
- Micro Irrigation Management: Technological Advances and Their Applications
- Micro Irrigation Scheduling and Practices
- Performance Evaluation of Micro Irrigation Management: Principles and Practices
- Potential of Solar Energy and Emerging Technologies in Sustainable Micro Irrigation
- Principles and Management of Clogging in Micro Irrigation
- Sustainable Micro Irrigation Design Systems for Agricultural Crops: Methods and Practices

**Book Series: Innovations in Agricultural and Biological Engineering**
***Senior Editor-in-Chief: Megh R. Goyal, PhD, PE***

- Advanced Research Methods in Food Processing Technologies
- Advances in Food Process Engineering: Novel Processing, Preservation and Decontamination of Foods
- Advances in Green and Sustainable Nanomaterials: Applications in Energy, Biomedicine, Agriculture, and Environmental Science
- Advances in Sustainable Food Packaging Technology
- Analytical Methods for Milk and Milk Products, Volume 1: Sampling Methods, Chemical and Compositional Analysis
- Analytical Methods for Milk and Milk Products, Volume 2: Physicochemical Analysis of Concentrated, Coagulated and Fermented Products
- Biological and Chemical Hazards in Food and Food Products: Prevention, Practices, and Management
- Bioremediation and Phytoremediation Technologies in Sustainable Soil Management, Volume 1: Fundamental Aspects and Contaminated Sites
- Bioremediation and Phytoremediation Technologies in Sustainable Soil Management, Volume 2: Microbial Approaches and Recent Trends
- Bioremediation and Phytoremediation Technologies in Sustainable Soil Management, Volume 3: Inventive Techniques, Research Methods, and Case Studies
- Bioremediation and Phytoremediation Technologies in Sustainable Soil Management, Volume 4: Degradation of Pesticides and Polychlorinated Biphenyls

- Dairy Engineering: Advanced Technologies and Their Applications
- Developing Technologies in Food Science: Status, Applications, and Challenges
- Emerging Technologies in Agricultural Engineering
- Engineering Interventions in Agricultural Processing
- Engineering Interventions in Foods and Plants
- Engineering Practices for Agricultural Production and Water Conservation: An Interdisciplinary Approach
- Engineering Practices for Management of Soil Salinity: Agricultural, Physiological, and Adaptive Approaches
- Engineering Practices for Milk Products: Dairyceuticals, Novel Technologies, and Quality
- Enzyme Inactivation in Food Processing: Technologies, Materials, and Applications
- Field Practices for Wastewater Use in Agriculture: Future Trends and Use of Biological Systems
- Flood Assessment: Modeling and Parameterization
- Food Engineering: Emerging Issues, Modeling, and Applications
- Food Process Engineering: Emerging Trends in Research and Their Applications
- Food Processing and Preservation Technology: Advances, Methods, and Applications
- Food Technology: Applied Research and Production Techniques
- Functional Dairy Ingredients and Nutraceuticals: Physicochemical, Technological, and Therapeutic Aspects
- Handbook of Research on Food Processing and Preservation Technologies: Volume 1: Nonthermal and Innovative Food Processing Methods
- Handbook of Research on Food Processing and Preservation Technologies, Volume 2: Nonthermal Food Preservation and Novel Processing Strategies
- Handbook of Research on Food Processing and Preservation Technologies, Volume 3: Computer-Aided Food Processing and Quality Evaluation Techniques
- Handbook of Research on Food Processing and Preservation Technologies, Volume 4: Design and Development of Specific Foods, Packaging Systems, and Food Safety
- Handbook of Research on Food Processing and Preservation Technologies, Volume 5: Emerging Techniques for Food Processing, Quality, and Safety Assurance

- Modeling Methods and Practices in Soil and Water Engineering
- Nanotechnology and Nanomaterial Applications in Food, Health, and Biomedical Sciences
- Nanotechnology Applications in Agricultural and Bioprocess Engineering: Farm to Table
- Nanotechnology Applications in Dairy Science: Packaging, Processing, and Preservation
- Nanotechnology Horizons in Food Process Engineering, Volume 1: Food Preservation, Food Packaging, and Sustainable Agriculture
- Nanotechnology Horizons in Food Process Engineering, Volume 2: Scope, Biomaterials, and Human Health
- Nanotechnology Horizons in Food Process Engineering, Volume 3: Trends, Nanomaterials, and Food Delivery
- Novel and Alternative Methods in Food Processing: Biotechnological, Physicochemical, and Mathematical Approaches
- Novel Dairy Processing Technologies: Techniques, Management, and Energy Conservation
- Novel Processing Methods for Plant-Based Health Foods: Extraction, Encapsulation and Health Benefits of Bioactive Compounds
- Novel Strategies to Improve Shelf-Life and Quality of Foods: Quality, Safety, and Health Aspects
- Phytochemicals and Medicinal Plants in Food Design: Strategies and Technologies for Improved Healthcare
- Processing of Fruits and Vegetables: From Farm to Fork
- Processing Technologies for Milk and Milk Products: Methods, Applications, and Energy Usage
- Quality Control in Fruit and Vegetable Processing: Methods and Strategies
- Scientific and Technical Terms in Bioengineering and Biological Engineering
- Soil and Water Engineering: Principles and Applications of Modeling
- Soil Salinity Management in Agriculture: Technological Advances and Applications
- State-of-the-Art Technologies in Food Science: Human Health, Emerging Issues and Specialty Topics

- Sustainable Biological Systems for Agriculture: Emerging Issues in Nanotechnology, Biofertilizers, Wastewater, and Farm Machines
- Sustainable Nanomaterials for Biomedical Engineering: Impacts, Challenges, and Future Prospects
- Sustainable Nanomaterials for Biosystems Engineering: Trends in Renewable Energy, Environment, and Agriculture
- Technological Interventions in Dairy Science: Innovative Approaches in Processing, Preservation, and Analysis of Milk Products
- Technological Interventions in Management of Irrigated Agriculture
- Technological Interventions in the Processing of Fruits and Vegetables
- Technological Processes for Marine Foods, From Water to Fork: Bioactive Compounds, Industrial Applications, and Genomics
- The Chemistry of Milk and Milk Products: Physicochemical Properties, Therapeutic Characteristics, and Processing Methods

# ABOUT SENIOR-EDITOR-IN-CHIEF

**Megh R. Goyal, PhD, PE**
*Retired Professor in Agricultural and Biomedical Engineering, University of Puerto Rico, Mayaguez Campus; Senior Acquisitions Editor, Biomedical Engineering and Agricultural Science, Apple Academic Press, Inc.*

Megh R. Goyal, PhD, PE, is currently a Retired Professor of Agricultural and Biomedical Engineering from the General Engineering Department at the College of Engineering at the University of Puerto Rico–Mayaguez Campus (UPRM); and Senior Acquisitions Editor and Senior Technical Editor-in-Chief for Agricultural and Biomedical Engineering for Apple Academic Press Inc.

During his long career, he has worked as a Soil Conservation Inspector; Research Assistant at Haryana Agricultural University and Ohio State University; Research Agricultural Engineer/Professor at the Department of Agricultural Engineering of UPRM; and Professor of Agricultural and Biomedical Engineering in the General Engineering Department of UPRM. He spent a one-year sabbatical leave in 2002–2003 at the Biomedical Engineering Department of Florida International University, Miami, USA.

Dr. Goyal was the first agricultural engineer to receive the professional license in agricultural engineering from the College of Engineers and Surveyors of Puerto Rico. In 2005, he was proclaimed the "Father of Irrigation Engineering in Puerto Rico for the Twentieth Century" by the American Society of Agricultural and Biological Engineers, Puerto Rico Section, for his pioneering work on micro irrigation, evapotranspiration, agroclimatology, and soil and water engineering.

During his professional career of 52 years, he has received many awards, including Scientist of the Year, Membership Grand Prize for the American Society of Agricultural Engineers Campaign, Felix Castro Rodriguez Academic Excellence Award, Man of Drip Irrigation by the Mayor of Municipalities of Mayaguez/Caguas/Ponce and Senate/Secretary of Agriculture of ELA, Puerto Rico, and many others. He has been recognized as one of the experts "who rendered meritorious service for the development of [the] irrigation sector in India" by the Water Technology Centre of Tamil Nadu Agricultural University in Coimbatore, India, and ASABE who bestowed on him the 2018 Netafim Microirrigation Award.

Dr. Goyal has authored more than 200 journal articles and edited more than 100 books. AAP has published many of his books, including *Management of Drip/Trickle or Micro Irrigation; Evapotranspiration: Principles and Applications for Water Management*; ten-volume set on *Research Advances in Sustainable Micro Irrigation*. Dr. Goyal has also developed several book series with AAP, including Innovations in Agricultural & Biological Engineering (with over 60 titles in the series to date), Innovations and Challenges in Micro Irrigation; and Innovations in Plant Science for Better Health: From Soil to Fork.

Dr. Goyal received his BSc degree in Engineering from Punjab Agricultural University, Ludhiana, India, and his MSc and PhD degrees from the Ohio State University, Columbus, Ohio, USA. He also earned a Master of Divinity degree from the Puerto Rico Evangelical Seminary, Hato Rey, Puerto Rico, USA.

Readers may contact him at goyalmegh@gmail.com.

# ABOUT THE EDITORS

**Junaid Ahmad Malik, PhD**
*Lecturer, Department of Zoology, Government Degree College, Bijbehara, Kashmir (Jammu and Kashmir), India*

Junaid Ahmad Malik, PhD, is a Lecturer in the Department of Zoology at Government Degree College, Bijbehara, Kashmir (J&K), India, and is actively involved with teaching and research activities. He has more than eight years of research experience. His areas of interest are ecology, soil macrofauna, wildlife biology, conservation biology, etc. Dr. Malik has published more than 20 research papers in various national and international peer-reviewed journals. He has authored four books, 16 book chapters, edited seven books, and more than 10 popular editorial articles. He is also serving as editor and reviewer of several journals with a reasonable repute. He has participated in several state, national, and international conferences, seminars, workshops, and symposia, and more than 20 conference papers are to his credit. He is a life member of SBBS (Society for Bioinformatics and Biological Sciences).

Dr. Malik received his BSc (2008) in Science from the University of Kashmir, Srinagar, Jammu and Kashmir; MSc (2010) in Zoology from Barkatullah University, Bhopal, Madhya Pradesh; and PhD (2015) in Zoology from the same university. He completed his BEd program in 2017 from the University of Kashmir, Srinagar, Jammu and Kashmir. He started his career as a Lecturer in the School Education Department, Government of Jammu and Kashmir for 2 years.

**Satish Kumar, PhD**
*Assistant Professor (Food Science and Technology), College of Horticulture and Forestry, Thunag-Mandi, Dr. Y.S. Parmar University of Horticulture and Forestry, Nauni-Solan – 173230, Himachal Pradesh, India*

Satish Kumar, PhD, is an Assistant Professor (Food Science and Technology) in the College of Horticulture and Forestry at Dr. Y.S. Parmar University of Horticulture and Forestry in Himachal Pradesh, India. He is currently working in the area of characterization of edible nano-emulsions for reducing post-harvest losses of fresh horticultural produce. He is also associated with the development of novel animal food analogs (milk and meat analogs) and protocols for utilization of waste generated from the agricultural and horticultural processing industry by cradle-to-cradle regenerative conceptualization.

Dr. Kumar served at Lovely Professional University, Phagwara-Punjab, India, as Assistant Professor (Food Technology and Nutrition). He has successfully guided 19 MSc scholars of Food Technology and Nutrition. He has participated in several national and international conferences, seminars, workshops, training, symposiums, etc. He is serving as an editorial board member for *Archives of Food Science and Nutrition Research* and *Global Journal of Food Science and Technology.*

He has to his credit 52 research papers, 12 book chapters, five conference papers and five abstracts. He has published papers in journals including Trends in Food Science and Technology and Critical Reviews in Food Science and Nutrition. His cumulative NAAS score is 290 and impact factor is 150. He is serving as an external examiner for various Indian state agricultural universities. He is also serving as editor and reviewer for several peer-reviewed journals of international repute. He is the Joint Secretary in Functional Food Association of India (FFAI) [Registered under HP Societies Registration Act, 2006]

Dr. Kumar holds a BSc (2011) in Horticulture, MSc (2013), and PhD (2018) in Food Technology from Dr. Y. S. Parmar University of Horticulture and Forestry Nauni-Solan (Himachal Pradesh), India. He has qualified ICAR, ASRB-NET (2014), and holds two course diplomas on enhancing soft skills and personality development (2018) and dairy and food process and products technology (2019) from the Indian Institute of Technology, Kanpur (Ministry of HRD, Government of India).

Readers may contact him at: satishsharma1666@gmail.com

**Ritesh Balaso Watharkar, PhD**
*Assistant Professor, College of Management, Food and Agriculture Management Department, MIT Art, Design and Technology University, Loni Kalbhor, Pune 412201, Maharashtra, India*

Ritesh Balaso Watharkar, PhD, is an Assistant Professor in the College of Management, Food and Agriculture Management Department at MIT Art, Design and Technology University, Loni Kalbhor, Pune, Maharashtra, India. He has worked as a Senior Visiting Scholar under the UNESCO Great Wall China Fellowship Program 2017–2018, and as Assistant Professor at K. K. Wagh Food Tech College (2011–2013) and Shramshakti College of Food Technology, affiliated with Mahatma Phule Agricultural University (2019–2020).

During his professional career, he has received awards such as Senior Research Fellow under the UNESCO/China Great Wall Fellowship Program and worked one year at the School of Biosystem Engineering and Food Science Zhejiang University, Hangzhou, China. He was also awarded for the best popular article from Krishidut Maxine, Maharashtra, India. Currently, he is a life member of the Association of Food Scientists and Technologists, India.

Dr. Watharkar has published more than 10 journal articles in reputed journals and attended more than 10 national and international conferences. He also published more than 10 popular articles in the Marathi Literacy Maxine.

He received his BTech degree in Food Technology from Dr. Balasaheb Sawant Konkan Agriculture University, Dapoli, Maharashtra, India; his MTech in Food Processing and Engineering degree is from the Karunya University, India; his PhD in Food Engineering and Technology degree from Tezpur University (A Central University), Assam, India.

For more information, contact at: watharkarritesh2019@gmail.com.

# CONTENTS

# CONTRIBUTORS

**Manab B. Bera**
Professor, Department of Food Engineering and Technology, Sant Longowal Institute of Engineering and Technology, Longowal – 148106, Punjab, India, Mobile: +91-8699841328,
E-mail: drmbbera@yahoo.com

**B. B. Borse**
Senior Principal Scientist and Professor AcSIR, Department of Spices and Flavor Science, CSIR-Central Food Technological Research Institute, Mysuru – 570020, Karnataka, India,
Mobile: +91-7259376170, E-mail: borsebb@cftri.res.in

**Anil Kumar Chauhan**
Professor, Department of Dairy Science and Food Technology, Institute of Agricultural Sciences, BHU, Varanasi – 221005, Uttar Pradesh, India, Mobile: +91-9450658188,
E-mail: dranilchauhan@rediffmail.com

**Jyotika Dhankhar**
Assistant Professor, Department of Food Technology, Maharishi Dayanand University, Rohtak – 124001, Haryana, India, Mobile: +91-9416529500, E-mail: jyotika.ft@mdurohtak.ac.in

**Himjyoti Dutta**
Assistant Professor, Department of Food Technology, Mizoram University, Aizawl – 796004, Mizoram, India, Mobile: +91-7042484759, E-mail: duttahimjyoti@gmail.com

**Rimpi Foujdar**
PhD Research Scholar, Department of Food Engineering and Technology, Sant Longowal Institute of Engineering and Technology, Longowal – 148106, Punjab, India, Mobile: +91-8968649677,
E-mail: foujdar.r.phd@gmail.com

**Megh R. Goyal**
Senior Editor-in-Chief (Agriculture and Biomedical Engineering) for AAP, Retired Professor in Agricultural and Biomedical Engineering, University of Puerto Rico-Mayaguez, Mayaguez – PR-USA, USA, Mobile: 001-787-536-0039, E-mail: goyalmegh@gmail.com

**Kuldeep Gupta**
Research Associate, Department of Molecular Biology and Biotechnology, Tezpur University (A Central University), Napaam – 784028, Tezpur, Assam, India, Mobile: +91-8486743328,
E-mail: guptak@tezu.ernet.in

**Sheena Haorongbam**
Assistant Professor, Department of Agribusiness Management and Food Technology, North-Eastern Hill University, Tura Campus, Tura – 794002, Meghalaya, India,
Mobile: +91-8413035334, E-mail: sheenahaorongbam@gmail.com

**Sujata F. Harlapur**
Professor, Department of Chemistry, Vemana Institute of Technology, Bangalore – 560034, Karnataka, India, Mobile: +91-9448908643, E-mail: sujataharlapur@vemanait.edu.in

**Suneeta Harlapur**
Associate Professor, Department of ECE, Vemana Institute of Technology, Bangalore – 560034, Karnataka, India, Mobile: +91-9900113905, E-mail: sunitahaveri@gmail.com

**Shazia Hasan**
PhD Research Scholar, Department of Chemistry, Aligarh Muslim University, Aligarh – 202002, Uttar Pradesh, India, Mobile: +91-9415428870, E-mail: shazneyazi@gmail.com

**Pratibha Jadhav**
PhD Research Fellow, Engineering Physics and Engineering Materials, Institute of Chemical Technology Mumbai Marathwada Campus Jalna, Beej Sheetal Innovations Center Pvt. Ltd. BT-6/7, Biotechnology Park, Additional MIDC Area, Aurangabad Road, Jalna – 431203, Maharashtra, India, Mobile: +91-7249338303, E-mail: pratibhajadhav5323@gmail.com

**Girish M. Joshi**
Associate Professor, Engineering Physics and Engineering Materials, Institute of Chemical Technology Mumbai Marathwada Campus Jalna, Beej Sheetal Innovations Center Pvt. Ltd. BT-6/7, Biotechnology Park, Additional MIDC Area, Aurangabad Road, Jalna – 431203, Maharashtra, India, Mobile: +91-8838660102, E-mails: varadgm@gmail.com; gm.joshi@marj.ictmumbai.edu.in

**Rajashri Kulal**
Project Assistant, Department of Spices and Flavor Science, CSIR-Central Food Technological Research Institute, Mysuru – 570020, Karnataka, India, Mobile: +91-9535905549, E-mail: rraja872@gmail.com

**Sandeep Kumar**
MTech Student, Department of Food Engineering and Technology, Sant Longowal Institute of Engineering and Technology, Longowal – 148106, Punjab, India, Mobile: +91-8426964399, E-mail: sandeepkumargathania@gmail.com

**Preeti Kundu**
PhD Candidate, Department of Food Technology, Maharishi Dayanand University, Rohtak – 124001, Haryana, India, Mobile: +91-8222003476, E-mail: preetikundu03@gmail.com

**Nangwanbiang Kurbah**
PhD Research Scholar, Department of Agribusiness Management and Food Technology, North-Eastern Hill University, Tura Campus, Tura – 794002, Meghalaya, India, Mobile: +91-8837296751, E-mail: knangwanbiang@gmail.com

**Manabendra Mandal**
Professor, Department of Molecular Biology and Biotechnology, Tezpur University (A Central University), Napaam, Tezpur – 784028, Assam, India, Mobile: +91-3712-267007, Fax: +91-3712-267005/267006, E-mail: mandal@tezu.ernet.in

**Rita Mehla**
Project Fellow, Dairy Chemistry Division, National Dairy Research Institute, Karnal – 132001, Haryana, India, Mobile: +91-8295344469, E-mail: ritamehla24@gmail.com

**Gargi D. Patil**
MTech Student, Engineering Physics and Engineering Materials, Institute of Chemical Technology, Mumbai Marathwada Campus Jalna, Beej Sheetal Innovations Center Pvt. Ltd. BT-6/7, Biotechnology Park, Additional MIDC Area, Aurangabad Road, Jalna – 431203, Maharashtra, India, Mobile: +91-9767322751, E-mail: che18gd.patil@stumarj.ictmumbai.edu.in

**Smith V. Patil**
MTech Student, Engineering Physics and Engineering Materials, Institute of Chemical Technology Mumbai Marathwada Campus Jalna, Beej Sheetal Innovations Center Pvt. Ltd. BT-6/7, Biotechnology Park, Additional MIDC Area, Aurangabad Road, Jalna – 431203, Maharashtra, India, Mobile: +91-7875002052, E-mail: che18cv.patil@stumarj.ictmumbai.edu.in

**Preetha Radhakrishnan**
Associate Professor, Department of Food Process Engineering, School of Bioengineering., SRM Institute of Science and Technology, Kattankulathur – 603203, Tamil Nadu, India, Mobile: +91-9176610351, E-mail: preethar@srmist.edu.in

**Shruthy Ramesh**
PhD Research Scholar, Department of Food Process Engineering, School of Bioengineering, SRM Institute of Science and Technology, Kattankulathur – 603203, Tamil Nadu, India, Mobile: +91-9544065350, E-mail: shruthyr@srmist.edu.in

**Muzamil Ahmad Rather**
PhD Research Scholar, Department of Molecular Biology and Biotechnology, Tezpur University (A Central University), Napaam – 784028, Tezpur, Assam, India, Mobile: +91-6005145124, E-mail: muzamil786.mr@gmail.com

**Abdul Rauf**
Professor, Department of Chemistry, Aligarh Muslim University, Aligarh – 202002, Uttar Pradesh, India, Mobile: +91-9412545345, E-mail: abduloafchem@gmail.com

# ABBREVIATIONS

| | |
|---|---|
| μm | micrometer |
| AATCC | American Association of Textile Chemists and Colorists |
| ABR | antibiotic resistance |
| AFM | atomic force microscopy |
| Ag | silver |
| AgNPs | silver nanoparticles |
| $Al_2O_3$ | aluminum oxide |
| AlOOH | aluminum oxyhydroxide |
| AMR | antimicrobial resistance |
| ATP | adenine triphosphate |
| AuNPs | gold nanoparticles |
| BET | Brunauer–Emmett–Teller |
| C30B | cloisite30b |
| $CaCO_3$ | calcium carbonate |
| CaO | calcium oxide |
| CCNF | carrot cellulose nano-fibrils |
| CMC | carboxymethyl cellulose |
| CNC | cellulose nanocrystal |
| CNF | cellulose nanofibrils |
| CNP | chitosan nanoparticle |
| CPC | cetylpyridinum chloride |
| CPSC | Consumer Product Safety Commission |
| CS | chitosan |
| CTAB | hexadecyltrimethylammonium bromide |
| CTAC | hexadecyltrimethylammonium chloride |
| $Cu_2O$ | copper (I) oxide |
| CVD | chemical vapor deposition |
| Da | Dalton |
| DDT | dichlorodiphenyltrichloroethane |
| DEAMTT | diethyl-4-(4-amino-5-mercapto-4h-1,2,4-triazol-3-yl) phenyl phosphonate |
| DLS | dynamic light scattering |
| DNA | deoxyribonucleic acid |
| DOAN | 9,10-dibromooctadecanoic acid capped AgNPs |

| | |
|---|---|
| DTNB | 5,5'-dithio-bis(2- nitrobenzoic acid) |
| *E. coli* | *Escherichia coli* |
| EBT | Eriochrome black T |
| EDS | energy-dispersive X-ray spectroscopy |
| EFSA | European Food Safety Authority |
| ELISA | enzyme-linked immunosorbent assay |
| EOB | elongation at break |
| EPA | Environmental Protection Agency |
| EPM | extracellular polymer matrix |
| EPS | exopolymeric substances |
| EU | European Union |
| FAO | Food and Agriculture Organization |
| FASP | flame-assisted spray pyrolysis |
| FDA | Food and Drug Administration |
| FSANZ | Food Standards of Australia and New Zealand |
| FSCJ | Florida State College at Jacksonville |
| FSP | flame spray pyrolysis |
| FTIR | Fourier transform infrared spectroscopy |
| GRAS | generally recognized as safe |
| GSH | glutathione |
| $H_2O_2$ | hydrogen peroxide |
| $H_2S$ | hydrogen sulfide |
| HCL | hydrochloric acid |
| ICP | intracranial pressure |
| ISO | International Organization for Standardization |
| LASiS | laser ablation synthesis in solution |
| LbL | layer-by-layer |
| LDH | lactate dehydrogenase |
| LDPE | low-density polyethylene |
| LIBS | laser-induced breakdown spectroscopy |
| LiCl/DMAc | LiCl/N,N-dimethylacetamide |
| LS | lipopolysaccharide |
| MB | methylene blue |
| MBC | minimum bactericidal concentrations |
| MCT | medium chain triglycerides |
| MgO | magnesium oxide |
| MIC | minimum inhibitory concentrations |
| MMT | montmorillonite |
| MPa | megapascal |

| | |
|---|---|
| MT | million tons |
| MV | millivolt |
| MWCNTs | multi-wall carbon nanotubes |
| NaFeEDTA | sodium ferric ethylenediaminetetraacetate |
| Nm | nanometer |
| NMMO | N-methyl morpholine-N-oxide |
| NMR | nuclear magnetic resonance |
| NMs | nanomaterials |
| NNPs | neem nano-particles |
| NP | nanoparticle |
| O/W | oil in water |
| OECD | Organization for Economic Co-operation and Development |
| OMMT | organically modified montmorillonite |
| ONP | oleic acid capped AgNPs |
| ONP | organic nanoparticle |
| OSHA | Occupational Safety and Health Administration |
| OTR | oxygen transmission rate |
| PANI | polyaniline |
| Pb NPs | lead NPs |
| PBAT | poly (Butylene Adipate Co-terephthalate) |
| PBS | poly (butylene succinate) |
| PC-g | pectin-g |
| PCL | poly(ε-caprolactone) |
| PDI | polydispersity index |
| PE | polyethylene |
| PEG | polyethylene glycol |
| PEI | polyethyleneimine |
| PET | polyethylene terephthalate |
| PGA | poly (glycolic acid) |
| PHA | polyhydroxy-alkanoate |
| PHB | polyhydroxy butyrate |
| PHBV | poly(3-hydroxybutyrate-co-3-hydroxyvalerate) |
| PLA | polylactic acid |
| PLGA | poly lactic-co-glycolic acid |
| PLLA | poly (L-lactic acid) |
| PMMA | polymethyl methacrylate |
| PPy | polypyrrole |
| PS | polystyrene |
| Pt NPs | platinum nanoparticles |

| | |
|---|---|
| PTh | polythiophene |
| PVA | polyvinyl alcohol |
| PVP | polyvinylpyrrolidone |
| QS | quorum sensing |
| RDA | recommended dietary allowance |
| ROS | reactive oxygen species |
| *S. aureus* | *Staphylococcus aureus* |
| SAED | selected area electron diffraction |
| SCOBY | symbiotic culture of bacteria and yeast |
| SD | sawdust |
| SDR | spinning disc reactor |
| SEM | scanning electron microscopy |
| SEPS9 | sepiolites9 |
| $SiO_2$ | silicon dioxide |
| SNAS | systemic nickel allergy syndrome |
| SNP | stearic acid capped AgNPs |
| SWCNT's | single-wall carbon nanotubes |
| TEA | triethylamine |
| TEM | transmission electron microscopy |
| TEMPO | 2,2,6,6-tetramethylpiperidin-1-oxyl |
| $TiO_2$ | titanium dioxide |
| TNT | titanium dioxide nanotube |
| tRNA | transport ribonucleic acid |
| TS | tensile strength |
| USD | United States Dollar |
| USFDA | United States Food and Drug Administration |
| UV | ultraviolet |
| UV-Vis | UV-Vis spectroscopy |
| W/O | water in oil |
| WHO | World Health Organization |
| WVP | water vapor permeability |
| WVTR | water vapor transmission rate |
| XPS | X-ray photoelectron spectroscopy |
| XRD | X-ray diffraction |
| ZnO | zinc oxide |
| $ZnO_2$ | zinc dioxide |

# PREFACE

Nanotechnology has advanced to new heights in recent years. Because of the rapid advancement of nanotechnology, a variety of nanomaterials (NMs) have found applications in a variety of fields. Although nanotechnology's suitability in numerous science fields (e.g., medicine, genetics, biology, bioengineering, mechanics, chemistry, etc.), has been recognized and accepted upon a couple of years, its versatility and application in the food industry is comparatively modern in comparison to the biomedical industry.

In the last decades, the food industry has invested a significant amount of money in nanotechnology science in order to discover new technologies for creating new flavors, extending food shelf life, and improving food protection and nutritional value. Nanotechnology's use in the food industry will result in intelligent nutrient delivery systems, "smart" foods, contaminant detection nanodevices/nanosensors, advanced food processing, antimicrobial chemicals, encapsulation, and green nanomaterials.

Nanotechnology's advancement in the food industry has resulted in revolutionary food packaging that can track spoilage and deter pathogenic microorganisms. In the not-too-distant future, immersive "smart" diets could "personalize" menus (new taste, ingredients, flavor, and color) to each consumer's nutritional preference or nutritional requirements. A functionalized food can mitigate disease risks to improve health. The distribution system's job is to regulate the rate at which the usable content is released as well as the circumstances that cause it to be released. The delivery mechanism is also responsible for preventing biological or chemical deterioration of the functionalized products. The distribution system can encapsulate substances such as (nano/micro) emulsification, transparent liquids, agglomerates, biomaterials, and other products.

Nanoencapsulation of food materials is among the most exciting fields of nanotechnology research. Nanoencapsulation enables bioactive substances to be encapsulated in tiny capsules (diameters in the micro- and nanoscales) to shield them from various destructive agents (such as pH variations, heat, and moistness) at the appropriate time and location. Probiotics, nutrients, fatty acids, enzymes, biologically active proteins, metabolites, polymers, resins, and carbohydrates are among the products currently used in nanoencapsulation methods. Flavors have a major impact on customer loyalty.

Since flavors are highly fragile and vulnerable to environmental factors, encapsulation is a feasible strategy for preserving them in the best possible conditions. Encapsulation techniques include melt infusion, coacervation, thermal treatment, extruding, and emulsification.

The fact that most of the intensive farming practices adopted during the industrial revolution, although profitable, are considered counterproductive in the long run, further complicates the situation. About 30% of the chemicals used by traditional methods are currently used by the crop. Spray drift (wind), volatilization, leaching into groundwater, and water leakage from drainage or rainfall all contribute to the loss of fertilizer and pesticides added in the field through traditional methods. Due to the high cost of products and their long-term destiny in the climate, the inefficiency of traditional field applications of fertilizers and pesticides is becoming a growing concern. Agriculture is a major cause of contamination due to pesticides that have leached into waterways. It is also a major cause of contaminants used in runoff that end up in rivers, wetlands, and estuaries. Eutrophication in wetlands, rivers, streams, and estuaries can be caused by fertilizer runoff containing nitrogen, potassium, and phosphorus.

Nanotechnology's influence is now being felt in medicine, astronomy, molecular biology, and farming, despite the fact that it is only in its infancy. Nanoencapsulation technology is transforming the way organic chemicals are used of the future in agriculture. In order to ensure that enough active agents are transported to where it is required, conventional means of administering agrochemicals in the field frequently result in intoxication. Excess pollutants wind up in rivers, causing environmental issues. Since it can be applied more effectively to a desired target region, nanoencapsulation allows for a reduction in chemical dose. The book entitled *Nanotechnology Horizons in Food Process Engineering: Volume 1: Food Preservation, Food Packaging and Sustainable Agriculture* consists of 11 chapters, such as:

- Chapter 1 describes the general role of nanotechnology for food and its preservation. The chapter discusses some useful techniques for potential applications in the food processing industry. It also deliberates the role of enzyme immobilization and nanofiltration.
- Chapter 2 delineates the importance of nanotechnology in food fortification and food processing. It also discusses the role of various nano-based packaging materials for food preservation.
- Chapter 3 debates various nanoscale formulations for foods and beverages like nanoemulsion, nanomicelles, nanoconjugation, and nanobiosensors.

- Chapter 4 talks about the potential of major polysaccharide nanoparticles (NPs) in foods, like cellulose, glycogen, pectin, and chitin. The role of various enzymes and hydrolases has also been covered.
- Chapter 5 focuses on the advent of nanotechnology through nanoencapsulation in probiotics, bioactive compounds, and additives in the preservation of foods.
- Chapter 6 explores how polymer nanoblends are preferable for food packaging applications. It discusses the role of protein nanoblends and organic as well as inorganic nanofillers for application in the food processing industry.
- Chapter 7 focuses on how actually nanoparticle as an individual contributes to increasing its physical properties. It overviews how actually the nano fillers influence on mechanical, barrier, and antimicrobial properties of composites. Various effects due to nano entity loading and the alteration in their properties are addressed. At last, the toxicological evaluation of nanoparticles on health and environment is discussed.
- Chapter 8 deals with several methods of fortification of different micronutrients and vitamins. Some of the fortificants and methods of their incorporation in commonly used nanoformulations have also been discussed.
- Chapter 9 describes the history, types, and different methods of synthesis of nanoparticles. The chapter also discusses biosynthesis of NPs, and their scope as antimicrobial and antibiofilm agents.
- Chapter 10 explores the utilization of neem nanoparticles as an elective antibacterial nanobiotics. It further discusses the advent of nanotechnological approaches in medicine and agriculture especially silver nanoparticles (AgNPs).
- Chapter 11 focuses on how metal nanomaterials may be used to remove organic compounds from environmentally dangerous chemicals. This chapter briefly overviews employment of different nanomaterials such as metal, metal oxide nanoparticles, metal composites and capped nanostructures for the successful elimination of biologically toxic materials catalytically and photo-catalytically. Eventually, recommendations for eliminating of such toxic materials are made based on the current practices of nanomaterials.

Since the field is still emerging and new programs are being identified on a routine basis, this book is not intended to be a systematic treatise. However, it is meant to be a helpful guide on some of the most recent nanotechnology concepts and techniques in food and agriculture. It will also serve as a forum for forming alliances, formulating plans, and generating new concepts and ways to address some of the most difficult problems facing food and agriculture in a rapidly changing environment. The book provides an in-depth look at the emerging status of nanotechnology in the food processing industry. It explains the benefits and drawbacks of each methodology to aid scientists in improving and developing food product sourcing and food hygiene monitoring methods.

***—Editors***

# PART I

# NANOTECHNOLOGY APPLICATIONS IN FOOD PRODUCTS

CHAPTER 1

# SCOPE OF NANOTECHNOLOGY FOR FOOD PRESERVATION

RAJASHRI KULAL and B. B. BORSE


## ABSTRACT

Nanotechnology is a precise, considerable science and technology rising within the food industry. It is involved in processing, packaging, storage, transportation, functionality, detection of food pathogens, extended shelf life, and other safety aspects of food. Nanostructured substances derived from metal oxides, inorganic metals, and their composites with bioactive components are used in food to extend shelf life and increase food quality. It can even bring down microbial contamination with the assistance of nano delivery systems and provide fresh and healthier food. In this chapter, nanotechnology focuses on different strategies of food processing and preservation techniques in conjugation with protection issues and governing measures on nano-processed food.


## 1.1 INTRODUCTION

Since the traditional period, man has been attempting to ameliorate and formulate advanced food protection techniques. From the primitive times, man has been trying to protect the food by reserving them in caves to shield it from being ruined to the chilling procedure of the 21$^{st}$ century. They also find cellar and cold streams for the preservation of food [24]. Aging, smoking, roasting, canning, steaming, fermenting, salting, sun drying, curing, bottling, carbonation, irradiation of food use of chemicals, and preservatives are the

Nanotechnology Horizons in Food Process Engineering, Volume 1. Megh R. Goyal, PhD, PE, Junaid Ahmad Malik, PhD, Satish Kumar, PhD & Ritesh B. Watharkar, PhD (Eds.)

few methods that are unremarkably utilized by men on an everyday premise. All these methods slow down the duplication of the hazardous microorganisms however none of those methods were applied with integrated scientific logic behind them. Techniques like preservation by using vacuum and osmotic inhibition are normally utilized methods for the protection of food. However, these techniques could not protect the food for a prolonged time. There was a necessity for a perpetual and additional dependable or determined solution for the protection of food.

Nanotechnology is relying on the prefix "nano," a Greek word signifying "predominate or dwarf." It is control and comprehension of matter at a measurement of approx. 1–100 nanometers. To be progressively explicit, nanotechnology is characterized because of the production, designing, and application of structure, devices, and systems through restricting the size state of the material at the nanometer ($10^{-9}$) scale wherever extraordinary phenomena empower novel application [116]. This innovation has altered the health care, textile, information technology, and energy sectors, etc. [78]. A few products empowered by nanotechnology are already in the market, those are antibacterial dressing, odor, and stain-repellent fabrics, light-diffracting cosmetics, transparent sunscreen lotions, scratch-free paints, long-lasting paints, soil repellent coating, etc. [98]. However, this technology has also been used in the food industry with considerable respect for food safety and quality. Preserving the concept of the healthful populace in mind, scientists have been attempting to discover imaginative innovation in improving the food safety and quality. The interruption of this technology within the realm of nutrition has caused the planning and advancement of thermal stabilized novel food, higher solubility, and bioavailability [125]. Implementation of nano-techniques in food science intrinsically has many approaches; because of its potentials, it may be applied in the enhancement of the food safety and quality. It serves a valuable role within the subject of food science, from boosting shelf life to elaborated food preservation to tracking pollutants or pathogens to the incorporation of bactericides or health dietary supplements [106]. The shelf life of foods will be extended by implementing this technology, reducing the amount of time when food deteriorates. As a result, more nutritious food would reach the population, improving people's wellbeing and potentially reducing the problem of food shortages. Several kinds of "nanosystems," which include nanosensors, nanofibers, solid nanoparticles (NPs), nanocomposites, nanotubes, nanocapsules, nano barcodes are the major nanomaterials (NMs) which discover their usage in the food preservation, processing, and packaging sectors [36]. This chapter discusses how nanotechnology is used in the food production and storage industry.

## 1.2 NANOTECHNOLOGY IN FOOD PROCESSING

The study of behavior and properties associated with nanostructures is known as nanoscience [62]. Organizing food structures at the nanoscale will change their inherent attributes which will, thus aid modulation in the quality of food at the macroscale with standardization and preparation. It has a significant effect on food structure and qualities including stability, taste, flavor release factor, and other elemental composition [66]. To enhance the food quality at the nanoscale, comprehensive knowledge of the connection between the properties of food and its structure is essential [3]. The food quality and structure are considerably altered throughout the food process and it depends upon the different processing conditions such as temperature, pressure, and mechanical power. Processing primarily influences the composition of food specially protein, starch, and lipids [117]. Food nanotechnology can help to protect bioactive components embodied in nanocarrier systems from harmful environmental factors, as well as improve food safety, security, taste, color, and shelf life. Nanotechnology may help being developed more beneficial food with lower salt, fat, and sugar to overcome foodborne disease [12].

### *1.2.1 NANOFILTRATION*

Nanofiltration, which is commonly used in wastewater treatment and water softening, is a combination of reverse osmosis and ultrafiltration. The aqueous solution of organic solutes with a molecular weight of 100–1,000 Da and a pressure of 1–4 MPa is concentrated or purified through its pore size scale of 0.5–1 nm [9, 123]. The nanofiltration membrane carries a negative charge on its surface, which attracts positively charged ions due to the Donnan effect [143]. Most of the nanofiltration composition is polymer substrates and polyamide are used as a thin-film membrane layer [9, 64].

### *1.2.2 ENZYME IMMOBILIZATION*

Enzyme immobilization is the interaction of the enzyme with the carrier. Size, isoelectric point, glycosylation, presence of hydrophobic and hydrophilic region is the key factor to be considered during the enzyme immobilization [68]. Generally, enzymes are used in the food process to achieve good nutritional value and improve the flavor [44]. Nanocarriers can act as an enzyme that helps to provide better activity with improved shelf life. As compared to traditional macroscale support components, this improves cost-effectiveness due to higher dispersion of enzymes in food matrices and a large surface-to-volume ratio [114]. The olive oil hydrolyzed by nano silicon dioxide ($SiO_2$) showed stability, reusability, and adaptability [8]. Other authors have proved the elevation of lipase enzymatic activity and its thermostability. The aggregation and denaturation occurring upon heating were avoided while the enzyme changed into encapsulated nanogels of cholesterol-bearing pullulan [124].

### *1.2.3 TEXTURE IMPROVEMENT*

The structure and size of the food impact the usefulness of nourishments that customers need along with texture, taste, and stability. It will play a significant role in handling and regulating the composition and size of the grain. Those incorporate more beneficial nourishments (low salt, low fat) with alluring sensory properties; elements with enhanced properties; and also, the capacity for the eradication of certain ingredients without compromising the stability, parenthetically within the emulsion, and in smart-aid for food process to take away food allergens like peanut protein [105]. Downsizing molecules of food size to nanosized crystals makes it particular for the surface area. Smaller particles improve dietary stability and spreadability, which would aid in the development of low-fat, nutritious foods. Water-in-oil-in-water (multiple emulsion) can circulate the lipids uniformly to diminish additional thickeners and stabilizers to accomplish a required texture [54]. Nanotubes are explored as the viscosifying agents and gelation. Nanotubes dependent on the milk protein – lactalbumin, for example, have a few advantages over conventional gelling and viscosifying operators [56]:

- They have a high length-to-distance ratio and have firmness, making them useful viscosifying agents. It improves the quality of food by requiring less protein as a result of their use.
- At low concentrations, it has a high storage modulus, implying that less protein is needed to construct a gel.
- From a clear gel that looks good in a food environment.
- Because of the physical contact, their implementation allows for reversible gel growth.
- It is possible to dismantle tubes in a controlled manner, which may be used to cause a controlled decline in gel pressure.

### 1.2.4 COLOR, FLAVOR, AROMA IMPROVEMENT

Food sensory qualities such as color, texture, flavor, and quality have improved as a result of the use of nanotechnology in the food industry [118]. Using nanoemulsion technologies, researchers discovered that – carotene can be used to control the color of food systems [30]. Changes in food viscosity influence the flavor perception. As viscosity increases, the flavor release profile decreases and as a consequence result in lower viscosity of food intensity. It modulates the flavor perception and retards the tasting signs of taste buds [140]. The size of oil droplet dispersion is an important emulsion feature that regulates aroma substance release. In the case of hydrophobic esters, an analysis to ascertain the effect of droplet size on aroma release profile revealed that the release of aroma compounds increases with a decrease in emulsion droplet size while preserving the oil content of the emulsion. Nanoencapsulation could also be employed to recover toxic components in natural lipid blends, and these nanocapsules can also be used in instant coffee as fragrance enhancers [50].

## 1.3 OTHER NANOTECHNIQUES IN FOOD PRESERVATION

### 1.3.1 FOOD PACKAGING

Packaging procedures are used to ensure that the food's consistency is maintained and that it is safe to eat. Packaging defends food from natural shocks and movements, microbial attack, and temperature changes, and it also protects food by scrounging oxygen as well as other food destroying variables. To reduce waste, biodegradable packaging materials are favored, so the concept of nanotechnology in the food industry has been adopted [111].

#### 1.3.1.1 NANOCOATINGS

An ultra-thin layer of thickness $<1$–100 nm on the surface is referred to as nanocoatings [38]. Nanocoatings do not change the topography of the surface and provide a flat surface in the same way as paint does. Furthermore, they are resistant to abrasion and wear. It functions as a gas buffer, has hydrophobic/hydrophilic or oleophobic properties, improves corrosion resistance, and improves conductive/insulating properties [102]. Multifunctional coating called smart coatings assures thermal insulation, self-healing function, and controlled release of active ingredients. Based on required function nanocoating uses antimicrobial coatings with silver, titanium oxide, zinc oxide (ZnO), or bioactive components such as an extract or essential oil from spices and herbs, polysaccharides, proteins, bacteriocins such as nisin, medicines, and organic acid for packing of food. Nano coating shows better processing properties and high-performance material compared to conventional coatings in case of quick-drying, high flexibility, high water vapor permeability (WVP), will not expand after contact with water [141]. Many types of nanocoating are referred in the applications of food packaging as nanocoating sandwiched like multilayer laminated packaging films, high barrier properties, an edible coating which acts as moisture, gas, and lipid barrier [19], covering of glass containers with organosilanes, nano surface biocides, etc. For processing meat products, replacing polyvinylidene fluoride coated films and oxide evaporated films with nano silica-coated films on PET (polyethylene terephthalate), nylon provides moisture and oxygen barrier, lowers manufacturing costs, and extends shelf life while retaining fragrance. These nanocoatings are eco-friendly and no formation of dioxins during incineration and have excellent optical and mechanical properties [129]. Intelligent type nanocoatings have been developed by the scientist which indicates any sort of contamination during storage. The ability to continuously and effectively track the gas content, oxygen content, and excess moisture present in the headspace of a food package and provide protection and consistency of the food after processing has been shown by detection of a system of gas content and non-invasive [4, 36]. Nanostructure of magnesium oxide (MgO) and polylactic acid (PLA) biopolymer blended packaging material has potential against bacterial films [139]. Zerovalent iron particles were synthesized to scavenge the oxygen molecule from the food [49]. However, nanocoatings require specialized equipment and complex methods, for example, plasma spray techniques and chemical vapor deposition (CVD).

#### *1.3.1.2 NANO LAMINATES*

Nanolaminates are a thin film with two or more food-grade layers. It can be developed by the deposition of material on the surface through layer-by-layer (LbL) assembly methods. The deposition can be enabled by electrostatic attraction, which makes the charged substance settled on a positively charged surface. This LbL technology permits exact control of thickness (1–100 nm) and properties of nanolaminates [149]. Nano laminates are used as coating rather than self-standing films since they are very delicate [77]. The application of nanolaminates to the food is done by dipping into arranged solutions or by spraying the solution containing substances on the food surface [149]. The LbL technique can be achieved by electrostatic bonding, covalent bonding, hydrogen bonding, charge-transfer interaction, and hydrophobic interaction between them [15]. Ionic strength, pH, molecular weight, concentration, and temperature of the solution, terminal layer, number of layers deposited, adsorption time, washing, and drying time are the factors that affect the physicochemical properties and formation of nanolaminates [76]. The different layers are made from polysaccharides, polyelectrolytes such as protein, charged lipids like surfactants and phospholipids, and colloidal particles such as micelle, droplets, and vesicles. These techniques may be utilized to encapsulate the antimicrobial, enzymes, color, flavor, anti-browning agents, and antioxidants into the films, thereby increasing the quality as well as shelf life of laminated food [149]. Nano laminates made from alginates and lysozyme-chitosan-rectorite complex have shown antimicrobial activity and preserve the pork for 3 days [67]. The nanolaminate film of PET with nanocapsules or nanoemulsion containing carvacrol improved the antifungal properties of packing material [42]. The use of alginate, chitosan (CS), carrageenan in nanolaminates helps to preserve the fresh-cut, pears, and mangoes [93, 94, 133] and a similar approach was applied to improve the cheese shelf life [95].

### *1.3.1.3 NANOCOMPOSITES*

Nanocomposites are made from a mixture of polymers and nanoparticles which helps to improve the property of the polymer by consolidating with it [159]. Nanocomposite gives highly versatile chemical functionality; thus, it is utilized for the development of high barrier properties [109]. These nanocomposites are helping foods to remain fresh and without any microbial contamination for a sustainable amount of period. Table 1.1 shows the use of certain nanocomposites as antimicrobial agents in food packaging. Usually, these act as a gas barrier and reduce the loss of carbon dioxide gas in the case of carbonated drinks [109]. Hence, it prolongs the stability of the product. The idea of nanocomposites development was encouraged by the biomineralization process where an organic substance (lipid, protein) reacts with an inorganic substance (calcium carbonate) and develops a material with more toughness [136]. Nano clay is used as a nanocomposite to create a gas barrier for bottles. Aluminum silicates are naturally occurring nano clay referred phyllosilicates that are stable, eco-friendly, and less expensive [32]. Aegis, Impen, and Durethan are commercially available nano clays. Aegis and Impen are oxygen scavengers that strengthen the barrier property of carbonated beverages by arresting carbon dioxide gas. The paperboard bottle for fruit juice is stiffened with Durethan [48]. Nanocomposite incorporated with ZnO and pediocin has antimicrobial activity [95]. Silver-lined nanomaterials (NMs) are often noted for their antibacterial activity, as silver adheres to the cell wall and improves permeability by degrading lipopolysaccharide (LS), causing permanent changes in bacterial DNA. PEG-covered garlic oil nanocomposites are found effective in the prevention of pests during the storage of packed food [100]. Bio-nanocomposites made up of starch and cellulose residues, polycaprolactone, and polyhydroxy butyrate (PHB) have shown to be effective as packaging laying materials [7]. Gatekeeper IN Fresh, a commercialized nanocomposite scavenges the ethylene gas produced by fruits and vegetables and delays the ripening process [57]. In the food packaging industry, enzyme immobilization is considered as an effective type of nanocomposite. These enzymes are used in food packaging processes and are formulated into nano clays [17].

**TABLE 1.1** Use of Nanocomposites as Antimicrobial Agents in Food Packaging

| Nanocomposite | Microorganism | Advantage | References |
|---|---|---|---|
| Silver nanoparticles, silver clays | *E. coli, Listeria* | Inhibition zone | [119] |
| Silver nanoparticle, kaolin, titanium dioxide, + hot air | Yeast and molds, *P. citrinum* | Decreases decaying of Myrica (bayberries) | [148] |
| Zinc oxide, nisin | *Salmonella enterica* | 3–4 log reduction in liquid egg | [73] |
| Titanium dioxide, zinc oxide, silver nanoparticles | Mesophilic | Increased shelf life of orange juice up to 28 days | [41] |
| Copper and zinc oxide | *S. aureus, C. albicans, E. coli* | 1–2 log reduction | [135] |
| Titanium dioxide | *E. coli* | Inhibition zone in meat extrudates | [142] |
| Silver montmorillonite | *Pseudomonas* spp. | >2log reduction | [70] |

#### 1.3.1.4 ACTIVE PACKAGING AND NANOENCAPSULATION

Active packaging consists of a specific substance that can release or absorb the component towards or against the circumferential environment of the packed food. The active food packaging NMs are produced for antimicrobial packaging applications [128]. Bioactive compounds can be integrated into packaging material to improve the viability and functionality of food packages by the process of encapsulation [55]. Food suppliers should select a successful antimicrobial substance-based approach to assure the quality of the packed product [43]. Nanoencapsulation allows immediate connection of NMs at the time of food consumption for example generally $SiO_2$ is applied as an aroma transporter in various food items [34]. Polyethylene (PE) blended nanopowders like anatase $TiO_2$, rutile $TiO_2$, silver, kaolin packaging material was investigated to preserve the fruits like strawberries [153]. The coating products had a strong bactericidal activity against *E. coli* and were recommended as a potential eco-friendly active packaging substance. Besides, antimicrobial nanoemulsion prepared by encapsulating citral essential oil by the use of ultrasonic power system, this results in the effective delivery of antimicrobial agents in food systems [88]. Active packaging built on metal and metal oxide NMs can be a powerful weapon for nanocomposites in terms of antimicrobial capacity [84]. Gram-positive and gram-negative pathogens such as *Salmonella typhimurium*, *E. coli,* and *L. monocytogenes* could be inhibited in cooked and raw chicken when benzoic acid encapsulated in poly lactic-co-glycolic acid (PLGA) particles [115]. In some cases, antimicrobial compounds are entrapped within the nanocavities of bulk polymer which has to migrate to the outer surface and come in contact with pathogens [128, 131]. Heat-sensitive compounds cannot be used in the active packaging as they inactivate during the packaging process. In the case of heat-sensitive compounds, using nanoencapsulated active compounds before incorporating them into the polymer is the safest option [58]. The various factors such as pH, heat, enzymatic activity, and relative humidity regulate the liberation of active compounds towards package headspace [6, 61].

### 1.3.2 NANOSIZED ADDITIVES

Nanosized additives could be intentionally integrated to change the food properties like color, taste, texture, and shelf life to fortify the food and enhance the bioavailability. Nanosized ingredients may employ as an index of freshness and quality of food or to that traceability [23, 60, 113]. The opportunities of nanotechnology in nutritional supplements, functional foods with nanosized additives like antioxidants, preservatives, antimicrobials, and vitamins are presently available for bioavailability, improved taste, and absorption [99]. To boost shrimp production, water-based nanosized chitin and chitin-based materials are used as additives (color and texture) [21]. Fat-soluble bio-actives require nanosized delivery systems to disperse bioactive into water systems. Solid lipid nanoparticles, nanodispersion, microemulsion liposomes are the promising delivery systems which increase the solubility of nano additives in water [71]. The nanosized microbial compounds enhance the stability or shelf life of food and protect it from deterioration. This was reported by the scientists, the antimicrobial activity of nanoparticles act as growth inhibitors or killing agents [31]. The essential oil nanoparticles and metal and metal oxide nanoparticles have been developed for its active antimicrobial property. These compounds damage the microbial cells by applying oxidative stress [112]. Among these compounds, essential oils are the common compounds. The nanosized essential oil has proven its antimicrobial activity, solubility, and stability. The nanosized chitosan-thyme essential oil has shown its inhibitory effect against *Bacillus cereus* and *Staphylococcus aureus* (*S. aureus*) [132].

### 1.3.3 NANOTECHNOLOGY IN FOOD FUNCTIONALITY

Food oxidation is caused by a chemical reaction between the different components of food and their surroundings. Thus, NMs were developed to manage or control the chemical reaction [89]. Nanotechnology can remove toxicants and improves nutritional supplements thus enhance the food functionality [69]. Antioxidants in combination with edible coating can retard the browning enzyme of cut fruits [122]. However, nano-ZnO coated active packaging decreases the polyphenol oxidase and pyrogallol peroxide and extend the fresh-cut Fuji apples shelf life [82]. Metal and metal oxide may cause ROS, oxidative stress, and leads to the unbalanced redox state of the cell [89]. These metal and metal oxides have shown its effectiveness against microbial contamination and increase the stability of food [52, 151]. The biopolymer membrane of DL-α-tocopherol with poly-ε-caprolactone has decreased the browning index of fresh-cut red delicious apples [161]. The nanomaterial such as $SiO_2$ can act as a flavor carrier and improve the flavor of nanofood products [34].

## 1.4 NANOTECHNOLOGY FOR FOOD SAFETY

Food safety is becoming a growing public health priority. Food safety's key goal is to ensure that the food does not have any negative effects on the consumer after consumption [108]. Recently, nanotechnology has numerous applications in food processing and safety as well as shelf-life extension, enhances nutraceutical value and diminish the packaging waste [150]. Nowadays, food safety has taken a major concern because of changed food habits and recipes. The contaminants, foodborne pathogens and toxins cause severe health issues. The detection of these pollutants and pathogens by conventional method takes more time and labor-intensive. Thus, advanced nanotechnology has sped up the tending of food safety problems in microbial contamination and has improved shelf-life, packaging techniques, and toxin detection [69]. NMs also used in the development of biosensors for the evaluation of microorganisms [69, 150].

### *1.4.1 USE OF NANO SENSORS*

Nanosensors are developed to identify the changes in color, flavor, and gasses produced due to spoilage in the food. Nanosensor comprises of sensing part and electronic data processing unit which detects changes in humidity, heat, light, gas, and chemicals further it converts into electrical signals [87]. These sensors are highly sensitive and selective compared to conventional sensors. The gas sensors made from metal such as palladium, platinum, gold [90], and conducting polymers. Conducting polymers governs magnetic, optical, and electrical properties and relates to pi-electron backbones. The changes in resonance show response pattern at the conducting polymer-based sensor due to the presence of the gas to be identified [120]. Different types of nanosensors were advanced to detect food spoiling substances and foodborne pathogens. Silver nano colloids [10], graphene oxide [89, 166], magnetic beads [63], carbon nanotubes [158], silver nanoparticles (AgNPs) [1], and plasmonic gold [43] are also used to identify the foodborne pathogens. Furthermore, nano barcodes made of synthetic DNA labeled with color-coded probes have been used to identify food contaminants [83]. The identification of *E. coli* in food samples is done by measuring and detecting light scattering across cells. These sensors function by binding with a known protein and a silicon chip that binds with other *E. coli* found in food [13]. Using G-liposomal protein nano-vesicles, an array-based immunosorbent assay system was demonstrated to detect *L. monocytogenes*, *E. coli*, and *Salmonella* spp. in pure and mixed cultures [25]. Furthermore, liposome-based techniques have been studied by different scientists to detect pathogens [126, 127, 137]. Nanosensors, such as nano cantilevers, use silicon-based materials to detect proteins and recognize pathogens that vibrate at different frequencies depending on their biomass [72]. The use of nanosensors in food packaging materials could help monitor physical, chemical, and biological changes during food processing. To detect chemicals, toxins, and foodborne pathogens, smart packaging with specialized nanosensors and nanodevices was developed [26, 80, 155]. This smart packaging improves the package's connectivity capabilities and the reliability of knowledge sharing during delivery [97]. Electronic tongue and electronic nose are examples of sensors that make use of a large number of chemical sensors connected to data processing systems [144]. The advantage of nanosensors with composition is listed in Table 1.2.

**TABLE 1.2** Advantage of Nano Sensors with Composition

| Nano Sensors with Composition | Application/Advantage | References |
|---|---|---|
| Nano biosensors | Used in food packaging to detect the bacteria and virus | [28] |
| Nano barcodes | Detect the quality of agricultural produce | [28] |
| Nano smart dust | To detect any sort of environmental pollution | [28] |
| Smart biosensors and biomimetic sensors | Detects and removes pathogens and act as pseudo cell surfaces | [28] |
| Array biosensors | Changes in color after the contact of spoiled food | [14] |
| Metal-based biosensors | Detect any sort of color change in the food | [75] |
| | Detect toxins (aflatoxin) in the milk | [96] |
| | Detect any changes in humidity, heat, light, chemicals, and gas. | [91] |
| | Detect gas produced due to spoilage. | [75] |
| Single-walled carbon nanotubes and DNA | Detect the pesticides present on the surface of fruits and vegetables | [134] |
| Polyaniline and carbon black | Detect presence of carcinogens in food. | [144] |

**TABLE 1.2** *(Continued)*

| Nano Sensors with Composition | Application/Advantage | References |
|---|---|---|
| | Detect foodborne pathogens | [160] |
| Abuse indicators | To detect the weather desired temperature has been achieved or not | [137] |
| Cerium oxide immunosensors | Used in food packaging to detect several toxins | [101] |
| Reflective interferometry | Used to detect *E. coli* contamination in packed foods | [147] |
| Carbon nanotubes coated with Liposomes and poly (3,4-ethylene dioxythiophene) | Used to detect cholera toxins | [146] |
| Micro conductometric immuno-sensor with gold nanoparticles | Used to detect *Salmonella*, gram-negative bacteria | [39] |

## 1.4.2 USE OF NANOPARTICLES

Nanoparticles serve several functions in food processing. It improves the color, stability, and flow property of a food [29]. Earlier, nanoparticles were used in drug delivery system but now it identifies its use in the food industry. Nanoparticles like titanium oxide, zinc oxide (ZnO) and silicate are utilized to decrease the oxygen flow into the packaging container, helps in decreasing the moisture leakage and preserve the food as a fresh for longer time [65]. Some nanoparticles bind specially and remove pathogens or chemical additives from food [103]. As a drying and anticaking agent, $SiO_2$ helps to absorb water from food [163]. Titanium dioxide ($TiO_2$) is a food coloring agent that also serves as a photocatalytic disinfectant. It is also used as a nutritional whitener in milk, cheese, and other dairy products [2]. It has been used as a barrier to UV protection in packaged foods. As it functions as an antibacterial, AgNPs defend the food against microbial outbreak. Nanosized silver particles act as a bacterial agent, which provides a bigger surface zone and dispersed easily and readily ionized [163]. Nanotechnology helps in controlling and managing food allergens [79, 110]. As exposed to immunoglobulin E, immunotherapy uses nanoparticles as an allergen delivery mechanism, allowing for easier administration, dose reduction, and allergen reduction [53]. To detect food allergens, a magnetic, nanoparticles-assisted aptamer-based fluorescence assay was investigated [162]. The heavy metal released from NMs has an adverse effect on human health [59]. The sol-gel and calcination process of MgO nanomaterial has demonstrated its ability against bacterial contamination and heavy metal removal from polluted water [20]. A study showed that silver penetrates inside the microbial structure and disrupts the ribosomal activity, thus causing an impediment to the synthesis of many enzymes. Compared to gram-positive microorganism, AgNPs are effective against gram-negative microorganisms. It is easy to penetrate the thin cell wall of gram-negative microorganism [163]. AgNPs have extended the shelf life of fruits and vegetables by consuming and decomposing ethylene [163]. Besides AgNPs, titanium, and zinc nanoparticles are also used in food packaging. Cadmium, selenium, copper, and copper oxide, MgO, ZnO, CS, telluride, and carbon nanotubes have been reported for its antibacterial activity. CS possesses a positive charge and it binds to a negatively charged cell wall, resulting in increased permeability and cell wall disruption [5]. Inorganic nanoceramic is used for deep frying in cooking oil. The oil with nanoparticles extends the food shelf life and makes its crispier [16]. Detection of foodborne pathogens with different nanoparticles and advantages of different nanoparticles in food packaging are listed in Tables 1.3 and 1.4.

**TABLE 1.3** Detection of Foodborne Pathogen Using Nanoparticles

| Nanoparticles | Pathogen Detected | References |
|---|---|---|
| Gold nanoparticles | *Salmonella enterica serotype Typhi* | [86] |
| | *Vibrio parahaemolyticus* | [152] |
| | *E. coli* | [145] |
| $ZrO_2$ nanoparticles | *E. coli* | [130] |
| FeON nanoparticles | *Listeria monocytes* | [154] |
| Magnetic nanoparticles | *Mycobacterium avium* | [157] |
| | *Salmonella typhimurium* and *Staphylococcus aureus* | [35] |
| | *E. coli* and *Staphylococcus epidermis* | [164] |
| Silicon/gold nanorod | *Respiratory syncytical virus* | [37] |
| | *E. coli, Salmonella enterica, Serotype typhi* | [51] |
| Immunomagnetic liposome nanoparticles | *Cronobacter sakazakii* | [126] |
| Liposome nanoparticles | *Salmonella typhimurium* | [165] |
| Gold nanoparticles conjugated with Aptamer | *Salmonella typhimurium* | [33, 107] |
| RuBpy doped silica | *E. coli* | [138, 156] |

**TABLE 1.4** Advantage of Nanoparticles in Food Processing and Preservation

| Nanoparticles Used | Applications/Advantages | References |
|---|---|---|
| Titanium dioxide | Used in milk, cheese, and other dairy products as food whitener | [2] |
| | Used as photocatalytic disinfecting agent | [163] |
| | Retards ripening, senescence, and decay, reduce browning activity | [81] |
| | Used in filtration | [22] |
| Silicon dioxide | It reduces moisture leakage in packaging | [29] |
| | It shows hygroscopic application in food as it absorbs moisture in food and act as and food colorant | [74] |
| | Act as drying agent, anticaking agent | [64] |
| Silver nanoparticles | Protect the food from microbial infestation | [5] |
| | It absorbs and decompose ethylene gas as well as extend the shelf life of fruits and vegetables | [16] |
| | Shows antimicrobial effect against *Staphylococcus aureus* and *E. coli* and reduces the growth of aerobic psychotropics, yeast, and molds | [41, 45–47, 162] |
| Zinc oxide | Reduces the oxygen flow from the container. | [5] |
| | Yeast and molds, *Salmonella, Lactobacillus plantarum* counts reduced effectively without changing food quality | [40, 41, 73] |
| Acrylic nanoparticles | Used in barrier packaging | [92, 121] |
| Carbon and nickel nanoparticles | Used in equipment coating as disinfecting | [11, 26] |
| Inorganic nanoceramics | Used for deep frying food. | [16] |
| Polymeric nanoparticles | They are bactericidal and effective in delivery systems | [16] |
| Silver oxide | Retards microbial spoilage | [165] |

## 1.5 SUMMARY

Nanotechnology is referred to as an "enabling technology" that promises to open up a slew of new science and technical possibilities [104]. Continuous advancement in using nanotechnology to improve the quality and quantities of food would catalyze the food processing industry from production to preservation, processing, packaging, and even waste disposal [18]. The use of nano-based substances, processes, and systems in manufacturing, packaging, shipping, and storage methods will help to preserve the nutritional value, flavor, scent, and texture of foods while still providing additional conveniences in near to natural form. Nanotechnology is opens up a universe of additional opportunities for the food industry; however, the incorporation of nanoparticles into the food chain will contribute to the development of harmful food pollutants and can damage human health. Thus, the transparency of environmental impact and safety issues to be the need when managing the improvement of nanotechnology in food systems. Nonetheless, in the future years, huge nanotechnology implementation in ensuring microbial food protection looks promising.

## KEYWORDS

- **nano fabrications**
- **nanosensors**
- **nanocoating**
- **nanoencapsulation**
- **nanofiltration**
- **nanolaminates**
- **nanoparticles**

## REFERENCES


1. Abbaspour, A., Norouz-Sarvestani, F., Noori, A., & Soltani, N., (2015). Aptamer-conjugated silver nanoparticles for electrochemical dual-aptamer-based sandwich detection of *Staphylococcus aureus*. *Biosens. Bioelectron, 68*, 149–155.
2. Acosta, E., (2009). Bioavailability of nanoparticles in nutrient and nutraceutical delivery. *Current Opinion in Colloid and Interface Science, 14*, 3–15.
3. Aguilera, J. M., & Lillford, P. J., (2008). Structure-property relationships in foods. Chapter 12. In: Aguilera, J. M., & Lillford, P. J., (eds.), *Food Materials Science* (pp. 229–253). New York – USA: Springer International Publishing.
4. Arndt, G. W., (2008). Leak testing. In: Yam, K. L., (ed.), *The Wiley Encyclopedia of Packaging Technology* (3rd edn., pp. 163–197). New York – USA. John Wiley and Sons, Inc.

5. Arshak, K., Adley, C., Moore, E., Cunniffe, C., Campion, M., & Harris, J., (2007). Characterization of polymer nanocomposite sensors for quantification of bacterial cultures. *Sensors and Actuators, B: Chemical, 126*(1), 226–231.
6. Ayala-Zavala, J. F., González-Aguilar, G. A., Ansorena, M. R., Alvarez-Párrilla, E., & De La Rosa, L., (2014). Nanotechnology tools to achieve food safety. Chapter 17. In: Bhat, R., & Gomez-Lopez, V. M., (eds.), *Practical Food Safety: Contemporary Issues and Future Directions* (pp. 341–353). Oxford – UK: Wiley-Blackwell.
7. Baeumner, A., (2004). Nanosensors identify pathogens in food. *Food Technology, 58*(8), 51–55.
8. Bai, Y. X., Li, Y. F., Yang, Y., & Yi, L. X., (2006). Covalent immobilization of triacylglycerol lipase onto functionalized nanoscale $SiO_2$ spheres. *Process Biochemistry, 41*(4), 770–777.
9. Baker, R. W., (2004). *Membrane Technology and Applications* (p. 460, 2nd edn.). England – UK: John Wiley and Sons Ltd.
10. Baranwal, A., Mahato, K., Srivastava, A., Maurya, P. K., & Chandra, P., (2016). Phytofabricated metallic nanoparticles and their clinical applications. *RSC Adv., 6*(107), 105996–6010.
11. Basavaraja, S., Balaji, S. D., Lagashetty, A., Rajasab, A. H., & Venkataraman, A., (2008). Extracellular biosynthesis of silver nanoparticles using the fungus *Fusarium semitectum*. *Mater. Res. Bull., 43*, 1164–1170.
12. Berekaa, M. M., (2015). Nanotechnology in food industry; advances in food processing, packaging and food safety. *Int. J. Curr. Microbiol. App. Sci, 4*(5), 345–357.
13. Bhattacharya, S., Jang, J., Yang, L., Akin, D., & Bashir, R., (2007). Biomems and nanotechnology based approaches for rapid detection of biological entities. *J. Rapid. Meth. Autom. Microbiol., 15*, 1–32.
14. Biswal, S. K., Nayak, A. K., Parida, U. K., & Nayak, P. L., (2012). Applications of nanotechnology in agriculture and food sciences. *International Journal of Science Innovations and Discoveries, 2*, 21–36.
15. Borges, J., & Mano, J. F., (2014). Molecular interactions driving the layer-by-layer assembly of multilayers. *Chemical Reviews, 114*(18), 8883–8942.
16. Bouwmeester, H., Dekkers, S., Noordam, M. Y., Hagens, W. I., Bulder, A. S., De Heer, C., Ten, V. S. E., et al., (2009). Review of health safety aspects of nanotechnologies in food production. *Regulatory Toxicology and Pharmacology, 53*(1), 52–62.
17. Burdo, O. G., (2005). Nanoscale effects in food-production technologies. *Journal of Engineering Physics and Thermophysics, 78*, 90–96.
18. Buttel, F. H., & Barker, R., (1985). Emerging agricultural technologies, public policy, and implications for third world agriculture: The case of biotechnology. *American Journal of Agricultural Economics, 67*(5), 1170–1175.
19. Cagri, A., Ustunol, Z., & Ryser, E. T., (2004). Antimicrobial edible films and coatings. *J. Food Prot., 67*, 833–848.
20. Cai, Y., Li, C., Wu, D., Wang, W., Tan, F., Wang, X., Wong, P. K., & Qiao, X., (2017). Highly active MgO nanoparticles for simultaneous bacterial inactivation and heavy metal removal from aqueous solution. *Chemical Engineering Journal, 312*, 158–166.
21. Chantarasataporn, P., Yoksan, R., Visessanguan, W., & Chirachanchai, S., (2013). Water-based nanosized chitin and chitosan as seafood additive through a case study of pacific white shrimp (*Litopenaeus vannamei*). *Food Hydrocolloids, 32*(2), 341–348.

22. Chau, C. F., Huei, W. S., & Yen, G. C., (2007). The development of regulations for food nanotechnology. *Tr. Food Sci. Technol., 18*, 269–280.
23. Chaudhry, Q., Scotter, M., Blackburn, J., Ross, B., Boxall, A., Castle, L., Aitken, R., & Watkins, R., (2008). Applications and implications of nanotechnologies for the food sector. *Food Additives and Contaminants, 25*(3), 241–258.
24. Chellaram, C., Murugaboopathi, G., John, A. A., Sivakumar, R., Ganesan, S., Krithika, S., & Priya, G., (2014). Significance of nanotechnology in food industry. *APCBEE Procedia, 8*, 109–113.
25. Chen, C. S., & Durst, R. A., (2006). Simultaneous detection of *Escherichia coli* O157:H7, *Salmonella* spp., and *Listeria monocytogenes* with an array-based immunosorbent assay using universal protein G-liposomal nanovesicles. *Talanta, 69* 232–238.
26. Chen, H., Zuo, X., Su, S., Tang, Z., Wu, A., Song, S., Zhang, D., & Fan, C., (2008). An Electrochemical Sensor for pesticide assays based on carbon nanotube-enhanced acetylcholinesterase activity. *Analyst, 133*, 1182–1186.
27. Cheng, Y., Liu, Y., & Huang, J., (2008). Rapid amperometric detection of coliforms based on MWNTs/Nafion composite film modified glass carbon electrode. *Talanta, 75*, 167–171.
28. Coles, D., & Frewer, L. J., (2013). Nanotechnology applied to European food production—A review of ethical and regulatory issues. *Trends in Food Science and Technology, 34*, 32–43.
29. Coma, V., (2008). Bioactive packaging technologies for extended shelf life of meat-based products. *Meat Science, 78*(2), 90–103.
30. Cushen, M., Kerry, J., Morris, M., Cruz-Romero, M., & Cummins, E., (2012). Nanotechnologies in the food industry–recent developments, risks and regulation. *Trends Food Sci. Technol., 24*, 30–46.
31. Dasgupta, N., Ranjan, S., Mundekkad, D., Ramalingam, C., Shanker, R., & Kumar, A., (2015). Nanotechnology in agro-food: From field to plate. *Food Res. Int., 69*, 381–400.
32. Davis, D., Guo, X., Musavi, L., Lin, C. S., Chen, S. H., & Wu, V. C. H., (2013). Gold nanoparticle-modified carbon electrode biosensor for the detection of *Listeria monocytogenes*. *Industrial Biotechnology, 9*, 31–36.
33. De Souza, R. J., Esparza, I., Ferrer, M., Sanz, M. L., Irache, J. M., & Gamazo, C., (2012). Nanoparticulate adjuvants and delivery systems for allergen immunotherapy. *J. Biomed. Biotechnol.*, 474605.
34. Dekkers, S., Krystek, P., Peters, R. J., Lankveld, D. P., Bokkers, B. G., Hoeven-Arentzen, P. H. V., Bouwmeester, H., & Oomen, A. G., (2011). Presence and risks of nanosilica in food products. *Nanotoxicology, 5*(3), 393–405.
35. Duan, N., Wu, S., Zhu, C., Ma, X., Wang, Z., Yu, Y., & Jiang, Y., (2012). Dual-color upconversion fluorescence and aptamer-functionalized magnetic nanoparticles-based bioassay for the simultaneous detection of *Salmonella typhimurium* and *Staphylococcus aureus*. *Anal. Chim. Acta, 723*, 1–6.
36. Duncan, T. V., (2011). Applications of nanotechnology in food packaging and food safety: Barrier materials, antimicrobials and sensors. *Journal of Colloid and Interface Science, 363*(1), 1–24.
37. Dungchai, W., Siangproh, W., Chaicumpa, W., Tongtawe, P., & Chailapakul, O., (2008). *S. typhi* determination using voltammetric amplification of nanoparticles: A highly sensitive strategy for metalloimmunoassay based on a copper-enhanced gold label. *Talanta, 2*, 727–732.

38. Egodage, D. P., Jayalath, H. T. S., & Samarasekara, A. M. P. B., (2017). Novel antimicrobial nano coated polypropylene based materials for food packaging systems. In: *Proceedings of the Moratuwa Engineering Research Conference (Mercon)* (pp. 122–134). Moratuwa, Sri Lanka.
39. El Ichi, S., Leon, F., Vossier, L., Marchandin, H., Errachid, A., Coste, J., Jaffrezic-Renault, N., & Fournier-Wirth, C., (2014). Microconductometric immunosensor for label-free and sensitive detection of gram-negative bacteria. *Biosensors and Bioelectronics, 54,* 378–384.
40. Emamifar, A., Kadivar, M., Shahedi, M., & Soleimanian-Zad, S., (2011). Effect of nanocomposite packaging containing Ag and ZnO on inactivation of *Lactobacillus plantarum* in orange juice. *Food Control, 22*(3, 4), 408–413.
41. Emamifar, A., Kadivar, M., Shahedi, M., & Soleimanian-Zad, S., (2010). Evaluation of nanocomposite packaging containing Ag and ZnO on shelf life of fresh orange juice. *Innovat. Food Sci. Emerg. Technol., 11*(4), 742–748.
42. Fabra, M. J., Flores-López, M. L., Cerqueira, M. A., De Rodriguez, D. J., Lagaron, J. M., & Vicente, A. A., (2016). Layer by-layer technique to developing functional nanolaminate films with antifungal activity. *Food and Bioprocess Technology, 9*(3), 471–480.
43. Fang, Z., Zhao, Y., Warner, R. D., & Johnson, S. K., (2017). Active and intelligent packaging in meat industry. *Trends Food Sci. Technol., 61,* 60–71.
44. Fernandes, P., (2010). Enzymes in food processing: A condensed overview on strategies for better biocatalysts. *Enzyme Res., 2010,* 19.
45. Fernandez, A., Picouet, P., & Lloret, E., (2010). Cellulose-silver nanoparticle hybrid materials to control spoilage-related microflora in absorbent pads located in trays of fresh-cut melon. *Int J Food Microbiol, 142*(1, 2), 222–228.
46. Fernandez, A., Picouet, P., & Lloret, E., (2010). Reduction of the spoilage related microflora in absorbent pads by silver nanotechnology during modified atmosphere packaging of beef meat. *J. Food Protect, 73*(12), 2263–2269.
47. Fernandez, A., Soriano, E., Lopez-Carballo, G., Picouet, P., Lloret, E., & Gavara, R., (2009). Preservation of aseptic conditions in absorbent pads by using silver nanotechnology. *Food Res. Int., 42*(8), 1105–1112.
48. Flanagan, J., & Singh, H., (2006). Microemulsions: A potential delivery system for bioactives in food. *Critical Reviews in Food Science and Nutrition, 46*(3), 221–237.
49. Foltynowicz, Z., Bardenshtein, A., S€Angerlaub, S., Antvorskov, H., & Kozak, W., (2017). Nanoscale, zero-valent iron particles for application as oxygen scavenger in food packaging. *Food Package Shelf Life, 11,* 74–83.
50. Freiberger, E. B., Kaufmann, K. C., Bona, E., Hermes De, A. P. H., Sayer, C., Leimann, F. V., & Gonçalves, O. H., (2015). Encapsulation of roasted coffee oil in biocompatible nanoparticles. *LWT-Food Sci. Technol., 64*(1), 381–389.
51. Fu, J., Park, B., Siragusa, G., Jones, L., Tripp, R., & Zhao, Y., (2008). An Au/Si hetero-nanorod-based biosensor for salmonella detection. *Nanotechnology, 15,* 155502.
52. Fu, P. P., Xia, Q., Hwang, H. M., Ray, P. C., & Yu, H., (2014). Mechanisms of nanotoxicity: Generation of reactive oxygen species. *J. Food Drug Anal., 22,* 64–75.
53. Gamazo, C., Gastaminza, G., Ferrer, M., Sanz, M. L., & Irache, J. M., (2014). Nanoparticle based-immunotherapy against allergy. *Immunotherapy, 6*(7), 885–897.

54. Garti, N., & Benichou, A., (2004). Recent developments in double emulsions for food applications. In: Friberg, S., Larsson, K., & Sjoblom, J., (eds.), *Food Emulsions* (pp. 352–412). Boca Raton – Florida: CRC Press.
55. Ghaani, M., Cozzolino, C. A., Castelli, G., & Farris, S., (2016). An overview of the intelligent packaging technologies in the food sector. *Trends Food Sci. Technol., 51,* 1–11.
56. Graveland-Bikker, J. F., & De Kruif, C. G., (2006). Unique milk protein based nanotubes: Food and nanotechnology meet. *Trends in Food Science & Technology, 17*(5), 196–203.
57. Gupta, S., & Moulik, S. P., (2008). Biocompatible microemulsions and their prospective uses in drug delivery. *Journal of Pharmaceutical Sciences, 97*(1), 22–45.
58. Hatzigrigoriou, N. B., & Papaspyrides, C. D., (2011). Nanotechnology in plastic food-contact materials. *J. Appl. Polym. Sci., 122*(6), 3720–3739.
59. He, X., Aker, W. G., Huang, M. J., Watts, J. D., & Hwang, H. M., (2015). Metal oxide nanomaterials in nanomedicine: Applications in photodynamic therapy and potential toxicity. *Curr. Top. Med. Chem., 15,* 1887–1900.
60. Hilty, F. M., Arnold, M., Hilbe, M., Teleki, A., Knijnenburg, J. T., Ehrensperger, F., Hurrell, R. F., et al., (2010). Iron from nanocompounds containing iron and zinc is highly bioavailable in rats without tissue accumulation. *Nature Nanotechnology, 5*(5), 374–380.
61. Ho, B. T., Joyce, D. C., & Bhandari, B. R., (2011). Release kinetics of ethylene gas from ethylene-a-cyclodextrin inclusion complexes. *Food Chemistry, 129,* 259–266.
62. Hochella, M. F., (2008). Nanogeoscience: From origins to cutting-edge applications. *Elements, 4,* 373–379.
63. Holzinger, M., Le Goff, A., & Cosnier, S., (2014). Nanomaterials for bio sensing applications: A review. *Frontier in Chemistry, 2,* 63.
64. Hong, S., Miller, M. D., & Bruening, M. L., (2006). Removal of dyes, sugars, and amino acids from NaCl solutions using multilayer polyelectrolyte nanofiltration membranes. *Industrial and Engineering Chemistry Research, 45,* 6284–6288.
65. Horner, S. R., Mace, C. R., Rothberg, L. J., & Miller, B. L., (2006). A proteomic biosensor for enteropathogenic *E. coli. Biosensors and Bioelectronics, 21*(8), 1659–1663.
66. Huang, Q., (2012). *Nanotechnology in the Food, Beverage and Nutraceutical Industries* (p. 280). Sawston, Cambridge.
67. Huang, W., Xu, H., Xue, Y., Huang, R., Deng, H., & Pan, S., (2012). Layer-by-layer immobilization of lysozyme–chitosan–organic rectorite composites on electrospun nanofibrous mats for pork preservation. *Food Research International, 48*(2), 784–791.
68. Hwang, E. T., & Gu, M. B., (2013). Enzyme stabilization by nano/microsized hybrid materials. *Engineering In Life Sciences, 13*(1), 49–61.
69. Inbaraj, B. S., & Chen, B. H., (2016). Nanomaterial-based sensors for detection of foodborne bacterial pathogens and toxins as well as pork adulteration in meat products. *J. Food Drug Anal., 24,* 15–28.
70. Incoronato, A. L., Buonocore, G. G., Conte, A., Lavorgna, M., & Del, N. M. A., (2010). Active systems based on silver montmorillonite nanoparticles embedded into bio-based polymer matrices for packaging applications. *Journal of Food Protection, 73,* 2256–2262.
71. Jafarizadeh-Malmiri, H., Sayyar, Z., Anarjan, N., & Berenjian, A., (2019). *Nanobiotechnology in Food: Concepts, Applications and Perspectives* (p. 420). The Gateway East – Singapore: Springer International Publishing.

72. Jain, K. K., (2003). Nanodiagnostics: Application of nanotechnology in molecular diagnostics. *Expert Rev. Mol. Diagn., 3*(2), 153–161.
73. Jin, T., & Gurtler, J. B., (2011). Inactivation of salmonella in liquid egg albumen by antimicrobial bottle coatings infused with allyl isothiocyanate, nisin and zinc oxide nanoparticles. *Journal of Applied Microbiology, 110*, 704–712.
74. Jones, N., Ray, B., Ranjit, K. T., & Manna, A. C., (2008). Antibacterial activity of ZnO nanoparticle suspensions on a broad spectrum of microorganisms. *FEMS Microbiology Letters, 279*, 71–76.
75. Kang, S., Pinault, M., Pfefferle, L. D., & Elimelech, M., (2007). Single-walled carbon nanotubes exhibit strong antimicrobial activity. *Langmuir, 23*(17), 8670–8673.
76. Klitzing, R. V., & Klitzing, V. R., (2006). Internal structure of polyelectrolyte multilayer assemblies. *Physical Chemistry Chemical Physics: PCCP, 8*(43), 5012–5033.
77. Kotov, N. A., (2003). Layer-by-layer assembly of nanoparticles and nanocolloids: Intermolecular interactions, structure and materials perspective. In: Decher, G., & Schlenoff, J. B., (eds.), *Multilayer Thin Films: Sequential Assembly of Nanocomposite Materials* (pp. 207–243). Weinheim – Germany: Wiley-VCH.
78. Kumar, O., & Rai, T., (2009). Nanotechnology in milk and its products. *Indian Dairyman,* 37–40.
79. Kumar, S., Dilbaghi, N., Barnela, M., Bhanjana, G., & Kumar, R., (2012). Biosensors as novel platforms for detection of food pathogens and allergens. *Bionano Science, 2*, 196–217.
80. Lerner, M. B., Goldsmith, B. R., & Mcmillon, R., (2011). A carbon nanotube immunosensor for salmonella. *AIP Adv., 1*, 042127.
81. Li, H., Li, F., Wang, L., Sheng, J., Xin, Z., Zhao, L., Xiao, H., Zheng, Y., & Hu, Q., (2009). Effect of nano-packing on preservation quality of Chinese jujube (*Ziziphus jujuba* Mill. Var. *Inermis* (Bunge) Rehd). *Food Chemistry, 114*(2), 547–552.
82. Li, X., Li, W., Jiang, Y., Ding, Y., Yun, J., Tang, Y., & Zhang, P., (2011). Effect of nano-Zno-coated active packaging on quality of fresh-cut 'fuji' apple. *Int. J. Food Sci. Technol., 46*, 1947–1955.
83. Li, Y., Tseng, Y. D., Kwon, S. Y., Despaux, L., Bunch, J. S., & Mceuen, P. L., (2004). Controlled assembly of dendrimer-like DNA. *Nat. Mater., 3*, 38–42.
84. Liau, S. Y., Read, D. C., Pugh, W. J., Furr, J. R., & Russell, A. D., (1997). Interaction of silver nitrate with readily identifiable groups: Relationship to the antibacterial action of silver ions. *Lett. Appl. Microbiol., 25*, 279–283.
85. Lin, S., Xu, M., Zhang, W., Hu, X., & Lin, K., (2017). Quantitative effects of amination degree on the magnetic iron oxide nanoparticles (MIONPs) using as adsorbents to remove aqueous heavy metal ions. *J. Hazard Mater., 335*, 47–55.
86. Lin, Y. H., Chen, S. H., Chuang, Y. C., Lu, Y. C., Shen, T. Y., Chang, C. A., & Lin, C. S., (2008). Disposable amperometric immunosensing strips fabricated by Au nanoparticles-modified screen-printed carbon electrodes for the detection of foodborne pathogen *Escherichia coli* O157:H7. *Biosens. Bioelectron, 23*(12), 1832–1837.
87. Lopez-Rubio, A., Gavara, R., & Lagaron, J. M. (2006). Bioactive packaging: Turning foods into healthier foods through biomaterials. *Trends in Food Science and Technology, 17*(10), 567–575.
88. Lu, W. C., Huang, D. W., Wang, C. C., Yeh, C. H., Tsai, J. C., Huang, Y. T., & Li, P. H., (2018). Preparation, characterization, and antimicrobial activity of nanoemulsions incorporating citral essential oil. *Journal of Food and Drug Analysis, 26*(1), 82–89.

89. Manke, A., Wang, L., & Rojanasakul, Y., (2013). Mechanisms of nanoparticle-induced oxidative stress and toxicity. *Biomed Res. Int., 15*, 942916.
90. Mannino, S., & Scampicchio, M., (2007). Nanotechnology and food quality control. *Veterinary Research Communications, 31*(1), 149–151.
91. Mao, X., Huang, J., Leung, M. F., Du, Z., Ma, L., Huang, Z., Li, P., & Gu, L., (2006). Novel core-shell nanoparticles and their application in high-capacity immobilization of enzymes. *Applied Biochemistry and Biotechnology, 135*(3), 229–239.
92. Mcmurray, T. A., Dunlop, P. S. M., & Byrne, J. A., (2006). The photocatalytic degradation of atrazine on nanoparticulate $TiO_2$ films. *J. Photochem. Photobiol. A, 182*, 43–51.
93. Medeiros, B. G. D. S., Pinheiro, A. C., Carneiro-Da-Cunha, M. G., & Vicente, A. A., (2012). Development and characterization of a nano multilayer coating of pectin and chitosan–evaluation of its gas barrier properties and application on 'Tommy Atkins' mangoes. *Journal of Food Engineering, 110*(3), 457–464.
94. Medeiros, B. G., De, S., Pinheiro, A. C., Teixeira, J. A., Vicente, A. A., & Carneiro-Da-Cunha, M. G., (2012). Polysaccharide/protein nanomultilayer coatings: Construction, characterization and evaluation of their effect on "Rocha" pear (*Pyrus communis* L.) shelf-life. *Food and Bioprocess Technology, 5*(6), 2435–2445.
95. Medeiros, B. G. D. S., Souza, M. P., Pinheiro, A. C., Bourbon, A. I., Cerqueira, M. A., Vicente, A. A., & Carneiro-Da-Cunha, M. G., (2014). Physical characterization of an alginate/lysozyme nano-laminate coating and its evaluation on 'coalho' cheese shelf life. *Food and Bioprocess Technology, 7*(4), 1088–1098.
96. Meetoo, D. D., (2011). Nanotechnology and the food sector: From the farm to the table. *Emirates Journal of Food and Agriculture, 23*(5), 387–407.
97. Mihindukulasuriya, S. D. F., & Lim, L. T., (2014). Nanotechnology development in food packaging: A review. *Trends Food Sci. Technology, 40*, 149–167.
98. Miller, G., (2008). *Nanotechnology- the New Threat to Food.* Global Research; https://www.globalresearch.ca/nanotechnology-the-new-threat-to-food/10755 (accessed on 11 February 2022).
99. Momin, J. K., Jayakumar, C., & Prajapati, J. B., (2013). Potential of nanotechnology in functional foods. *Emirates J. Food Agricult, 25*(1), 10–19.
100. Moraru, I., Panchapakesan, C. P., Huang, Q., Takhistov, P., Liu, S., & Kokini, J. L., (2003). Nanotechnology: A new frontier in food science. *Food Technology, 57*(12), 24–29.
101. Mousavi, S. R., & Rezaei, M., (2011). Nanotechnology in agriculture and food production. *Journal of Applied Environmental and Biological Science, 1*(10), 414–419.
102. Müller, K., Bugnicourt, E., Latorre, M., Jorda, M., Echegoyen, S. Y., Lagaron, J. M., Miesbauer, O., et al., (2017). Review on the processing and properties of polymer nanocomposites and nanocoatings and their applications in the packaging, automotive and solar energy fields. *Nanomaterials, 7*(4), 74.
103. Nam, J. M., Thaxton, C. S., & Mirkin, C. A., (2003). Nanoparticle-based bio-bar codes for the ultrasensitive detection of proteins. *Science, 301*(5641), 1884–1886.
104. NAS, (2006). *A Matter of Size: Triennial Review of the National Nanotechnology Initiative*. National Academy of Sciences, Washington DC; National Academies Press.
105. Neethirajan, S., & Jayas, D. S., (2011). Nanotechnology for the food and bioprocessing industries. *Food and Bioprocess Technology, 4*(1), 39–47.
106. Neo, Y. P., Ray, S., Jin, J., Gizdavic-Nikolaidis, M., Nieuwoudt, M. K., Liu, D., & Quek, S. Y., (2013). Encapsulation of food grade antioxidant in natural biopolymer by

electrospinning technique: A physicochemical study based on zein–gallic acid system. *Food Chemistry, 136*(2), 1013–1021.
107. Oh, S. Y., Heo, N. S., Shukla, S., Cho, H. J., Vilian, A. E., Kim, J., Lee, S. Y., et al., (2017). Development Of gold nanoparticle-aptamer-based LSPR sensing chips for the rapid detection of *Salmonella typhimurium* in pork meat. *Scientific Reports, 7*(1), 1–10.
108. Pal, M., (2017). Nanotechnology: A new approach in food packaging. *J. Food E Microbiol. Safety Hyg., 2*, 121.
109. Pandey, S., Zaidib, M. G. H., & Gururani, S. K., (2013). Recent developments in clay-polymer nano composites. *Scientific Journal of Review, 2*(11), 296–328.
110. Pilolli, R., Monaci, L., & Visconti, A., (2013). Advances in biosensor development based on integrating nanotechnology and applied to food-allergen management. *Trends Anal. Chem., 47*, 12–26.
111. Pradhan, N., Singh, S., Ojha, N., Shrivastava, A., Barla, A., Rai, V., & Bose, S., (2015). Facets of nanotechnology as seen in food processing, packaging, and preservation industry. *Biomed Research International, 2015.*
112. Prakash, B., Kujur, A., Yadav, A., Kumar, A., Singh, P. P., & Dubey, N. K., (2018). Nanoencapsulation: An efficient technology to boost the antimicrobial potential of plant essential oils in food system. *Food Control, 89*, 1–11.
113. Puddu, M., Paunescu, D., Stark, W. J., & Grass, R. N., (2014). Magnetically recoverable, thermostable, hydrophobic DNA/silica encapsulates and their application as invisible oil tags. *ACS Nano, 8*, 2677–2685.
114. Rashidi, L., & Khosravi-Darani, K., (2011). The applications of nanotechnology in the food industry. *Crit. Rev. Food Sci. Nutr., 51*(8), 723–730.
115. Ravichandran, M., Hettiarachchy, N. S., Ganesh, V., Ricke, S. C., & Singh, S., (2011). Enhancement of antimicrobial activities of naturally occurring phenolic compounds by nanoscale delivery against *Listeria monocytogenes*, *Escherichia coli* O157: H7 and *Salmonella typhimurium* in broth and chicken meat system. *J. Food Saf., 31*(4), 462–471.
116. Ravichandran, R., & Kala, P. S., (2006). Nanoscience and nanotechnology: Perspectives and overview. *School Sci.*, 43–49.
117. Raynes, J. K., Carver, J. A., Gras, S. L., & Gerrard, J. A., (2014). Protein nanostructures in food–should we be worried?. *Trends Food Sci. Technol., 37*, 42–50.
118. Restuccia, D., Spizzirri, U. G., Parisi, O. I., Cirillo, G., Curcio, M., Iemma, F., Puoci, F., et al., (2010). New EU regulation aspects and global market of active and intelligent packaging for food industry applications. *Food Control, 21*, 1425–1435.
119. Rhim, J. W., Hong, S. I., Park, H. M., & Ng, P. K. W., (2006). Preparation and characterization of chitosan-based nanocomposite films with antimicrobial activity. *Journal of Agricultural and Food Chemistry, 54*, 5814–5822.
120. Ribeiro, H. S., Chu, B. S., Ichikawa, S., & Nakajima, M., (2008). Preparation of nanodispersions containing β-carotene by solvent displacement method. *Food Hydrocolloids, 22*(1), 12–17.
121. Robinson, D. K. R., & Morrison, M. J., (2010). *Nanotechnologies for Food Packaging: Reporting the Science and Technology Research Trends: Report for the Observatory NANO* (pp. 1–20).
122. Rojas-Grau, M. A., Soliva-Fortuny, R., & Martin-Belloso, O., (2009). Edible coatings to incorporate active ingredients to fresh-cut fruits: A review. *Trends Food Sci. Technol., 20*, 438–447.

123. Salehi, F., Razavi, S. M. A., & Elahi, M., (2011). Purifying anion exchange resin regeneration effluent using polyamide nanofiltration membrane. *Desalination, 278*, 31–35.
124. Sawada, S. I., & Akiyoshi, K., (2010). Nano-encapsulation of lipase by self-assembled nanogels: Induction of high enzyme activity and thermal stabilization. *Macromolecular Bioscience, 10*(4), 353–358.
125. Semo, E., Kesselman, E., Danino, D., & Livney, Y. D., (2007). Casein micelle as a natural nano-capsular vehicle for nutraceuticals. *Food Hydrocolloids, 21*(5, 6), 936–942.
126. Shukla, S., Lee, G., Song, X., Park, S., & Kim, M., (2016). Immunoliposome based immunomagnetic concentration and separation assay for rapid detection of cronobacter sakazakii. *Biosens Bioelectron, 77*, 986–994.
127. Shukla, S., Leem, H., & Kim, M., (2011). Development of a liposome based immunochromatographic strip assay for the detection of *Salmonella. Anal. Bioanal. Chem., 401*, 2581–2590.
128. Silvestre, C., Duraccio, D., & Cimmino, S., (2011). Food packaging based on polymer nanomaterials. *Prog. Polym. Sci., 36*(12), 1766–1782.
129. Smolander, M., & Chaudhry, Q., (2010). Nanotechnologies in food packaging. Chapter 6. In: Chaudhry, Q., Castle, L., & Watkins, R., (eds.), *Nanotechnologies in Food* (pp. 86–101). London – UK: RSC Publishing.
130. Solanki, P. R., Kaushik, A., Chavhan, P. M., Maheshwari, S. N., & Malhotra, B. D., (2009). Nanostructured zirconium oxide based genosensor for *Escherichia coli* detection. *Electrochem. Commun., 11*(12), 2272–2277.
131. Sonawane, S. S., Rathod, A. P., Wasewar, K. L., & Varma, M. N., (2011). Nanocomposites for food packaging applications. *Res. J. Chem. Environ., 15*(2), 721–723.
132. Sotelo-Boyás, M., Correa-Pacheco, Z., Bautista-Baños, S., & Gómez, Y. G. Y., (2017). Release study and inhibitory activity of thyme essential oil-loaded chitosan nanoparticles and nanocapsules against foodborne bacteria. *Int. J. Biol. Macromol., 103*, 409–414.
133. Souza, M. P., Vaz, A. F. M., Cerqueira, M. A., Texeira, J. A., Vicente, A. A., & Carneiro-Da-Cunha, M. G., (2015). Effect of an edible nanomultilayer coating by electrostatic self-assembly on the shelf life of fresh-cut mangoes. *Food and Bioprocess Technology, 8*(3), 647–654.
134. Sozer, N., & Kokini, J. L., (2009). Nanotechnology and its applications in the food sector. *Trends in Biotechnology, 27*(2), 82–89.
135. Stanić, V., Dimitrijević, S., Antić-Stanković, J., Mitrić, M., Jokić, B., Plećaš, I. B., & Raičević, S., (2010). Synthesis, characterization and antimicrobial activity of copper and zinc-doped hydroxyapatite nanopowders. *Applied Surface Science, 256*(20), 6083–6089.
136. Stucky, G. D., (1997). Oral presentation. *WTEC Workshop on R&D Status and Trends in Nanoparticles, Nanostructured Materials, and Nanodevices in the United States.* Rosslyn – VA.
137. Su, H. C., Zhang, M., Bosze, W., Lim, J. H., & Myung, N. V., (2013). Metal nanoparticles and DNA co-functionalized single-walled carbon nanotube gas sensors. *Nanotechnology, 24*(50), 505502.
138. Su, S. L., & Li, Y., (2004). Quantum dot biolabeling coupled with immunomagnetic separation for detection of *Escherichia coli* O157:H7. *Anal. Chem., 76*(16), 4806–4810.
139. Swaroop, C., & Shukla, M., (2018). Nano-magnesium oxide reinforced polylactic acid biofilms for food packaging applications. *Int. J. Biol. Macromol., 113*, 729–736.

140. Taylor, A., Mcclements, D., & Decker, E., (2009). Measurement and simulation of flavor release from foods. In: McClements, D. J., & Decker, E., (eds.), *Designing Functional Foods: Measuring and Controlling Food Structure Breakdown and Nutrient Absorption* (pp. 294–313). Cambridge – UK: Woodhead Publishing.
141. Vasile, C., (2018). Polymeric nanocomposites and nanocoatings for food packaging: A review. *Materials, 11*(10), 18–34.
142. Verran, J., Packer, A., Kelly, P., & Whitehead, K. A., (2010). Titanium coating of stainless steel as an aid to improved cleanability. *International Journal of Food Microbiology, 141*, 134–139.
143. Vezzani, D., & Bandini, S., (2002). Donnan equilibrium and dielectric exclusion for characterization of nanofiltration membranes. *Desalination, 149*(1–3), 477–483.
144. Vidhyalakshmi, R., Bhakyaraj, R., & Subhasree, R. S., (2009). Encapsulation 'the future of probiotics'—A review. *Advances in Biological Research, 3*(3, 4), 96–103.
145. Vikesland, P. J., & Wigginton, K. R., (2010). Nanomaterial enabled biosensors for pathogen monitoring a review. *Environ. Sci. Technol., 10*, 3656–3669.
146. Viswanathan, S., Wu, L. C., Huang, M. R., & Ho, J. A., (2006). Electrochemical immunosensor for cholera toxin using liposomes and poly(3,4-ethylene dioxythiophene)-coated carbon nanotubes. *Anal. Chem., 78*(4), 1115–1121.
147. Wanekaya, A. K., Chen, W., Myung, N. V., & Mulchandani, A., (2006). Nanowire-based electrochemical biosensors. *Electroanalysis, 18*(6), 533–550.
148. Wang, K., Jin, P., Shang, H., Li, H., Xu, F., Hu, Q., & Zheng, Y., (2010). A combination of hot air treatment and nano-packing reduces fruit decay and maintains quality in postharvest Chinese bayberries. *Journal of the Science of Food and Agriculture, 90*(14), 2427–2432.
149. Weiss, J., Takhistov, P., & Mcclements, D. J., (2006). Functional materials in food nanotechnology. scientific summary. *J. Food Sci., 71*(9), R107–R116.
150. Wesley, S. J., Raja, P., Sunder, R. A. A., & Tiroutchelvamae, D., (2014). Review on nanotechnology applications in food packaging and safety. *Int. J. Eng. Res., 3*, 645–651.
151. Wu, H., Yin, J. J., Wamer, W. G., Zeng, M., & Lo, Y. M., (2014). Reactive oxygen species-related activities of nano-iron metal and nano-iron oxides. *J. Food Drug Anal., 22*, 86–94.
152. Yakes, B. J., Lipert, R. J., Bannantine, J. P., & Porter, M. D., (2008). Detection of *Mycobacterium avium* subsp. *paratuberculosis* by a sonicate immunoassay based on surface-enhanced Raman scattering. *Clin. Vaccine Immunol., 15*(2), 227–234.
153. Yang, F. M., Li, H. M., Li, F., Xin, Z. H., Zhao, L. Y., Zheng, Y. H., & Hu, Q. H., (2010). Effect of nano-packing on preservation quality of fresh strawberry (*Fragaria ananassa* duch. CV *Feng Xiang*) during storage at 4°C. *Journal of Food Science, 75*(3), C236–C240.
154. Yang, H., Qu, L., Wimbrow, A. N., Jiang, X., & Sun, Y., (2007). Rapid detection of *Listeria monocytogenes* by nanoparticle-based immunomagnetic separation and real-time PCR. *Int. J. Food Microbiol., 118*(2), 132–138.
155. Yang, J. Y., Li, Y., Chen, S. M., & Lin, K. C., (2011). Fabrication of a cholesterol biosensor based on cholesterol oxidase and multiwall carbon nanotube hybrid composites. *Int. J. Electrochem. Sci., 6*, 2223 2234.
156. Yang, L., & Li, Y., (2006). Quantum dot bioconjugates for simultaneous detection of *Escherichia coli* O157:H7 and *Salmonella typhimurium*. *Analyst, 131*, 394–401.

157. Yang, L., & Li, Y., (2005). Quantum dots as fluorescent labels for quantitative detection of *Salmonella typhimurium* in chicken carcass wash water. *J. Food Prot., 68*(6), 1241–1245.
158. Yang, M., Peng, Z., Ning, Y., Chen, Y., Zhou, Q., & Deng, L., (2013). Highly specific and cost-efficient detection of *Salmonella paratyphi* combining aptamers with single-walled carbon nanotubes. *Sensors, 13,* 6865–6881.
159. Yotova, L., Yaneva, S., & Marinkova, D., (2013). Biomimetic nanosensors for determination of toxic compounds in food and agricultural products (review). *Journal of Chemical Technology and Metallurgy, 48*(3), 215–227.
160. Yuan, Y., Gao, Y., Zhao, J., & Mao, L., (2008). Characterization and stability evaluation of β-carotene nanoemulsions prepared by high pressure homogenization under various emulsifying conditions. *Food Research International, 41*(1), 61–68.
161. Zambrano-Zaragoza, M. L., Mercado-Silva, E., Gutierrez- Cortez, E., Cornejo-Villegas, M. A., & Quintanar-Guerrero, D., (2014). The effect of nanocoatings with a-tocopherol and xanthan gum on shelf-life and browning index of fresh-cut "red delicious" apples. *Innov. Food Sci. Emerg. Technol., 22,* 188–196.
162. Zhang, Y., Wu, Q., Sun, M., Zhang, J., Mo, S., Wang, J., Wei, X., & Bai, J., (2018). Magnetic-assisted aptamer-based fluorescent assay for allergen detection in food matrix. *Sensors and Actuators B: Chemical, 263,* 43–49.
163. Zhao, R., Torley, P., & Halley, P. J., (2008). Emerging biodegradable materials: Starch- and protein-based bio-nanocomposites. *Journal of Materials Science, 43*(9), 3058–3071.
164. Zhao, X., Hilliard, L., Mechery, S., Wang, Y., Bagwe, R., & Jin, S., (2004). A rapid bioassay for single bacterial cell quantitation using bioconjugated nanoparticles. *Proc. Natl. Acad. Sci., 42,* 15027–15032.
165. Zhou, L., Lv, S., He, G., He, Q., & Shi, B. I., (2011). Effect of PE/AG2O nanopackaging on the quality of apple slices. *J. Food Qual., 34*(3), 171–176.
166. Zuo, P., Li, X., Dominguez, D. C., & Ye, B. C., (2013). A PDMS/paper/glass hybrid microfluidic biochip integrated with aptamer functionalized graphene oxide nano-biosensors for one-step multiplexed pathogen detection. *Lab Chip, 13,* 3921–3928.

# CHAPTER 2

# EMERGING NANOTECHNOLOGY FOR FOOD PRESERVATION

SHRUTHY RAMESH and PREETHA RADHAKRISHNAN

## ABSTRACT

Nowadays, nanotechnology plays a major role in the food industry. During food processing as well as storage, the degradation or inactivation of the bioactive compounds in food is a common issue. The above-said issues can be overcome to a certain extent by introducing nanoencapsulation techniques, including preparation of nanoemulsion. Similarly, nanocomposite-based packaging technology is another emerging area in food preservation and storage. This chapter summarizes the importance of the nanotechnology in the food industry, nanotechnology for food processing, various nanotechnology-based techniques for food preservation, the role of nano-based packaging material for food storage/preservation, applications of nanotechnology in food preservation, safety issues of nano-based techniques for food preservation, and future aspects of nanotechnology for food preservation.

## 2.1 INTRODUCTION

The term 'nanotechnology' means the process or technology which can make nanostructural/nanoscale products ranging in size from 1 to 100 nm [68]. Nanotechnology has applications in biomedicine, agriculture, cosmetics, food, clothing, and many more [48]. In food processing sector, nanotechnology has a prominent role mainly in food preservation [25]. In addition to that nanotechnology plays an important role in packaging, designing of sensors and encapsulation techniques in food sector [9]. In food

Nanotechnology Horizons in Food Process Engineering, Volume 1. Megh R. Goyal, PhD, PE, Junaid Ahmad Malik, PhD, Satish Kumar, PhD & Ritesh B. Watharkar, PhD (Eds.)

industry, nanotechnology has applied for achieving a specific or desired property by incorporating NMs to particular food products [56]. Various studies highlighted to assure food safety with the help of nanotechnology. For example, nano-biocomposites as food packaging show enhanced shelf life and antimicrobial properties, nano additives improved the nutritional value without affecting the taste or quality and so on [19, 27, 33, 60]. The production and processing of nanostructured materials in the food industry is facing challenges, especially in its cost-effectiveness and safety issues [1]. The authorities assisting or evaluating the regulation and guidelines for the safe use of nano-based materials in food are EPA, FDA, and Directorate of European Health and Consumer Protection [27].

This chapter summarizes the importance of the nanotechnology in the food industry, nano technological innovations for food processing, nanotechnology-based food preservation techniques, nano-based packaging material for food preservation, safety issues of nano-based techniques for food preservation, applications of nanotechnology in food preservation and future aspects of nanotechnology for food preservation.

## 2.2 IMPORTANCE OF NANOTECHNOLOGY IN FOOD INDUSTRY

Nanotechnology influenced various areas in food sectors such as packaging, food preservation, and food additives [9] (Figure 2.1). By considering consumer preferences, these aspects will ensure/provide the quality and safety of the food products [27]. The commonly accepted nanomaterials (NMs) used in the food industry include organic, inorganic, and combination of both. Organic nanoparticles (ONP), such as metals and metal oxides and inorganic, refer to the nanoparticles from the natural products (directly from the earth). An example of an organic and inorganic combination is the mix of metal nanoparticles and clay nanoparticles [50]. Gold nanoparticles (AuNPs) and silver nanoparticles (AgNPs) are commonly used nanoparticles; whereas, AgNPs can be prominently used in the food industry due to their antimicrobial property [50]. In the food industry, the use of food-grade titanium dioxide ($TiO_2$) nanoparticles is notable/common as a food additive, disinfectant agent, and a flavor enhancer [50].

Nanotechnology has been applied for the improvement of the flow properties, nutritional quality, color, flavor, and shelf life of process food. Indeed, nanotechnology can help to produce food items that have lower fat, salt, and sugar to conquer lifestyle diseases. Food additives such as $SiO_2$ and $TiO_2$

of limited/required amount have been permitted recently [27]. In an early report, it was mentioned that during hydrolysis of olive oil, for dynamic stability, adaptation, and reuse, the covalently immobilized porcine triacylglycerol lipase and $SiO_2$ nanoparticles along with reactive aldehyde group was added [8]. Nano/microstructure assemblies of nanoparticles have been designed for the encapsulation of food ingredients, functional food additives and nutritional supplements [6]. The application of nanosized additives can enhance taste, absorption, stability of antimicrobials and antioxidants, and the bioavailability of nutrients such as vitamins [43].

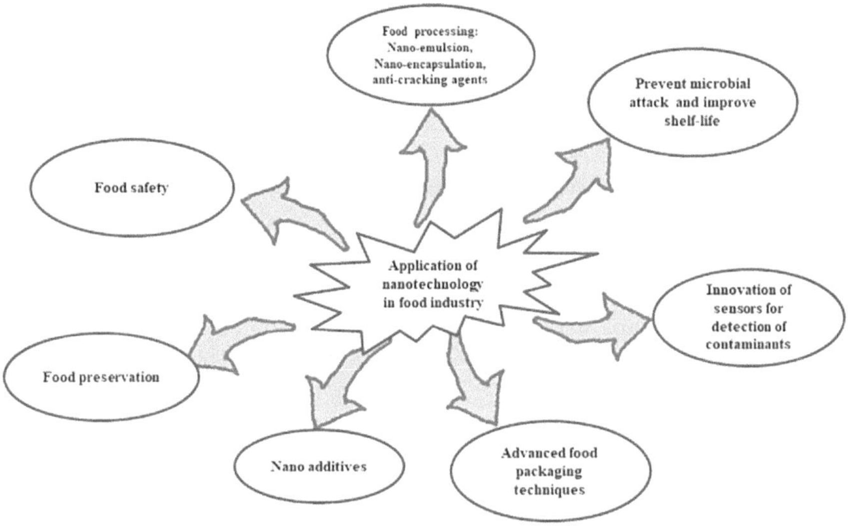


**FIGURE 2.1** Application of nanotechnology in the food industry.

The encapsulation using nanomaterial is more efficient than conventional methods. The main benefits of nanoencapsulation are the enhancement of stability, ease of handling, volatile ingredient retention, enhanced solidity, controlled delivery of active ingredients, taste/flavor locking or masking ability, pH triggered controlled release, enhanced efficiency, bio-availability, and long-lasting organoleptic perception [15].

In a previous study, it was described that the encapsulation of desired active or volatile compounds and the masking of the unwanted compounds and odors of the food sample results in extended shelf-life/food preservation [46]. The basic six steps of preparing nano-capsules are emulsification, precipitation, emulsion-diffusion method, and polymer coating [40]. One of the examples of nanoencapsulation is the incorporation of zein fiber with gallic acid with the help of electrospinning and the study claims this as a novel method of encapsulation [46]. Lipids can be protected by zein fiber-based nanoencapsulation technique and help to prevent the lipid from degradation and also helps to perform the targeted delivery. Above said new technique of nanoencapsulation was performed in food packaging industry [64].

The difference between the normal emulsion and nanoemulsion is that the latter does not make any appearance change of the food samples [46]. The preservation of the milk was reported by nanoencapsulation technique using α-tocopherol in fat droplets [46]. Some of the commercially available nanoencapsulated products are canola, active oil with fortified phytosterols, fortified fruit juice with encapsulated vitamins iron, threonine, and lycopene (High Vive, USA) [3].

Nowadays, nanotechnology plays a vital role in improving the appearance, texture, and taste of food. Nanoencapsulation is mainly used for enhancing the taste of the food. The nanoencapsulation was also reported for the preparation of the reactive plant pigments. Arroyo et al. [4] explained improved photostability and thermal stability of cyanide-3-O glucoside from soybean by nanoencapsulation technique. The bioactive complexes of proteins, lipids, carbohydrates, and vitamins cannot resist under acidic conditions and enzymatic action (inside stomach and duodenum), encapsulation technique can overcome the above-said problem. The nanoparticle coated edible capsules can strengthen the delivery system of vitamins and micronutrients [4]. Moreover, these nanoparticles coated edible capsules can obstacle the gas interchange, act as anti-browning agents, and are also used to prevent the loss of colors, flavor, and antioxidant activity [4]. The nanosized composites and inorganic additives such as calcium, iron, silver, magnesium, silica, and selenium were reported as additives and preservatives for the improvement of taste and flavors [4].

The current food product market is mainly focusing on improved/ enhanced shelf-life food items such as fast food, confectionery products, canned/tinned meat, and vegetables, and so on. Nowadays, the trend in the food industry is edible coated food items. The edible coatings are mainly done for baked products, chocolate, and dairy products such as cheese, butter, and paneer. Moreover, the coated food products provide enhanced flavor, color, antioxidant, and antimicrobial properties to the food items [7]. Furthermore, in the food industry, nanofillers helps to remove the lactose content from milk, removing color without affecting the flavor and other organic content from various juices and wines [45]. Another remarkable property of the nano filter is that it can eliminate bacteria from water and milk, without the boiling process [57].

Nanoemulsions are droplet size in the range of 2–200 nm and having narrow size distributions. The nanoemulsions are mainly used for various applications in food industries, including encapsulation of bioactive compounds, especially as an efficient delivery system of bioactive lipophilic compounds including omega-three fatty acids, vitamins, flavors, preservatives, and nutraceutical [59]. For example, a study describes nanoemulsion – ice cream with a lower fat enhances its sensory and rheological properties [70]. Furthermore, there are many nanotechnology-based food items available commercially. Some of the common food items available are fortified fruit juice, shakes, active oils, health supplements (omega-3, vitamin-E, Lycovit), food additives, and food supplements [28, 59].

## 2.3 NANOTECHNOLOGY IN FOOD PROCESSING

Nanotechnology plays an important role in food processing. The application and advantages of nanotechnology during food processing are shown in Table 2.1. These engineered NMs show unique properties, thereby improving food quality also. Inorganic metal nanoparticles such as copper, iron, and zinc oxides (ZnOs) are generally recognized as safe (GRAS) by FDA as dietary supplements for animal feed [23].

**TABLE 2.1** Applications and Advantages of Nanotechnology in Food Processing

| Technique | Applications | Advantages | References |
|---|---|---|---|
| Nano additives | Nano-sized particular additives | For improved antibacterial/antioxidant properties. | [65] |
| Nanostructures | Nanosized ingredients (organic/inorganic) | Improved texture and flavor, enhanced bioavailability, increased antibacterial property. | [58] |
| Nanocomposites | Packaging of food, flavor, beverages, and so on. | Improves shelf life, enhances quality, and ensures safety. | [60] |
| | Packaging of fish, meat, and poultry products | Improves antibacterial property, increases the shelf life. | [33] |
| Nanocapsules | Standard ingredient as oil for nutraceutical. | Improved bioavailability. | [59] |

**TABLE 2.1** *(Continued)*

| Technique | Applications | Advantages | References |
|---|---|---|---|
| Nanoemulsion | Proteins, carbohydrates, oil | For maintaining the texture of food, weight gain supplement, reduce sedimentation. | [52] |
| | Delivery of nutrients | Targeted delivery of nutrients, Increased bioavailability of nutrients. | [59] |
| Nanoencapsulation | Boosters | The regular boosters are fortified by bioactive compounds | [59] |
| | Nanotea, slim shakes and so on. | Nano-clusters of products will take place and will enhance the flavor. | [2] |
| | Nanocochlertes | Helps to improve the quality of food. | [2] |
| | Colloidosomes | Delivery of vitamins and minerals in the food | [70] |
| | Nanoliposomes | Odor entrapment, get rid of unwanted compounds in food. | [5] |
| Nanolithography | Nanoparticle based intelligent reactive layer for food packaging. | For tractability and authentication, to prevent adulteration. | [59] |
| Nanosensors | Detection of contamination | Precise detection of contaminants and time consuming. | [63] |
| | Detection of rancidity and off-flavor of dairy products, freshness of meat and fish, aging, and spoilage of wine, spoilage of alcoholic/nonalcoholic beverages | Instant detection, ease of handling, high accuracy. | [73] |

### *2.3.1 NANOENCAPSULATION*

Nanoencapsulation means entrapping inside a protective layer that can preserve bioactive molecules, and it can release bioactive molecules at proper time and condition [5]. The significant property of encapsulation is that it can preserve both the physical and chemical properties of the preferred bioactive compounds. The bioactive compounds commonly used for the encapsulation are antioxidants, vitamins, lipids, and protein [2]. The main advantages of the nanoencapsulation during food processing are the increased surface area, improved solubility, optical transparency, solubilization property, incompatibility preventing, and toxicity reduction. The nanoencapsulation process can be divided into three modes; liquid-liquid nanoencapsulation, solid-liquid nanoencapsulation, and solid-solid nanoencapsulation. The two methods mainly preferred for nanoencapsulation are size reduction approach (top-down approach) and precipitation method (bottom-up process). Figure 2.2 shows the classification and methods of nanoencapsulation.

Micronization is the process which reduces the size of the solid particles. Micronization is linked to nanoencapsulation in the food industry, and it is accomplished using a new method known as a supercritical fluid extractor. High-pressure homogenization and sonication are other approaches for size reduction from microstructure to nano size. During nanoencapsulation, hot/cold high-pressure homogenization is performed for enhancing the properties. The bottom-up technique as precipitation is the new trend nowadays for nanoencapsulated food preservation. Due to precipitation technique the nanoencapsulated product improves its stability and functional properties [2].

Nanoemulsion is a mixture of oil-emulsion-water and continuous phase as water. The particle diameter of the oil droplet will be less than 100 nm. The main advantages of nanoemulsion are kinetically stable; having a higher specific area, optically clear, requires fewer surfactants and less viscous [10]. The commonly used method for nanoemulsion preparation is high-pressure homogenization. During the homogenization process, the mixture of oil-water-surfactant is passing through the high pressure of 10,000–20,000 psi. The process creates turbulence, and cavitation tends to reduce the particle size [30, 49]. Atul and Niraj [5] described that the nanosized emulsion of oil can improve the activities of nisin, lauric arginate, and lysozyme [20]. There are many other methods for nanoencapsulation, including double-layer nanoencapsulation by electrostatic deposition and multi-assembled nanoencapsulation [52, 54].

The nano-emulsions were reported for various applications in food industries, including encapsulation of bioactive compounds, food processing, storage protection, and efficient delivery system of bioactive lipophilic compounds including omega-three fatty acids, vitamins, flavors, preservatives, and nutraceuticals [41]. For example, the nano-size emulsion is based on ice cream with a lower fat [70].

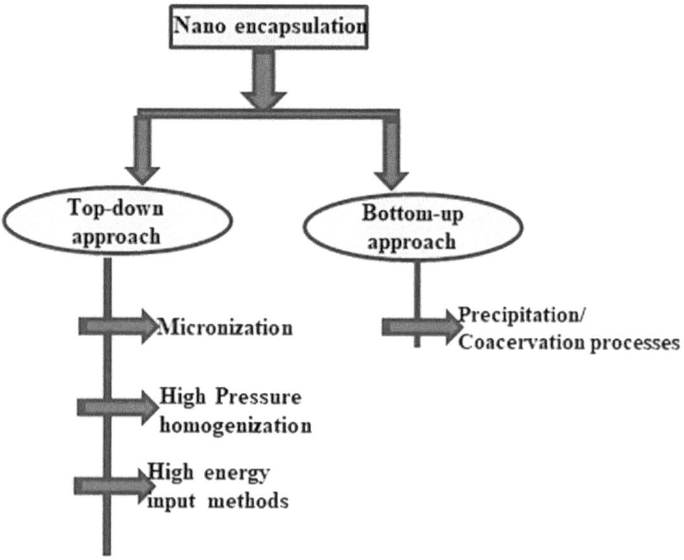


**FIGURE 2.2** Classification of nanoencapsulation technique.

### 2.3.2 OTHER NANOTECHNOLOGY-BASED TECHNIQUES FOR FOOD PRESERVATION

Nanotechnology plays an essential role for the improvement of the quality and taste of food. In addition to those various sensors are used to detect the color change, spoilage, humidity, toxin content, pesticide content, and food-borne pathogens [11, 17, 58, 63, 69, 74]. Metal NMs such as gold, platinum, and palladium are used in sensors for food analysis, especially for the detection of gas production due to the spoilage of food items. Nanosensors can also be used to detect aflatoxins, humidity, color change of food inside the package, and detect chemicals into electrical signals. E-Nose has been made for the identification and detection of various odors and the concentration of odorants. E-nose are commonly used to analyze food samples:

- for the detection of rancidity and off-flavor of dairy products;
- for the detection of freshness and shelf life of fish and red meat;
- for detecting the aging of wine, spoilage in wine and so on. Many studies reported the importance of E-Nose for the screening of contamination due microbes by analyzing the metabolic changes.

The major advantage of E-Nose is its ease of use, instant result, and accuracy in analysis of food items [63]. The ZnO nanowire was reported in E-Nose to detect the flavor and freshness of the food [16].

## 2.4 THE ROLE OF NANO-BASED PACKAGING MATERIAL FOR FOOD PRESERVATION

Nowadays, nanotechnology-based packaging has gained wide attention in the food industry. Nanocomposite-based packaging provides better barrier properties and prevent entry of light, oxygen, UV rays and change moisture content for food items. Nanoparticles are also excellent in strength, durability, elasticity, thermal stability, lighter weight, and flammability reduction [53]. Chaudhary et al. [15] reported nanoparticles-based food packaging materials with an improved antimicrobial property for enhanced shelf life, barrier, and mechanical properties. In addition to that, the nanoparticles were used as carriers for the delivery of flavors, antioxidants, anti-browning agents, enzymes, and other materials to extend the shelf life of the food items [14, 37, 53].

Food packaging using inorganic NMs made up of metal oxides and metals such as silver, $TiO_2$, ZnO, iron, magnesium oxide (MgO), silicon dioxide ($SiO_2$), and carbon nanoparticles are also used as antimicrobial agents [42, 57, 71]. Antimicrobial property is essential for food packaging composites; $TiO_2$ is widely used as an antimicrobial agent which produces highly reactive oxygen species (ROS), which are toxic to the bacteria. The silver dioxide is also commonly used as an antimicrobial agent for the elimination of *E. coli* and *Bacillus cereus* spores [57]. There are various other nanoparticles-based food packaging reported with antimicrobial activity, for example, MgO [72], $Cu_2O$ [72], ZnO [24] and CS [51] carbon nanotubes were reported for water purification and associated applications [15, 29].

The main advantage of biodegradable packages is its recyclability, and biodegradability [60]. Once the dynamic packages are disposed to the environment, the microorganisms will degrade the packages aerobically and produce carbon dioxide, water, and biomass; thereby, developing fewer environmental problems [33, 34, 39]. Active packaging also will provide better keeping quality [61]. The principle behind active packaging is the incorporation of particular compounds that impart antimicrobial/antioxidant or intrinsic characterization [33]. The limited/focused/controlled delivery of active agents from the packaging material improves the shelf life of the food variety [38]. But the usage of artificial anti-oxidative agents, including butylated hydroxyl toluene, and trysters in active packaging additions are restricted due to the toxicity resulting from their migration into the food products [38]. Hence the usage of synthetic additives is replaced now, and the usage of essential oil from plants, spices, and tocopherols has become a trend [60].

In the case of packaged meat and meat products, the essential oil incorporated packages inhibit the microbial growth responsible for spoilage and thereby improve the keeping quality of food products [61]. Water vapor removal, temperature regulator, oxygen scavenging, ethanol release, and shelf healing are the critical functions of active packaging. Oxygen scavengers utilized as a part of packaging material hinder oxidative responses by expelling oxygen and can be added explicitly to the packaging area. Antimicrobial composites control the growth of spoilage and pathogenic microorganisms, which act as growth inhibitor and killing agent. Ag (Silver) nanoparticles are the commonly used antimicrobial agents in nanocomposites, and they are having large surface area for the interaction with microbial cells, resulting in an improved bacterial effect (Figure 2.3). Silver particles breakdown lipopolysaccharides (LSs) in the bacterial cell wall and penetrate into the cell and finally destruct the microbial DNA [15].

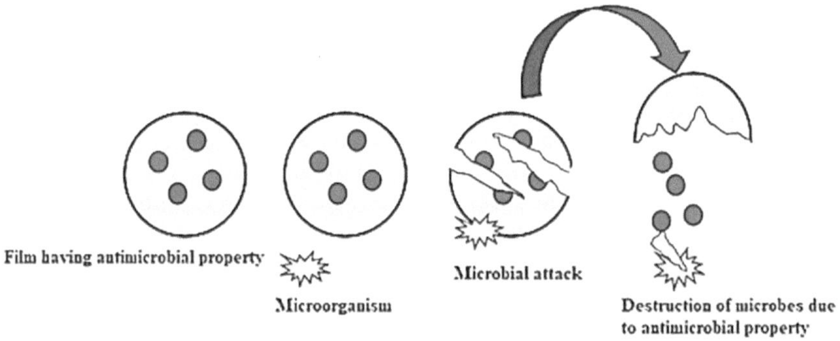


**FIGURE 2.3** Mechanism of the destruction of microorganisms by silver nanoparticle-based composites having antimicrobial property.

$TiO_2$ as an antimicrobial agent in the nanocomposite which inactivates several food-related pathogens by peroxidation of polyunsaturated phospholipids present in the cell wall membrane of the pathogen. $ZnO_2$ is also acting as an antimicrobial agent [66] (Figure 2.4). Silver zeolite is another important material in the food industry. Its antimicrobial activity is mainly due to ROS production by silver, which results in cell death [32, 35]. Compared to silver nanocomposite and zeolite-based material, the silver NMs show better antimicrobial activity for the long-term packaging of food [32].

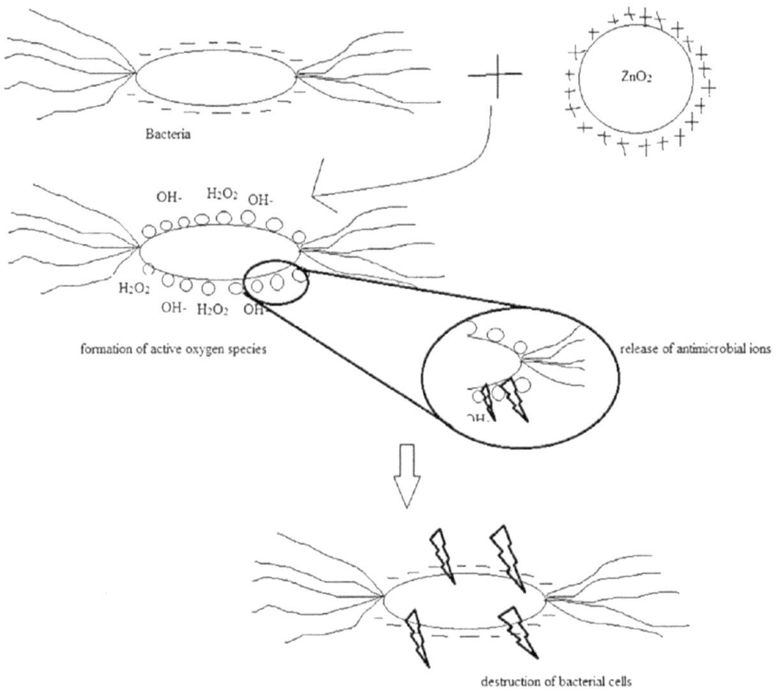


**FIGURE 2.4** Destruction of bacteria by $ZnO_2$ as antimicrobial agent in active packaging.

Another fastest developing field is nanocomposite-based packaging, which is a better option for traditional packaging [65]. Over the most recent decades, the utilization of synthetic polymers for packaging has expanded because they are simple to process, adaptable, cheap, and lightweight [62]. The traditional plastic polymers are not biodegradable, so they cause fundamental environmental issues. At the same time, biodegradable packages are ecofriendly, and they enhance the shelf life of the food items. Reinforcement of biopolymer provides better mechanical and physical property for the packaging material [31, 60].

Nanocomposites are having a significant role in food preservation. They are made up of a combination of nanoparticles and polymers. Many studies described that the nanocomposites are also involved in food preservation [33, 61]. The main advantage of nanocomposites is their versatility towards the chemical functionality, improved barrier properties (thermal properties), enhanced antimicrobial properties, and improved gas barrier properties [21]. Nano laminates also has lot of applications in food industry. Nano laminate is having two or more layers of nanomaterial and all chemically and physically bonded [22]. Nano lamination can be incorporated with anti-browning agents, antioxidants, enzymes, and antimicrobials and thereby improve the shelf life/quality of food [21].

## 2.5 SAFETY ISSUES OF NANO-BASED TECHNIQUES FOR FOOD PRESERVATION

The unique property of nanomaterial is the surface area which makes chemically active than the bulk materials. So, they can participate in biological reactions and leads to harmful side effects on humans and the environment [36]. The nanoparticle's size is another impact because of its absorption ability through the dermal-cell; it crosses through the biological barriers, and this leads the depositions in tissues, mainly in the central nervous system, leading to disorders in humans. Recently, the studies show $TiO_2$ nanoparticles promoted "tumor-like" progressions in human beings [26].

The nanosized particles, such as metal oxides and others, can be translocated into the food items and cause health defects. Those are highly reactive substances and can easily cross the membrane and capillaries, resulting in toxicity [65, 67]. Various regulatory authorities, such as the Food Standards of Australia and New Zealand (FSANZ), USFDA (United States Food and Drug Administration), control the regulations and usage of the nano additives and related products in food and agriculture [12]. The size, surface area, the dose and structure of the nanoparticle are also significant in its toxicity. Preliminary investigations indicated the increased use leads to poisonous quality to liver, kidneys. Therefore, nanotoxicity studies play an important role in providing clear guidelines and roadmaps for reducing the risks in the optimal use of NMs [47, 55].

## 2.6 FUTURE ASPECTS OF NANOTECHNOLOGY FOR FOOD PRESERVATION

Nanosystem for food preservation is still a developing area of research. Researchers are trying to develop many efficient methods for improved bioavailability, taste, and appearances [59]. The most important aspect of the nanotechnology in food processing is the development of new food textures with the help of immobilizing enzymes, aggregated, and nano fibrillated protein [13, 18]. The new trends and forthcoming trends of nanotechnology in food processing are smart feed for animals, targeted delivery of nutrients as supplements and nano-bio labels for tracking the information of food [44].

## 2.7 SUMMARY

The nanotechnology has a lot of scope in the field of food processing and preservation. It can be used for the enhancement of flavor and texture, encapsulation of various active volatile compounds, improvement of shelf life of food by incorporating nanomaterials in active packaging, nanoemulsion for stability of bioactive compounds in functional foods, and so on. Many food industries are promoting nanotechnology for food preservation. But, the awareness and the designing of the novel food preservation technique based on non-toxic and eco-friendly material is mandatory, which makes it more attractive for consumers. Rather than using chemical substitutes, the designing/discovery of natural and renewable resources as nanosystems for food preservation can make wonders, as well as these findings, can assure healthy, safe, and quality food preservation technology. The efficient use of nano-based technology for food preservation provides consumer safety and superior quality food.

## ACKNOWLEDGMENTS

The authors acknowledge the School of Bioengineering, SRM Institute of Science and Technology (SRMIST). They express thanks to Prof. C. Muthamizchelvan, Pro-Vice-Chancellor (Engineering and Technology), and Dr. M. Vairamani, Dean, School of Bioengineering.

## KEYWORDS

- **bionanocomposites**
- **food preservation**
- **generally recognized as safe**
- **nanoemulsion**
- **nanoencapsulation**
- **reactive oxygen species**

## REFERENCES


1. Adabi, M., Naghibzadeh, M., Adabi, M., Zarrinfard, M. A., Esnaashari, S. S., Seifalian, A. M., Faridi-Majidi, R., et al., (2017). Biocompatibility and nanostructured materials: Applications in nanomedicine. *Artificial Cells, Nanomedicine, and Biotechnology, 45*(4), 833–842.
2. Alejandro, P., Asensio, C. M., Labot, J. M., Allemandi, D. A., & Palma, S. D., (2016). Nanoencapsulation in the food industry: Manufacture, application and characterization. *Journal of Food Bioengineering and Nanoprocessing, 1*(1), 56–79.
3. Alfadul, S. M., & Elneshwy, A. A., (2010). Use of nanotechnology in food processing, packaging and safety: Review. *African Journal of Food, Agriculture, Nutrition and Development, 10*(6), 6–24.
4. Arroyo, B. J., Santos, A. P., De Melo, E. A., Campos, A., Lins, L., & Boyano-Orozco, L. C., (2019). Bioactive compounds and their potential use as ingredients for food and its application in food packaging. Chapter 8. In: Campos, M. R. S., (ed.) *Bioactive Compounds* (pp. 143–156). Amsterdam – Netherlands: Elsevier Inc.
5. Atul, R. K., & Niraj, V., (2014). Nanoencapsulation in the food industry: Technology of the future. In: Gaonkar, A. G., Vasisht, N., Khare, A. R., & Sobel, R., (eds.), *Microencapsulation in the Food Industry* (pp. 127–142). Illinois – USA: Khare Consulting, Palatine, Bio Delivery Sciences International, Raleigh, North Carolina, USA.
6. Augustin, M. A., & Hemar, Y., (2009). Nano- and micro-structured assemblies for encapsulation of food ingredients. *Chem. Soc. Rev., 38*(4), 902–912.
7. Azeredo, H. M. C., Henrique, C. L. M., Wood, D., Williams, T. G., Roberto, J. A. B., & Tara, H. M., (2009). Nanocomposite edible films from mango puree reinforced with cellulose nanofibers. *J. Food Sci., 74*, 31–35.
8. Bai, Y. X., Li, Y. F., Yang, Y., & Yi, L. X., (2006). Covalent immobilization of triacylglycerol lipase onto functionalized nanoscale $SiO_2$ spheres. *Process Biochem., 125*(4), 574–582.
9. Baranwal, A., Srivastava, P., Kumar, V. K., Bajpai, P. K., & Maurya, P. C., (2018). Prospects of nanostructured materials and their composites as antimicrobial agents. *Front Microbiol., 9*, 420–447.
10. Behnam, D., (2004). *Method for Producing an Active Ingredient Concentrate, and an Active Ingredient Concentrate*. US Patent Application 20040081670A1.
11. Biswal, S. K., Nayak, A. K., Parida, U. K., & Nayak, P. L., (2012). Applications of nanotechnology in agriculture and food sciences. *International Journal of Science Innovations and Discoveries, 2*(1), 21–36.
12. Bowman, D. M., & Hodge, G. A., (2006). Nanotechnology: Mapping the wild regulatory frontier. *Futures, 38*, 1060–1073.
13. Cerqueira, M. A., Pinheiro, A. C., Ramos, O. L., Silva, H., Bourbon, A. I., & Vicente, A. A., (2017). Advances in food nanotechnology. In: Busquets, R., (ed.), *Emerging Nanotechnologies in Food Science: Micro and Nano Technologies* (pp. 11–38). Boston – Massachusetts: Elsevier Inc.
14. Cha, D., & Chinnan, M., (2004). Biopolymer-based antimicrobial packaging: A review. *Crit. Rev. Food Sci. Nutr., 44*, 223 237.
15. Chaudhry, Q., Castle, L., & Watkins, R., (2010). *Nanotechnologies in Food* (p. 181). Royal Society of Chemistry Publishers, Cambridge, UK.

16. Chouhan, R. S., VivekBabu, K., Kumar, M. A., Neeta, N. S., Thakur, M. S., Amitha, R. B. E., Pasha, A., et al., (2006). Detection of methyl parathion using immuno-chemiluminescence based image analysis using charge-coupled device. *Biosens. Bioelectron., 21*, 1264–1271.
17. Coles, D., & Frewer, L. J., (2013). Nanotechnology applied to European food production—A review of ethical and regulatory issues. *Trends in Food Science and Technology, 34*(1), 32–43.
18. Couch, L. M., Wien, M., Brown, J. L., & Davidson, P., (2016). Food nanotechnology: Proposed uses, safety concerns and regulations. *Agro. Food Ind. Hitech, 27,* 36–39.
19. Das, M. P., Livingstone, J. R., Veluswamy, P., & Das, J., (2017). Exploration of we Delia chinensis leaf-assisted silver nanoparticles for antioxidant, antibacterial and *in vitro* cytotoxic applications. *J. Food Drug Anal., 26*(4), 917–925.
20. Dasgupta, N., & Ranjan, S., (2018). Nanotechnology in food sector. Chapter 1. In: Dasgupta, N., & Ranjan, S., (eds.), *An Introduction to Food Grade Nanoemulsions* (Vol. 13, pp. 1–18). The Gateway East – Singapore: Springer International Publishers.
21. Dasgupta, N., Ranjan, S., Mundekkad, D., Ramalingam, C., Shanker, R., & Kumar, A., (2015). Nanotechnology in agro-food: From field to plate. *Food Res. Int., 69*, 381–400.
22. Decher, G., & Schlenoff, J. B., (2003). *Multilayer Thin Films: Sequential Assembly of Nanocomposite Materials* (p. 543). Germany; Wiley Blackwell Publishers.
23. Dorier, M., Béal, D., Marie-Desvergne, C., Dubosson, M., Barreau, F., & Houdeau, E., (2017). Continuous *in vitro* exposure of intestinal epithelial cells to E171 food additive causes oxidative stress, inducing oxidation of DNA bases but no endoplasmic reticulum stress. *Nanotoxicology, 11*, 751–761.
24. Emamifar, A., Kadivar, M., Shahidi, M., & Soleimanian-Zad, S., (2011). Effect of nanocomposite packaging containing Ag and ZnO on inactivation of *Lactobacillus plantarum* in orange juice. *Food Control, 22*, 408 413.
25. Fu, P. P., (2014). Introduction to the special issue: Nanomaterials – toxicology and medical applications. *J. Food Drug Anal., 22*, 1, 2.
26. Hajipour, M. J., Fromm, K. M., Ashkarran, A. A., De Aberasturi, D. J., De Larramendi, I. R., Rojo, T., Serpooshan, V., et al., (2012). Antibacterial properties of nanoparticles. *Trends Biotechnol., 30*(1), 599–611.
27. He, X., Aker, W. G., & Hwang, H. M., (2014). An *in vivo* study on the photoenhanced toxicities of S-doped $TiO_2$ nanoparticles to zebrafish embryos (*Danio rerio*) in terms of malformation, mortality, rheotaxis dysfunction, and DNA damage *Nanotoxicology, 8*(1), 185–195.
28. He, X., & Hwang, H. M., (*2016*). Nanotechnology in food science: Functionality, applicability, and safety assessment. *J. Food Drug Anal., 24*(3), 671–681.
29. He, X., Aker, W. G., Pelaez, M., Lin, Y., Dionysiou, D. D., & Hwang, H. M., (2016). Assessment of nitrogenefluorine-codoped $TiO_2$ under visible light for degradation of BPA: Implication for field remediation. *J. Photochem. Photobiol. Chem., 314*(6), 81–92.
30. Huang, J. Y., Li, X., & Zhou, W., (2013). Safety assessment of nanocomposite for food packaging application. *Trends in Food Science & Technology, 45*(2), 187–199.
31. Huang, Q., Yu, H., & Ru, Q., (2010). Bioavailability and delivery of nutraceuticals using nanotechnology. *J. Food Sci., 75*, 50–57.
32. Inoue, Y., Hoshino, M., Takahashi, H., Noguchi, T., Murata, T., Kanzaki, Y., Hamashima, H., & Sasatsu, M., (2005). Bactericidal activity of Ag-zeolite mediated by reactive oxygen species under aerated conditions. *Journal of Inorganic Biochemistry, 50*, 86–87.

33. Jancy, S., Shruthy, R., & Preetha, R., (2020). Fabrication of packaging film reinforced with cellulose nanoparticles synthesized from jack fruit non-edible part using response surface methodology. *International Journal of Biological Macromolecules, 142*, 63–72.
34. Jin, T., & Zhang, H., (2008). Biodegradable polylactic acid polymer with nisin for use in antimicrobial food packaging. *J. Food Sci., 73*, 127–134.
35. Kawahara, K., Tsuruda, K., Morishita, M., & Uchida, M., (2000). Antibacterial effect of silver zeolite on oral bacteria under anaerobic condition. *Dent. Mater., 6*, 452–465.
36. Klaine, S. J., Alvarez, P. J. J., Batley, G. E., Fernandes, T. F., Handy, R. D., Lyon, D. Y., Mahendra, S., et al., (2008). Nanomaterials in the environment: Behavior, fate, bioavailability and effects. *Environmental Toxicology and Chemistry, 27*(9), 1825–1851.
37. LaCoste, A., Schaich, K., Zumbrunnen, D., & Yam, K., (2005). Advanced controlled release packaging through smart blending. *Package Technol. Sci., 18*, 77 87.
38. Lopez-de-Dicastillo, R., Gavara, P., & Hernandez-Munoz, (2011). Food applications of active packaging EVOH films containing cyclodextrins for the preferential scavenging of undesirable compounds *J. Food Eng., 104*, 380–386.
39. Lopez-Rubio, A., Almenar, E., Hernandez-Muñoz, P., Lagarón, J. M., Catalá, R., & Gavara, R. (2004). Overview of active polymer-based packaging technologies for food applications. *Food Reviews International, 20*(4), 357–387.
40. Maynard, A. D., Aitken, R. J., Butz, T., Colvin, V., Donaldson, K., Oberdörster, G., Philbert, M. A., et al., (2006). Safe handling of nanotechnology. *Nature, 444*(7117), 267–269.
41. McClements, D. J., & Rao, J., (2011). Food-grade nanoemulsions: Formulation, fabrication, properties, performance, biological fate, and potential toxicity. *Critical Reviews in Food Science and Nutrition, 51*(4), 285–330.
42. Mihindukulasuriya, S. D. F., & Lim, L. T., (2014). Nanotechnology development in food packaging: A review. *Trends Food Sci. Technol., 40*, 149–167.
43. Momin, J. K., Jayakumar, C., & Prajapati, J. B., (2013). Potential of nanotechnology in functional foods. *Emirates J. Food Agricult, 25*(1), 10–19.
44. Muthu, T., Rajakumar, G., & Chung, I. M. (2018). Nanotechnology: Current uses and future application in food industry. *3 Biotech, 8*(1), 74–95.
45. Naoto, S., Hiroshi, O., Mitsutoshi, N., & Mitsutoshi, N., (2009). Micro- and nanotechnology for food processing. *(Food Safety Series), Resource: Engineering and Technology for a Sustainable World, 16*, 1–19.
46. Neha, P., Singh, S., Ojha, N., Shrivastava, A., Barla, A., Rai, V., & Bose, S., (2015). Facets of nanotechnology as seen in food processing, packaging, and preservation industry. *BioMed Research International, 4*, 1–17.
47. Othman, S. H., (2014). Bio-nanocomposite materials for food packaging applications: Types of biopolymer and nanosized filler. *Agric. Agric. Sci. Proc., 2*, 296–303.
48. Pathkoti, K., Manubolu, M., & Hwang, H. M., (2017). Nanostructures: Current uses and future applications in food science. *J. Food Drug Anal., 25*(2), 245–253.
49. Pereda, M., Marcovich, N. E., & Ansorena, M. R., (2018). Nanotechnology in food packaging applications: Barrier materials, antimicrobial agents, sensors, and safety assessment. In: Martínez, L., Kharissova, O., & Kharisov, B., (eds.), *Handbook of Ecomaterials* (p. 338). The Gateway East – Singapore: Springer International Publishers.
50. Peters, R. J., Bouwmeester, H., Gottardo, S., Amenta, V., Arena, M., & Brandhoff, P., (2016). Nanomaterials for products and application in agriculture, feed and food. *Trends Food Sci. Technol., 54*, 155–164.

51. Qi, L., Xu, Z., Jiang, X., Hu, C., & Zou, X., (2004). Preparation and antibacterial activity of chitosan nanoparticles. *Carbohydr. Res., 339*, 2693–2700.
52. Quintanilla-Carvajal, M. X., Camacho-D´ıaz, B. H., & Meraz-Torres, L. S., (2010). Nanoencapsulation: A new trend in food engineering processing. *Food Engineering Reviews, 2*(1), 39–50.
53. Rai, M., Ingle, A. P., Gupta, I., Pandit, R., Paralikar, P., Gade, A., Chaud, M. V., & Dos, S. C. A., (2019). Smart nanopackaging for the enhancement of food shelf life. *Environ. Chem. Lett., 17*, 277–290.
54. Salvia-Trujillo, L., Rojas-Graü, M. A., Soliva-Fortuny, R., & Martín-Belloso, O., (2015). Use of antimicrobial nanoemulsions as edible coatings: Impact on safety and quality attributes of fresh-cut Fuji apples. *Postharvest Biol. Technol., 105*, 8–16.
55. Sanders, K., Degn, L. L., Mundy, W. R., Zucker, R. M., Dreher, K., Zhao, B., Roberts, J. E., & Boyes, W. K., (2012). *In vitro* phototoxicity and hazard identification of nano-scale titanium dioxide. *Toxicol. Appl. Pharmacol., 258*, 226–236.
56. Sandoval, B., (2009). Perspectives on FDA's regulation of nanotechnology: Emerging challenges and potential solutions. *Compr. Rev. Food Sci. Food Saf., 8*(1), 375–393.
57. Sekhon, B. S., (2010). Food nanotechnology – An overview. *Nanotechnol. Sci. Appl., 3*, 1–15.
58. Senturk, A., Yalcın, B., & Otles, S., (2013). Nanotechnology as a food perspective. *Journal of Nanomaterials & Molecular Nanotechnology, 6*(2), 1–19.
59. Shivraj, N. H., Baskar, V., Selvaraj, D., Nile, A., Xiao, J., & Kai, G., (2020). Nanotechnologies in food science: Applications, recent trends, and future perspectives. *Nano-Micro Letters, 12*(45), 1–34.
60. Shruthy, R., & Preetha, R., (2019). Cellulose nanoparticles from agro-industrial waste for the development of active packaging. *Applied Surface Science., 484*, 1274–1281.
61. Shruthy, R., Jancy, S., & Preetha, R., (2020). Cellulose nanoparticles synthesized from potato peel for the development of active packaging film for enhancement of shelf-life of raw prawns (*Penaeus monodon*) during frozen storage. *International Journal of Food Science and Technology*, 22–31.
62. Sorrentino, A., Gorrasi, G., & Vittoria, V., (2007). Potential perspectives of bio-nanocomposites for food packaging applications. *Trends in Food Science & Technology, 18*(2), 84–95.
63. Su, H. C., Zhang, M., Bosze, W., Lim, J. H., & Myung, N. V., (2013). Metal nanoparticles and DNA co-functionalized single-walled carbon nanotube gas sensors. *Nanotechnology, 24*(50), 1–18.
64. Thomas, K., & Sayre, P., (2005). Research strategies for safety evaluation of nanomaterials, part I: Evaluating the human health implications of exposure to nanoscale materials. *Toxicological Sciences, 87*(2), 316–321.
65. Tiede, K., Boxall, A. B. A., Tear, S. P., Lewis, J., & David, H., (2008). Detection and characterization of engineered nanoparticles in food and the environment. *Food Addit. Contam., 25*, 795–821.
66. Valdiglesias, V., Costa, C., Sharma, C., Kilic, G., Pesaro, E., Teixeira, J. P., Dhawan, A., & Laffon, B., (2013). Comparative study on effects of two different types of titanium dioxide nanoparticles on human neuronal cells. *Food Chem. Toxicol., 57*,352–361.
67. Valiyaveettil, S., Teow, Y., Asharani, P. V., Prakash, M., & Valiyaveettil, S., (2011). Health impact and safety of engineered nanomaterials. *Chem. Commun., 47*, 7025–7038.

68. Vivek, K. B., Kamle, M., Shukla, S., Mahato, D. K., Chandra, P., Hwang, S. K., Kumar, P., et al., (2018). Prospects of using nanotechnology for food preservation, safety, and security. *Journal of Food and Drug Analysis, 26*(4), 1201–1214.
69. Wanekaya, A. K., Chen, W., Myung, N. V., & Mulchandani, A., (2006). Nanowire-based electrochemical biosensors. *Electroanalysis, 18*(6), 533–550.
70. Weiss, J., Decker, E. A., McClements, D. J., Kristbergsson, K., Helgason, T., & Awad, T., (2008). Solid lipid nanoparticles as delivery systems for bioactive food components. *Food Biophysics, 3*(2), 146–154.
71. Weiss, J., Takhistov, P., & Mcclements, D. J., (2006). Functional materials in food nanotechnology. *J. Food Sci., 71*, 107–116.
72. Yoon, K. Y., Byeon, J. H., Park, J. H., & Hwang, J., (2007). Susceptibility constants of *Escherichia coli* and *Bacillus subtilis* in silver and copper nanoparticles. *Sci. Total Environ, 373*, 572–575.
73. Yotova, L., Yaneva, S., & Marinkova, D., (2013). Biomimetic nanosensors for determination of toxic compounds in food and agricultural products (review). *Journal of Chemical Technology and Metallurgy, 48*(3), 215–227.
74. Yuan, Y., Gao, Y., Zhao, J., & Mao, L., (2008). Characterization and stability evaluation of β-carotene nanoemulsions prepared by high pressure homogenization under various emulsifying conditions. *Food Research International, 41*(1), 61–68.

# CHAPTER 3

# ROLE OF NANOMATERIALS FOR FOOD PROCESSING AND PRESERVATION

RIMPI FOUJDAR, SANDEEP KUMAR, ANIL KUMAR CHAUHAN, and MANAB B. BERA

## ABSTRACT

The application of bio-based nanoparticles (NPs) in food science and technology is an emerging sector. The combination of nanostructure and biomaterials provides numerous benign properties related to food safety and quality. The biomaterials-based NPs have demonstrated a potential impact on each stage of progression from food production to consumption, such as enhance productivity, shelf life, and nutrient retention either through nano-encapsulation or nanoemulsion improve quality by nano conjugation and smart nano-biodegradable packaging material. Further, these nanomaterials (NMs) are extensively used in the development of biosensors for the detection of food pathogens and chemicals. Despite having numerous merits of these NMs, a serious concern on its safety in food application remains and calls for authentic and reproducible biosafety analysis based on both in-vitro as well as in-vivo studies. Such studies with a proper regulatory aspect could help to reduce the risk assessment and ensure the bio-safety issues associated with its implementation in food processing and preservation technology.

## 3.1 INTRODUCTION

Food processing and packaging is an essential part of our daily life. Its applications are gradually increasing over time. Today's food preservation relies

Nanotechnology Horizons in Food Process Engineering, Volume 1. Megh R. Goyal, PhD, PE, Junaid Ahmad Malik, PhD, Satish Kumar, PhD & Ritesh B. Watharkar, PhD (Eds.)

more on food additives and packaging materials derived from the chemical or artificial route. Such practice has led to a rise in the degradation of both human health and the environment. In developed and developing countries, the incidence of food poisoning and other health-threatening diseases are still making headlines of the news. Consequently, innovative food processing and packaging materials are growing to accomplish the requirements of society. Bio-based NMs are designed or fabricated using different techniques and added during the food preparation or in the final product formulation to improve food quality and safety [37].

A wide range of bio-based nanostructured materials such as nano-emulsion, nanoconjugates, nanomicelles, metal NPs, and nanostructured biopolymers are available and used as novel food additives, preservatives, biosensors, and smart biodegradable packaging materials [3, 9, 34, 65]. The fate and transformation of these NPs in the human body as well as in the environment depend on their physicochemical properties. The alteration of each set of factors such as shape, size, or surface properties can result in a risk assessment of cell cytotoxicity [76]. However, biomaterials such as biomolecules (protein, carbohydrate, lipid, vitamins, and minerals), plant polyphenols, microorganisms, based NMs exhibit lower toxicity than chemically modified NMs [6, 10, 81, 102].

The *in-vitro* and *in-vivo* studies provide information about whether NMs show toxicity or not, what quantity of nanomaterial exhibit nontoxic to normal cells. Therefore, the safe application of bio-nanomaterial in the society needs a proper cell line study along with suitable regulatory legislation. Multi governmental agencies, including the FDA, EPA, OSHA (Occupational Safety and Health Administration), and CPSC (Consumer Product Safety Commission) in the United States, have been involved in giving direction on nanotechnology regulation [46].

The current scenario of the food market showed that the growth of nano-based food products is seen suspicious due to neophobia (tendency to avoid new food) and scientific gap of knowledge about their potential risk for human health and environment [91]. These challenges are a potential barrier to the commercialization of food nanotechnology products. Although, more than 500 NMs are commercially available [35]. Nano-encapsulated silver and titanium dioxide ($TiO_2$) are found maximum in use including application as food additives and food contact materials [35]. As a concern of future developments, it seems that chemically synthesized toxic silver, copper NPS will be replaced by bio-based nano-encapsulates and nanoconjugate and its application in food.

This chapter provides an understanding of the formulation of NMs, application in novel food production and preservation, cytotoxicity study as well as regulatory aspect related to bio-nano materials. It also covers the present and future market perspective of NMs application in food sectors.

## 3.2 THE NANOSCALE FORMULATION FOR FOODS AND BEVERAGES

Constituents of food such as; some protein, polysaccharides, phytochemicals, fats as well as vitamins have naturally occurred in the form of nano-sized particles. The application of nano-sized particles in food and beverages is not new. The properties of the food materials changes with change in length scale, and such changes in properties of materials have been observed when the size of the particles is reduced. The particles (nanoscale) exhibit unique functional, physical, and optical properties. One of the most well-known examples is the processing of milk, after homogenization, the particle size of milk reduces to less than 100 nm. Figure 3.1 illustrates how properties change with the reduction of particle size. As particle size reduced the surface area expands (Figure 3.1(a)), which helps to improve the penetration power and showed an effective delivery system. Figure 3.1(b) illustrates the unique optical properties of silver at the nanoscale. Similarly, the changes in color with the reduction of size in quantum dots (Figure 3.1(c)) fascinate its use in labeling purposes. Such novel technologies are being used to develop novel food with the new formulation of nano-sized particles which can alter its nutrients, color, flavor even its structure to fulfill the desirability of consumer's preferences.

### *3.2.1 NANOEMULSION*

Emulsion refers to the mixing of two immiscible liquids, typically has the particle size with a mean diameter in the range of 100 nm to 100 μm [65]. This type of dispersed solution is thermodynamically unstable because the free energy of separated phases (oil and water) is lower than emulsion itself. Nanoemulsions are more thermodynamically stable having smaller particle sizes (500 nm or less) than conventional emulsions [41]. They also have a tendency of gravitational separation and aggregation over time. The stability could be improved by the selection of suitable emulsifiers, temperatures as well as processing methods. They are classified based on two phases, dispersed, and continuous phase. Oil in water (O/W) where oil is dispersed in water as the continuous phase and another is water in oil (W/O) nanoemulsions where water is dispersed in oil as the continuous phase. Apart from it, multiple-nanoemulsion is also a type of nanoemulsion that recently comes into light due to their potential application in the food and pharmaceutical industries. This bi-continuous phase nanoemulsion is further divided into two categories that are oil in water in oil (O/W/O) and water in oil in water (W/O/W) type nanoemulsion [95].

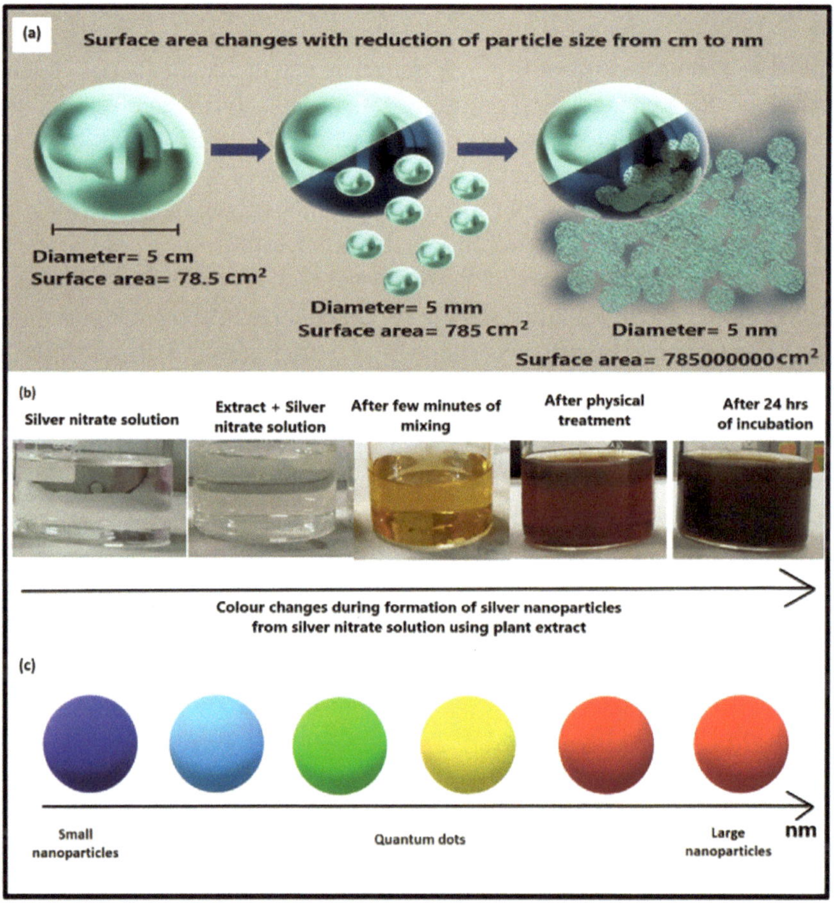


**FIGURE 3.1** Properties change with reduction of particles: (a) surface area increases [adapted and modified from: Nano. Gov. [72]]; (b) color changes with formation of nanoparticles (own work); (c) As particle size of quantum dot increases, the color shifted from dark blue to dark red [adapted and modified from: Wikipedia contributors [105]].

The selection of a type of nanoemulsion depends on its application in the food industry. The nanostructured multiple nanoemulsion or nanostructure multilayer emulsion provides an ability to carry several functional ingredients in a single delivery system [87]. It offers to release the functional component which is encapsulated inside the component of multiple nanoemulsion systems to a particular area with a response of specific environmental triggers [87]. The nanoemulsion can be prepared using two approaches: low and high energy approaches which are illustrated in Figure 3.2.

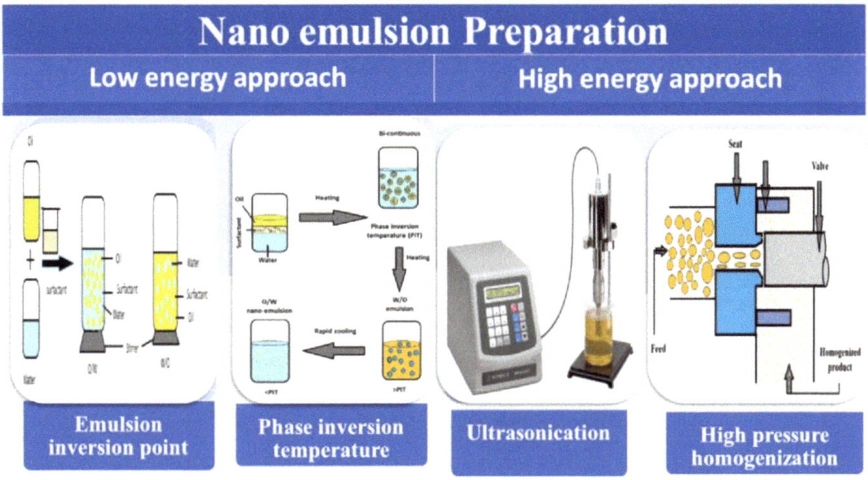


**FIGURE 3.2** Nanoemulsion preparation method.

Low energy methods require less power density ($10^7$–$10^9$ W/kg) whether high energy methods are based on high power density input ($10^3$–$10^5$ W/kg) [42]. Low energy methods comprise with mixing of components that are oil, water, and surfactant. Emulsion inversion point and phase inversion temperature are mainly two types that are associated with low energy methods [58]. In the first method, the composition has been mixed in two different ways; one is mixing of water and surfactant and then adding oil, the second one is mixing of oil and surfactant, and then adding water [58]. The way of mixing depends on the type of surfactant and liquid phase used during the preparation. In phase inversion temperature, the nanoemulsions are formed by changing the temperature, i.e., heating or cooling. The method depends on the solubility or molecular geometry of surfactant with changes in temperature. The driving force of phase inversion methods deviates from the temperature and physicochemical properties of surfactant [52].

The packing parameter is used to describe the molecular geometry of surfactant ($p = \alpha_T/\alpha_H$, where p is packing parameter, $\alpha_T$ and $\alpha_H$ indicate the cross-sectional areas of the hydrophilic head and lipophilic tail-group). The molecules of surfactant having a tendency to naturally associate together in water phase due to the hydrophobic effect and form monolayers with optimum curvature. The packing parameter determines the optimum curvature of surfactant monolayer as well as type of nanoemulsions:

- If, the packing parameter (P) is less than 1 ($P < 1$) indicates convex shape optimum curvature, which favors the formation of oil in water emulsion or nanoemulsion.
- Similarly, P is greater than 1 ($P > 1$) associates with concave shape optimum curvature and tend to favor the formation of water in oil (W/O) nanoemulsions.
- While, P =1 leads to zero curvature that means it neither favors the formation O/W nor W/O, instead leads to the development of bi-continuous system.

On the other hand, the second approach or high energy method includes ultrasonication and high-pressure homogenization. In ultrasonication, an electrical input is converted into mechanical and then it is converted into pressure fluctuations. Ultimately acoustic cavitation occurs which causes formation growth and the collapse of bubbles and helps to break the large molecules into smaller ones [43]. Similarly, high-pressure homogenization comprised of the passing of a mixture of components (oil, water, and surfactant) through a small gap (5–10 μm) [22]. This small gap led to increasing the pressure drop up to a few 1,000 bars, and the deformation of droplets occurs through the large value of shear forces result in the production of small droplets [22].

The high energy methods are able to produce high intensity disruptive forces which have the ability to develop very small droplets needed to form nanoemulsion. The extreme intense energy required is that the disruptive forces produced by mechanical device (homogenizer) which need to go beyond the restoring forces holding the droplet into spherical shapes [92]. The Laplace pressure law ($\Delta P = \gamma/2r$) is used to determine the restorative forces. The increased Laplace pressure leads to a decrease in droplets radius, simultaneously increase the interfacial tension ($\gamma$). Hence the droplets radii become extreme small in a high energy-based homogenizer.

Both approaches have merits and demerits. High energy methods are more efficient because the preparation of nanoemulsion does not restrict on surfactant choice. Also, a smaller amount of surfactant is required to produce stable nanoemulsion. High energy methods are more stable than low energy methods. However, high energy approaches required a high amount of energy input as compared to low energy approaches. High energy increases the temperature in the sample solution which leads to degradation of heat sensible products.

In contrast, low energy approaches are energy efficient whether the surfactant choice is restrictive. The amount of surfactant is required significantly higher as compared to high energy methods and often requires two surfactants for more efficient results [65].

The emulsion inversion point method mainly involves using simple equipment, operate at constant room temperature and moderate shear rates, therefore the researchers generally prefer this method to produce bio-based nanoemulsions to avoid the degradation of the thermal sensitive product. However, this method requires a large amount of surfactant which is relatively expensive and may cause toxicity [65]. Hence, ultrasonication can be used for the formulation of bio-based nanoemulsion as they can maintain the temperature at a constant level during the processing time using a thermal probe and water circulating bath. The study on the different types of bio-based nanoemulsion and their preparation method, as well as applications, is represented in Table 3.1.

### 3.2.2 NANO MICELLES

Nano micelles are core-shell nanostructures formed by spontaneous self-assembly of amphiphilic molecules with a hydrophilic surface and hydrophobic core or vice versa [67]. Both regions of hydrophobic and hydrophilic structures make them suitable to deliver the poor soluble and unstable compounds [67]. A strong cohesive force in between bioactive compound and core molecule segments along with cross-linking of the shell or core improves the stability of nano-micelles [38]. Using environment (like pH, temperature light, ultrasonication, etc.), responsive bio-polymeric micelles, prolonged circulation, and delivery at a specific site of active molecules can be successfully achieved [36].

The nano micelles with an improved delivery system using high intensity ultrasound waves (20–40 kHz) have been discovered. The ultrasonic waves are capable of accelerating the permeability and extravasation at the cancer cell which enhances the nutraceuticals uptake. The release behavior of bioactive molecules can be modified by influencing the amplitude, power density, pulse length, pulse duration, ultrasound frequency [50]. Dong et al. [21] produced ultrasonic-assisted Spinosad (fungicide) micelles stabilized with chitosan-isoleucine-polyethylene glycol methyl ether showed antifungal effect against *Fusarium oxysporum* than unencapsulated Spinosad. The shell was able to protect the Spinosad from photodegradation which showed a good barrier against the light.

Protein such as casein, polysaccharides like chitosan (CS), inulin biopolymers are widely used as a micelle's delivery vehicles due to their biocompatibility and able to improve the solubility of bioactive hydrophobic molecules [21, 89, 103]. Wang et al. [103] have reported self-assembled CS nano micelles were prepared by grafting deoxycholic acid and N-acetyl-L-cysteine which showed efficient delivery of quercetin by enhancing the solubility. An injectable micellar nano-formulation of quercetin exhibits good hemocompatibility.

### 3.2.3 NANO-CONJUGATION

Nano-conjugation refers to the involvement of either self-assembled or solid bond mediated biopolymeric NPs [34, 101]. A suitable functional group is required for nano-conjugation with several molecules or biomolecules. For instance, exposure of structures with aqueous phase results in the formation of self-assembled NPs with hydrophobic parts inside the core and hydrophilic portion towards the aqueous phase.

NPs surface coating with biopolymers provides several biological as well as food applications such as biosensing and imaging. It also helps in reducing the risk of disease, protecting the food against spoilage, improving nutritional and functional properties [34].

It prevents aggregation within the NPs in simulated adverse conditions [44]. On the contrary, the conventional conjugation between two molecules may have some limitations such as poorly solubility, large particle size, low encapsulation efficiency, unpredictability, and inefficient in particular applications. Nanoconjugates, nanoencapsulation, and nanocomposites are the typically reactive product of nano-conjugation (Figure 3.3(a–c)).

**TABLE 3.1** Different Types of Biomaterial-Based Nano-Emulsion, Preparation Method and Their Applications in Foods

| Bio-Materials | Type Nano-emulsion of Surfactants | Stabilizer | Methods Used | Processing Conditions | Nanoemulsion Type | Particle Size (nm) | PDI | Zeta Potential (mv) | Stability | Application in Food | Ref. |
|---|---|---|---|---|---|---|---|---|---|---|---|
| **1. High Energy Methods** | | | | | | | | | | | |
| Palm oil | Tween 80, polyglycerol poly-ricinoleate | Sucrase, Xanthan gum | Ultra-sonication | 40% amplitude and 180 s | O/W/O | 175.5 ± 9.8 | 0.232 ± 0.012 | – | – | Can be used as oil phase thickeners | [86] |
| Tuna Fish oil | Octenyl succinic anhydride (OSA) modified starch | – | Ultra-sonication | Coarse emulsion was prepared using homogenizer at 15,000 rpm for 3 min at room temperature and then sonicated at 40% applied power for 13 min | O/W | 226 | – | 34.2 ±1.026 | Showed negligible changes after 30 days of storage | Can be incorporated to produce dairy and beverage products | [5] |
| Cinnamon essential oil | Tween-80 | Pullulan polysaccharides | Homogenization and ultrasonication | Coarse emulsion was prepared by homogenization at 12,000 g for 4 min and then sonicated (30% power output of 225 watt) for 15 min | O/W | 162.1 | 0.193 | –35.4 | – | Exhibit excellent antimicrobial effect and help to increase the shelf life of strawberries during room storage.<br>Can be used in quality improvement of fruit storage | [16] |

**TABLE 3.1** *(Continued)*

| Bio-Materials | Type Nano-emulsion of Surfactants | Stabilizer | Methods Used | Processing Conditions | Nanoemulsion Type | Particle Size (nm) | PDI | Zeta Potential (mv) | Stability | Application in Food | Ref. |
|---|---|---|---|---|---|---|---|---|---|---|---|
| **2. Low Energy Methods** | | | | | | | | | | | |
| *Bixa orellana L.* seeds | Polysorbate-80/ sorbitan monooleate | – | Magnetic stirrer | Homogenized at stirrer at 500 rpm for 1 hr | O/W | 180–220 | 0.432 ± 0.03 | –27.8 ± 2.0 | Stable up to 21 days | As food colorants | [12] |
| Green seed oil | – | – | Phase inversion | Self-emulsification | O/W | 16.44 ± 0.01 | – | – | – | Act as anti-oxidant and antibacterial | [107] |
| DHA and EPA | – | – | Emulsion phase inversion | – | O/W | – | – | – | – | Development of functional apple juices | [110] |

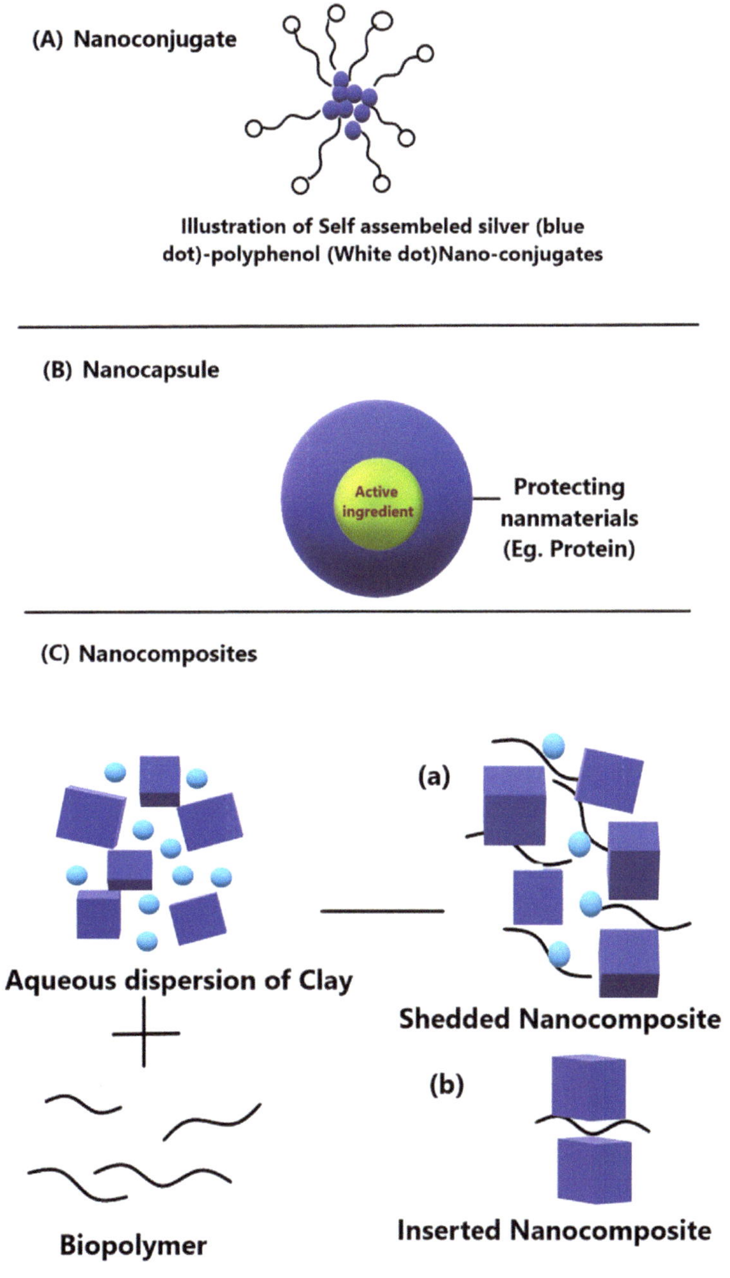


**FIGURE 3.3** Illustration of: (a) nanoconjugate (adapted and modified from: Tran and Tran, [101]); (b) nano-encapsulate; and (c) nanocomposite (adapted and modified from: Perotti et al. [82]).

#### 3.2.3.1 BIO-NANOCONJUGATE, NANO-ENCAPSULATE, AND NANOCOMPOSITES

The bio-nanoconjugate is a reactive product formed by conjugation of biomolecules within NMs (such as silver-polyphenol, polysaccharides-protein, protein-phenol, polysaccharide-phenol nanoconjugate) [34]. The materials used in conjugation can be organic or inorganic. The combination of two materials not only enhances the activity but also improves its physicochemical properties and thus expands its range of application in the food and nutritional field. Nano-encapsulation, on the other hand, refers to a capsule-like process of nano-conjugation. In nano-encapsulation active ingredient (bioactive compound) is surrounded with a core or protective nanomaterial (Figure 3.3(b)).

The encapsulation provides a protective layer to preserve the physicochemical and organoleptic properties of the actual products as well as improve the delectableness of volatile aroma compounds. Flavors and aromas are extensively used in the food industry for encapsulation. Examples are presented in Table 3.2. Formulation of nanoconjugates and nano-encapsulates comprised of the use of biomolecules such as polysaccharides, proteins, phenolic compounds, as well as some metal (silver, gold, copper, etc.). The methods encountered with the production of nanoconjugates and nano-encapsulates are Millard reaction, antisolvent precipitation, coacervation, self-assembly, and ionic gelation [30, 47, 99]. Electrostatic interaction and weak hydrogen bonds are mainly associated with these types of methods for conjugation and interaction. Moreover, free radical grafting is allied with covalent strong bond and help to produce stable nanoconjugates [94].

Both nanoparticles have numerous advantageous properties which aided its application in the food industry. Formulation within biocompatible NPs masks the flavor, aromas, and control interaction of active ingredients with the food matrix, ensures the availability at a target time, and a specific rate. Additionally, it protects from moisture, heat, chemical or biological degradation during processing and storage. This nano-formulation can be used to convert liquids into powders, makes the handling easier than liquid-based NPs such as nanoemulsion [101].

Bio-nanocomposite is other examples which are associated with nano-conjugation (Figure 3.3(c)). They are prepared using a different types of bio-based NPs like nanocrystals of cellulose and CS, biosynthesized metal nanoconjugates, nano clay, nanocarbon tubes through the percolated network within a medium [19, 80]. In the case of the polysaccharide-based nanocomposite, the appropriate percolation network creates large unwanted aggregation within the matrix which is the main issue for the fabrication [71]. This issue is mainly associated with strong hydrogen bonding interaction of nanocrystals with each other. Such interaction also plays a significant role in producing required network fillers inside the medium. Limitation can be avoided by dispersing the NPs into a polymer matrix to enhance the interfacial bond between polymeric NPs and preserving the strong hydrogen bonding interactions [24]. For instance, cellulose nanofibers and CS bio-nano composite were obtained through covalent linkages using polymer chain surface modification. It was found that covalent linkages of nanocomposites increase the tensile strength (TS) up to 150% [19]. A brief overview of the materials, methods, and application in the food industry of these nano conjugations-based products have been represented in Table 3.2.

## 3.3 NANOTECHNOLOGY TO IMPROVE FOOD PROCESSING AND PRESERVATION

### 3.3.1 BIO-NANO MATERIALS AS FOOD ADDITIVES

Bio-based NPs can be used as additives to enhance the physicochemical and bioactive properties of food materials. To boost the functional and bioactive properties, nano-formulation of nutraceuticals such as phytosterol, polyphenol, apigenin, etc., are incorporated [68, 69]. The incorporation of bio-based NPs into food materials aided the health benefits and help to reduce the risk of many diseases. In a study, a beverage containing cinnamon NPs enhance 40% of total phenolic content and 60% of antioxidant activity [69]. The physical stability and color of beverages were also improved. These observations suggested that formulated NPs could be a promising and novel process for ready to drink products with enriched antioxidant compounds and better stability. The nanoencapsulation of lycopene enhanced the solubility and facilitated its use in aqueous solution-based orange drink, also masked its tomato tase, and improved the homogeneity of fortified drinks [108].

A recent study on the development of functional bread fortifying with 5% microcapsules of garlic oil-based nanoemulsion using calcium alginate was reported [73]. The presence of microencapsulated garlic oil in fortified bread causes white crumb, this result indicated that the microencapsulation and nanoemulsion protected the color and oxidation of garlic oil, no change in cohesiveness and springiness over the storage period, and the sensory parameters also found better in functional bread as compare to market garlic bread. [73].

**TABLE 3.2** Overview of Biomaterials, Methods, and Applications in Food Industry of Nanoconjugation-based Products

| Nano-Conjugation | Type of NM | Methods Used | Particle Size (nm) | PDI | Zeta Potential (mv) | Encapsulation Efficiency | Application in Food | Advantageous Properties | References |
|---|---|---|---|---|---|---|---|---|---|
| Nano-conjugate | Zein-bovine serum albumin-caffeic acid | Antisolvent precipitation | 217.2 | – | – | 86.5 | As encapsulate material for resveratrol | Improved the photo and thermal stability and preserve the antioxidant activity | [30] |
| | Chitosan-tannin-lysozyme | Ion cross-linking and ultrasonication | – | – | – | – | As multi-functional nano-delivery systems | Prolonged the stability and exhibit antioxidant and antibacterial activity | [99] |
| | Gelatin and chia mucilage | Coacervation | 76.5 | – | – | 91.79 | As encapsulate material for Oregano essential oil | – | [47] |
| | Lignin-tall oily fatty acid-cellulose | Free radical grafting | <200 | – | – | – | As antimicrobial agent | – | [94] |
| Nano encapsulate | Garlic oil with chitosan and whey protein | – | 101 | – | –7.27 | – | Enhanced the shelf life of vacuum packaged sausage | Retard lipid oxidation and growth of spoilage bacteria; No significant changes in sensory | [27] |

**TABLE 3.2** *(Continued)*

| Nano-Conjugation | Type of NM | Methods Used | Particle Size (nm) | PDI | Zeta Potential (mv) | Encapsulation Efficiency | Application in Food | Advantageous Properties | References |
|---|---|---|---|---|---|---|---|---|---|
| | Sour tea extracts with Maltodextrin and whey protein concentrate | – | 139.03 | – | – | 67.30 | Increased the shelf life of chicken nuggets | – | [8] |
| Nanocomposite film | Gelatin based chitosan–tea polyphenol | Electrospray | 205 | – | – | 66.7 | Increased the stability of oxidizable oil over 14 days and can be used as edible packaging material | Improve barrier and mechanical properties; Exhibited antioxidant activity | [104] |
| | Carboxymethyl cellulose (CMC)-Guar gum-*Mentha* leaves extract-silver | – | – | – | – | – | Increase the shelf life, freshness, and weight loss of strawberry | Antimicrobial property against 10 food pathogenic bacteria | [54] |

gases and water vapors during cellular respiration. In the same manner, carbonated beverages release oxygen and carbon dioxide to avoid oxidation and decarbonization. The elimination of gases and water vapors varied with food and packaging materials. Consequently, these complications of packaging can be sorted out by the incorporation of various novel bio-nano materials into the packaging system.

In modern classification, the novel food packaging is categorized as "improve," "active" and "intelligent" packaging systems that are based on application and type of food material to be packed [45]. Biosynthesized silver, titanium, copper, gold, clay, and cellulose-based NPs are widely employed in the food packaging system to improve the quality (antioxidant and antimicrobial) and packaging properties (mechanical strength, barrier, water vapor transmission rate (WVTR)) of the food product. Each of these nanomaterials has a different chemical structure and characteristics which provide diverse functional applications in food packaging.

The starch and polysaccharides-based biopolymers are an attractive alternative approach to replace the existing packaging materials. The biodegradability nature of polysaccharides-based biopolymers, availability in large scale, and comparatively cheaper make them suitable for application [45]. However, high moisture absorbance and less capability to inhibit the microbial growth are major drawbacks of these biopolymers. Therefore, they are used as nanofillers to reinforce into bioplastics. One good example is the inclusion of nanocellulose in the form of reinforcement to thermoplastic starch [77]. The reinforcement improves 140% of TS, crystallinity, water barrier, and thermal properties [51]. Another example has reported on bacterial cellulose nanofibrils (CNF) based multifunctional nanocomposite film. The film was prepared by zein protein nanoparticle through environment-friendly routes (scalable paper making process) [61]. The strong interfacial adhesion and induced hydrogen bond via interaction between zein and cellulose cause remarkable tensile properties and improved thermal stability.

Integrating nano cellulose into silver NPs is another potential example of active food packaging to increase the shelf life of food products. The nanocomposite film can inhibit food pathogenic microbes. The capability of nanocomposite to improve the shelf life of strawberries was analyzed and results divulged a positive impact towards freshness and shelf life [54]. Beyond these aspects, biosynthesized silver NPs are one of the most attractive fillers due to excellent antimicrobial, antioxidant, and anticancer properties. They have the ability to gas sensing and catalytic activity. They not only act as ethylene absorbent but also decompose it, which helps to improve the shelf life of fruits and vegetables. Recently, an excellent application of green bottle fly biosynthesized AgNP-based carrageenan nanocomposite on decontamination of drinking water pathogen was reported [1]. In this, 30–100 nm biosynthesized NPs-based bio-nano composites were produced, which exhibited a uniform, transparent, elastic film with good mechanical properties and thermal stability (270°C). SEM results showed that silver NPs based nanocomposite are well dispersed within the polymeric matrix compared to plain (without silver NPs based nanocomposite) polymeric matrix, hence exhibited enhanced thermal stability [1].

#### *3.3.4.1 NANO-BIOSENSORS*

A combination of nanotechnology, chemistry, and biology leads to the development of nano-bio sensors. Nano-biosensors should consist of at least one bio-based compound as a substrate or sensing element. They provide information about microbial spoilage, adulterations, pollutants, and food freshness [53]. Many times, consumers are disappointed due to the food package. Since the food packaging system protects the food from the outside exposure, the consumers rely on expiry dates provided by producers or manufacturers. The information is mainly based on assumptions of food stored or transported. If the storage or transport conditions are desecrated during any period, the quality and food might get spoiled, which cannot be predicted by retailer/trader/consumer until the food package is made open [53]. In this sense, nano-bio sensors can alleviate the limitations of food packaging through the unique biochemical and electro-optical properties of nano-bio sensors. Such sensors can sense the presence of undesirable gases, aromas, chemical adulterants, toxins, food pathogens and also the fluctuations of environmental conditions, enhance the quality and safety of packaged food products, help the consumers to purchase fresh and healthy food products without any doubt [79].

The bio-based sensors in food packaging are generally based on colorimetric detection of analytes. In the intelligent food packaging system, fruits, and vegetables are coated with natural dye (such as anthocyanins, curcumin, and alizarin) as a pH indicator [2]. The change in color is associated with protonation or deprotonation of these compounds. Purple corn extract and silver NPs with CS film showed color variation at pH 3.0–10.0 [83].

The colorimetric-pH-based nanosensors are also able to monitor the deterioration of protein. The spoilage of proteins leads to the formation of alkaline nitrogen volatile compounds such as total volatile basic nitrogen, histamine, ammonia, putrescine, and cadaverine. [13]. Halo chromatic zein protein nanofibers-based nano-sensors fabricated with alizarin dye indicator provide high sensitivity alerts to sense the total volatile-based nitrogen in real-time for rainbow trout fish stored at 4°C [2]. In this, no change in color was detected during the starting of the 4th days of storage at 4°C, a light purple color was observed after 4th days, and the color turned into magenta during the 10th and 12th days of storage, which indicates the spoilage (Figure 3.4).

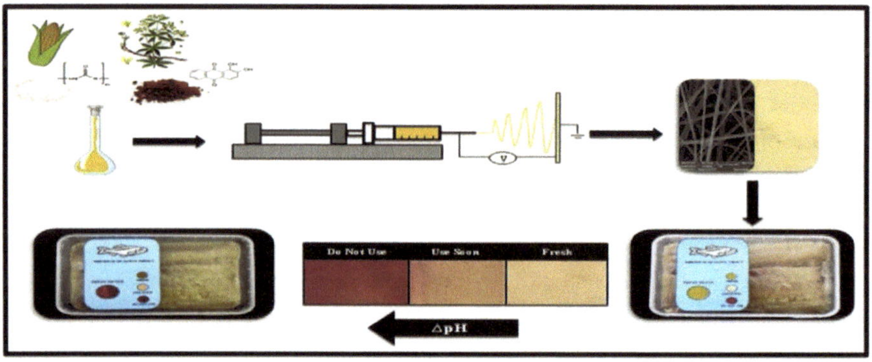


**FIGURE 3.4** Halochromatic nanosensors (left to right): (1) mixing of zein protein-alizarin dye; (2) nanofibers produced using electrospinning method; (3) nanostructure of nanofibers (right to left); (4) packaging of fresh trout fish into film with nanofibers; (5) changes of color from yellow to red as indication of spoilage.

*Source:* Reprinted with permission from: Aghaei et al. [2]. © 2020 Elsevier.

Biosynthesized silver NPs also exhibited colorimetric effect and sensed hydrogen sulfide ($H_2S$) produced by spoilt meat [109]. Since a strong bond between silver (Ag) and $H_2S$ exist and can form $Ag_2S$ to change the color from yellow to colorless within the package system.

On the other side, various chemically synthesized NMs already exploited as nano-sensors. For instance, copper NPs carbon-based tensile film can be used to determine the humidity [63]. The nanocomposite of iron NPs acts as a sensor to determine the moisture [100]. It was demonstrated that the titanium and zinc oxide (ZnO) NPs embedded with perylene-based fluorophores nanofibers are used to detect organic volatile compounds (ethanol, gaseous amine) [85].

These chemically derived NPs can be synthesized by bio-based materials such as plant extracts and microorganisms, which give a more beneficial impact on the development of nanosensors due to low toxicity and bioactive properties. However, reports on the mechanism of interaction between biosynthesized NPs with the environmental condition such as moisture, ethanol, gaseous amine, volatile compounds, oxygen, ethylene gas, and temperature are found to be limited during the writing of the chapter.

## 3.4 MECHANISMS INVOLVED IN NANOSTRUCTURE-MEDIATED TOXICITY

The toxicity of NPs on cancer cells or normal cells depends upon many factors such as shape, size, surface charges, compositions, and surface area [97]. The tiny particles improve the penetration power and easily move into the different body parts through several pathways [88] which cause toxicity towards the normal cell. They can either enter through cells by transcytosis mechanism or diffuse by the cell membrane [48]. They are also capable to penetrate the endothelial and epithelial barriers and enter into lymph and blood vessels which transport the particles through blood or lymph stream into different body parts such as the brain, heart, liver, kidneys, bone marrow, spleen, and nervous system [25, 78]. Combustion derived carbon-based NPs enhanced platelet aggregation cause thrombosis [84].

Experimental modeling of NPs on the body showed that it causes inflammation of the respiratory tracts, neurodegenerative diseases, myocardial infarction [62, 63]. It is clear from the study that NPs not only invade the organs but also enter into different cell organelles such as ribosomes, mitochondria, cytoplasm, etc.. They alter drastically the cell metabolism and cause DNA cuts, mutation, and apoptosis [11].

Gallic acid encapsulated with surfactant (polysorbate 80) NPs induces hemolysis activity of red blood cells resulted in membrane damage in a dose-dependent manner [6]. Also, excessive administration of emulsion-based NPs altered xenobiotic metabolism associated genes, accelerate ketone body production, and stearoyl-co-desaturase activity [81]. On the contrary, NPs have excellent potential to inhibit tumor cells. The mechanism of toxicity towards cancer cells is the same as a normal cell. Although silver NPs on cancer cell toxicity study has taken to throw light in the understanding of the mechanism of toxicity towards cancer cells (Figure 3.5).

The exact mechanism of metal NPs as an anticancer activity is not understood so far. Nevertheless, it is supposed that the generation of reactive oxygen species (ROS), inhibition of VEGF-made actions, arrest the Sub-G1 cell, up-regulation of caspase 3 expressions, and p53 protein, are the main mechanism associated with anticancer activity. Besides, the oxidation process enhances oxidative stress and initiates the oxidation of glutathione (GSH) to glutathione disulfide. GSH is an antioxidant that prevents cells from the damage of ROS. Moreover, silver NPs reduce the activity of enzymes that are responsible for DNA damage repair known as DNA-dependent protein kinase [10]. Similarly, biogenic metallic nanoparticle; the CS formulated curcumin NPs induced colorectal cell apoptosis by arresting the cell cycle at the G2/M phase [17].

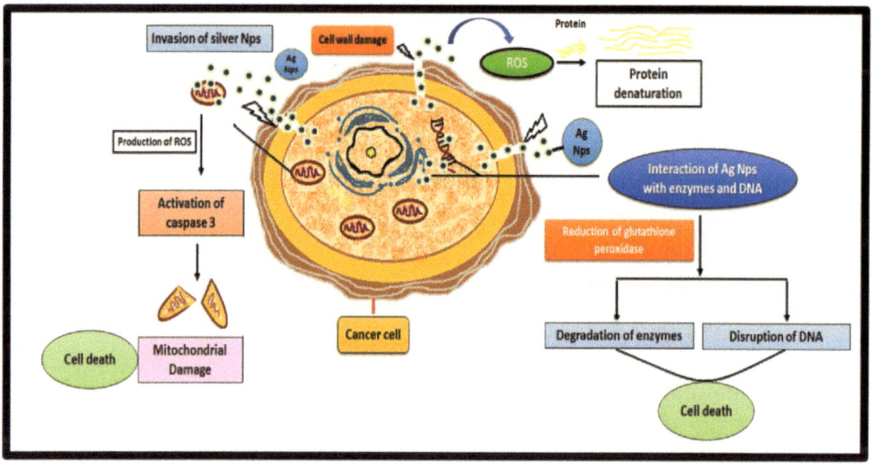


**FIGURE 3.5** Cytotoxicity mechanisms towards cancer cell by colloidal bio-based silver NPs.

*Source:* Adapted and modified from: Barbadi et al. [10].

Combustion or chemically derived NPs have higher toxicity towards normal cells as compare to bio-based NPs [10, 64]. Bioactive compound-based NMs are more bio-compatible and exhibit lower toxicity [6, 56, 81, 98, 102]. However, few studies also demonstrated toxicity towards the normal cell at high doses of bio-based NPs. These efforts give hope for the future and could bring a revolution on bio-based NPs for their safe application in food and pharmaceuticals for human beings, animals as well as the environment. In the next section, a brief discussion on the toxicity effect of bio-based NMs has given based on *in-vitro* as well as *in-vivo* studies.

### 3.4.1 IN-VIVO AND IN VITRO-TOXICITY OF BIONANOMATERIALS ON CANCER CELLS AS WELL AS NORMAL CELLS

As discussed earlier in this chapter, the toxicity of NPs towards normal cells is a major drawback which hinders its application in food as well as pharmaceutical sectors. Parenthetically, *in-vitro*, and *in-vivo* on cell lines study and experimental animals have remarkable importance for determining the toxicity. The *in-vitro* study makes it easier to understand the toxicity effect on each specific cell lines or tissues. In the case of *in-vivo* studies, they are very helpful to determine the toxicity level of individual body organs or the whole body. The *in-vitro* study is important during preliminary steps. In contrast, *in-vivo* studies are more reliable due to the use of live organisms which helps to investigate the ability of NPs to deliver the encapsulated bioactive compounds into the target cells, as well as provide information on the biocompatibility between NPs and host organs [3]. The toxicity of NMs varied with their concentration, penetration power, duration of interaction with living material, and stability in the biological system. Production and delivery of safe biocompatible NPs can only be based on understanding all aspects of the mechanism underlying toxicity as well as the interaction between all factors. In this regard, an experimental in vitro and in vivo study has taken to exemplify the effect of different factors on toxicity towards the normal cell as well as cancer cells. The toxicity effect of *in-vitro* as well as *in-vivo* study in both normal and cancer cells is presented in Tables 3.3 and 3.4.

The *in-vitro* study of several bio-based NMs showed that the toxicity towards normal cells depends on concentration. At lower concentrations, they do not exhibit any toxic effect while excessive administration showed toxicity [102]. The study also revealed that particle size has an important role to exhibit the cytotoxicity against cancer cells. Larger particles showed infrequent permeability across the Caco2-cell monolayer while small nano-size particles significantly improved the passive transport mechanism through the cell membrane [93]. The small-sized nanoemulsion (233 nm) showed 2.5 times higher efficiency of cellular uptake as compared to larger size-based nanoemulsion (350 nm).

The formulation also plays an important role in penetrability. For instance, lecithin-based nanoemulsion enhanced the permeability in cell monolayer despite having lesser differences in particles size than the combination of tween 20 and glycerol monooleate based nanoemulsion. It could be explained by the fact that the structure of phospholipid present in soy lecithin is similar to the phospholipid bilayer structure of the cell membrane hence improved permeability [93]. A few studies also reported that conjugation within NPs is more prone to inhibit the cancer cells as compared to individual bioactive compounds [17, 20, 98]. For example, curcumin-chitosan nanoconjugate showed IC50 value less than 10 μM whereas, the IC50 of free curcumin was 10 μM after 72 h of incubation on HT-29 cells. The conjugation enhanced the antioxidant activity which encourages cell apoptosis in the colorectal cancer cell [17]. Another recent example of an in-vivo study conferred that nano-encapsulated monoclonal antibody (Bevacizumab) with albumin was more effective to reduce the glycolysis and metabolic tumor volume of colorectal cancer than free monoclonal antibody [20].

Method of preparation of NPs was another important factor which also affects the cell cytotoxicity. It was found that the level of intratumoral monoclonal antibody was four times higher in animal tumor cells treated with the nanoencapsulation of monoclonal antibody prepared by the coacervation method as compared to the conventional formulation [20].

## 3.5 SAFETY ASPECTS OF NANOPARTICLE AS REGULATORY CONCERNS

The rapid development of nanotechnology applications in the food industry highlighted public concerns and calls for appropriate regulations and legislation to meet the safety requirement. Regulatory agencies such as ISO (2005), EFSA (2009 and 2011), FAO, and WHO (2013), and USFDA (2015) have issued several guidelines. These guidelines were mainly focusing on the potential risk assessment to food safety, management of food and agricultural products, application of food nanomaterial for animals [28, 31, 106]. Other agencies such as FSCJ (Florida State College at Jacksonville), EPA, OSHA, and CPSC have involved providing regulations related to the effective and safe use of NMs [46].

**TABLE 3.3** *In-vitro* Cytotoxicity Study of Biomaterials-Based Nanoparticles in Healthy Cells and Cancerous Cells

| Type of NM | Materials Used | Size (nm) | Zeta Potential (mv) | Dose | *In-Vitro* Cytotoxicity | | References |
|---|---|---|---|---|---|---|---|
| | | | | | Healthy Cells | Cancer Cells | |
| Nanoemulsion | Pea protein and Vitamin D | 170–350 | –25 | ≤0.0625 mg/ml | No toxic effect (cell viability > 80%) | – | [102] |
| | Rice bran protein, quercetin, and soybeans oil | 219.7 | –44.83 | 0.5–5 µM | Reduced cytotoxicity to normal cell than quercitin individual | – | [15] |
| Nano-encapsulate | Gallic acid and poly-lactic—co-glycolic acid (PLGA) | 225 | –22 | 20, 40, and 80 µg/ml | No toxicity to red blood cell lines | – | [6] |
| | Resveratrol and PLGA | – | – | – | No adverse effects on normal macro-phage cells | Induce apoptosis in prostate cancer cell lines | [74] |
| Nanoconjugate | Gold resveratrol | – | – | – | – | Enhanced cytotoxicity on Hep G 2 cell lines 9 times than free resveratrol at same concentration | [29] |
| | Curcumin, folate, and chitosan | 197.48 | –6.55 | – | Nearly no toxic to L-929 non-cancerous cell | Cytotoxic effect against MCF-7 breast cancer cell | [26] |
| | Silver, *Melia azedarach Linn* (Meliaceae or Neem) | 78 | –24.9 | 300 µg/ml | At lower concentra-tion does not exhibit any toxicity to the normal HBL cell | Cytotoxic effect against Hela cell lines | [98] |

**TABLE 3.4** *In-Vivo* Cytotoxicity Study of Biomaterials-Based Nanoparticles in Healthy Cells and Cancerous Cells

| NM | Type of NM | Particle Size (nm) | Zeta Potential | Dose | *In-Vivo* Cytotoxicity | | References |
|---|---|---|---|---|---|---|---|
| | | | | | Healthy Cell | Cancer Cell | |
| Nano-emulsion | *Citrus aurantium L.* bloom essential oils | 76.9 | –45 | 10 and 20 mg/kg/day | – | Induced overexpression of Cas-3 and trigger apoptic impacts on A-549 lung cancer cell line | [75] |
| | Vit E | 112 | +56 | 2 g/kg/day | No toxicity | – | [81] |
| Nano encapsulate | Curcuminoids | – | – | – | – | Cytotoxicity effects against lung, breast adenocarcinoma, hepatocellular carcinoma, and cervical carcinoma cells | [23] |
| | Albumin and Bevacizumab, PEG | 300 | – | 145 μg/mg | – | Reduction of glycolysis and tumor volume of HT-29 xenographic model of human colorectal cancer | [20] |
| Nano conjugate | Silver, *Melia azedarach Linn* (Meliaceae or Neem) | 78 | –24.9 | – | – | Showed enhanced cytotoxic effect against Dalton's ascites lymphoma (DAL) mice model than extract | [98] |
| | Silver, Longan fruit juice | 4–10 | – | 12.5–125 μg/ml | Did not exhibit any toxicity to normal healthy cells of Wistar albino rat | – | [56] |

Furthermore, USFDA indicates that all novel food materials (food packaging, additives, coloring agents, and dietary supplements) should go through the premarket review [32]. European Union (EU) makes strict mandatory guidelines for labeling and premarket safety assessment [88]. However, additional areas such as labeling waste management, and disposal are yet to be covered by the regulatory system [46]. Collaborative efforts are required to address the safety assessment of food nanotechnology concerning public health and policy-making guidance. Regulatory agencies must engage themselves to address safety issues, implement the risk regulation regarding nanotechnology based on collected outcomes from recent research studies.

## 3.6 PRESENT AND FUTURE ASPECTS OF NANOTECHNOLOGY IN THE MARKET AS CONSUMER PERCEPTION

The potentials of nanotechnology applications in food and pharma have been recognized. Nanotechnology has found its use in food processing, packaging, and storage applications. Approximate 150–600 nano-based food products and 400–500 nano-based food packaging applications are commercially available [18]. Nestle and Unilever have developed nanoemulsion incorporated ice cream with reduced fat content without compromising the fatty texture and flavors [90]. Commercially, nano encapsulated bioactive compounds such as vitamin and fatty acids are available for application in processing and preservation of food and beverages. Aqua nova, a German company developed nano micelles (30 nm) for encapsulation of vitamins C, E, and fatty acids to use in preservation.

Aqua nova sells the encapsulated products as "Nova Sol" name. One German magazine (Fleischwirtchaft) claimed about the advantageous effect of Nova sol. Nova sol helps in faster meat processing, improve the color stability, and also used as ready to eat liquid form [33]. A company of South Korea (Fine polymer Inc.) involve in making of fresh box food container using nanosilver (60–70 nm) and polyethylene (PE). They claimed that the incorporation of nanosilver provides an excellent antimicrobial property against several bacteria and fungi subsequently prolonging the freshness of food as compared to conventional food containers [14]. A list of commercially available nano products in food with advantageous properties are presented in Table 3.5.

**TABLE 3.5** Commercially Available Nano Products in Foods with Advantageous Properties

| Type of Food Application | Name of Product | Nano-Materials | Particle Size | Manufacture Company | Beneficial Properties | References |
|---|---|---|---|---|---|---|
| Preservatives | Aquasol preservative | Nanoencapsule of water-insoluble components | – | Aquanova | Enhance the preservative effect and improve the uptake of nutritional content | [57] |
| Additives | Bioral™<br>Omega 3<br>Nano-chelates | Nano-chelates | As small as 50 nm | Bio-delivery sciences international | Enhance the bioavailability of Omega 3 fatty acids to cakes, muffins, Pasta, soups, cookies and cereals | [57] |
| Functional beverages | Oat chocolate and oat vanilla nutritional drink | Iron nanoparticles | < 300 nm | Toddler health | Increase the reactivity as well as bioavailability of micronutrient nanoparticles | [57] |
| Contact materials | Teapot treated with silver nanoparticles | Silver nanoparticles | – | Song Sing Nanotechnology Co. | Exhibit antimicrobial activity | [4] |
| Contact materials | Nanosilver cutting board | Silver nanoparticles | – | A Do global | Exhibit antibacterial activity | [57] |
| Food packaging materials | Fresherlonger™ Plastic storage BagfresherLonger™Miracle food storage | Silver nanoparticles | 25 nm | Sharper Image$^R$Company, USA | Enhanced antimicrobial activity Food remain fresh 3 to 4 times longer | [14] |
| Food packaging materials | Sina Antibacterial Food storages | Silver nanoparticles, polypropylene, and silicon | – | Dai Dong Tien Corporation, Vietnam | Eliminate the bad smells; prevent the microorganism's growth; improve the shelf life of food | [14] |

Despite having all beneficial properties, nanotechnology-based food products are still waiting to come with fully commercialize. Because, the lack of knowledge about the benefits and safety of nanoproducts, unknown technology for consumers, and the absence of appropriate regulatory guidelines. However, it is estimated that the market of global nanotechnology will grow USD (United States Dollar) 112.49 billion during 2019–2023 [39]. This data offers an analysis of the consumer perceptions on nanofood and encourage in-depth study regarding the safety of nanotechnology.

## 3.7 SUMMARY

Bionanomaterials are currently lying in the investigative phase for food applications. Many new possibilities come with a wide range of applications. The use of bionanomaterials to improve the quality and safety of products will be supposed as progressive changes in existing technology. But its commercialization in the food market is limited due to the lack of data on nanopathology or toxicity and limited regulatory bodies. It is necessary to disseminate proper knowledge regarding the benefits and risks of such a novel food product to the consumer so that they can keep having a choice to use or not to use the nanotechnology-based products. Ultimately, the success of these developments will rely on consumer acceptances and regulatory guidelines. Nevertheless, the research outcomes in terms of the beneficial functionality of bionanomaterials in food products give hope for its potential use in developing safe nanotechnology-based food products.

## KEYWORDS

- **bionanomaterials**
- **cytotoxicity**
- **nano-antimicrobial**
- **nanoconjugates**
- **nanoparticles**
- **smart food packaging**

## REFERENCES


1. Abu-Saied, M. A., El-Nouby, M., Taha, T., El-Shafeey, M., Alshehri, A. G., Alamri, S., & Moustafa, M., (2020). Potential decontamination of drinking water pathogens through k-carrageenan integrated green bottle fly bio-synthesized silver nanoparticles. *Molecules, 25*(8), 1936–1940.
2. Aghaei, Z., Ghorani, B., Emadzadeh, B., Kadkhodaee, R., & Tucker, N., (2020). Protein-based halochromic electrospun nanosensor for monitoring trout fish freshness. *Food Control, 111*, 107065.
3. Akbari-Alavijeh, S., Shaddel, R., & Jafari, S. M., (2020). *In vivo* assays for evaluating the release of nanoencapsulated food ingredients. Chapter 5. In: Jafari, S., (ed.), *Release and Bioavailability of Nanoencapsulated Food Ingredients* (pp. 179–207). Cambridge – USA: Academic Press.
4. Alfadul, S. M., & Elneshwy, A. A., (2010). Use of nanotechnology in food processing, packaging and safety–review. *African Journal of Food, Agriculture, Nutrition and Development, 10*(6).
5. Ali, A., Rehman, A., Shehzad, Q., Khan, S., Karim, A., Afzal, N., Hussain, A., Yang, F., & Xia, W., (2020). Development and characterization of nanoemulsions incorporating tuna fish oil. *Int. J. Res. Agric. Sci., 7*(1), 2348–3997.
6. Alves, A. D. C. S., Mainardes, R. M., & Khalil, N. M., (2016). Nanoencapsulation of gallic acid and evaluation of its cytotoxicity and antioxidant activity. *Materials Science and Engineering: C, 60*, 126–134.
7. Bagheri, D. H., Bahrami, A., Jafari, S. M., & Williams, L., (2020). Micro/nanoencapsulation strategy to improve the efficiency of natural antimicrobials against *Listeria monocytogenes* in food products. *Critical Reviews in Food Science and Nutrition,* 1–19.
8. BahramiFeridoni, S., & KhademiShurmasti, D., (2020). Effect of the nanoencapsulated sour tea (*Hibiscus sabdariffa* L.) extract with carboxymethyl cellulose on quality and shelf life of chicken nugget. *Food Science & Nutrition, 8*(7), 3704–3715.
9. Bajpai, V. K., Kamle, M., Shukla, S., Mahato, D. K., Chandra, P., Hwang, S. K., Kumar, P., et al., (2018). Prospects of using nanotechnology for food preservation, safety, and security. *Journal of Food and Drug Analysis, 26*(4), 1201–1214.
10. Barabadi, H., Ovais, M., Shinwari, Z. K., & Saravanan, M., (2017). Anti-cancer green bionanomaterials: Present status and future prospects. *Green Chemistry Letters and Reviews, 10*(4), 285–314.
11. Barua, S., & Mitragotri, S., (2014). Challenges associated with penetration of nanoparticles across cell and tissue barriers: A review of current status and future prospects. *Nano Today, 9*(2), 223–243.
12. Bitencourt, A. P., Duarte, J. L., Oliveira, A. E., Cruz, R. A., Carvalho, J. C., Gomes, A. T., Ferreira, I. M., et al., (2018). Preparation of aqueous nanodispersions with annatto (*Bixa orellana* L.) extract using an organic solvent-free and low energy method. *Food Chemistry, 257*, 196–205.
13. Bulushi, I. A., Poole, S., Deeth, H. C., & Dykes, G. A., (2009). Biogenic amines in fish: Roles in intoxication, spoilage, and nitrosamine formation—A review. *Critical Reviews in Food Science and Nutrition, 49*(4), 369–377.

14. Bumbudsanpharoke, N., & Ko, S., (2015). Nano-food packaging: An overview of market, migration research, and safety regulations. *Journal of Food Science, 80*(5), R910–R923.
15. Chen, W., Ju, X., Aluko, R. E., Zou, Y., Wang, Z., Liu, M., & He, R., (2020). Rice bran protein-based nanoemulsion carrier for improving stability and bioavailability of quercetin. *Food Hydrocolloids,* 106042.
16. Chu, Y., Gao, C., Liu, X., Zhang, N., Xu, T., Feng, X., Yang, Y., Shen, X., & Tang, X., (2020). Improvement of storage quality of strawberries by pullulan coatings incorporated with cinnamon essential oil nanoemulsion. *LWT, 122*, 109054.
17. Chuah, L. H., Billa, N., Roberts, C. J., Burley, J. C., & Manickam, S., (2013). Curcumin-containing chitosan nanoparticles as a potential mucoadhesive delivery system to the colon. *Pharmaceutical Development and Technology, 18*(3), 591–599.
18. Chuah, L. H., Roberts, C. J., Billa, N., Abdullah, S., & Rosli, R., (2014). Cellular uptake and anticancer effects of mucoadhesive curcumin-containing chitosan nanoparticles. *Colloids and Surfaces B: Biointerfaces, 116*, 228–236.
19. De Mesquita, J. P., Donnici, C. L., Teixeira, I. F., & Pereira, F. V., (2012). Bio-based nanocomposites obtained through covalent linkage between chitosan and cellulose nanocrystals. *Carbohydrate polymers, 90*(1), 210–217.
20. De Redín, I. L., Expósito, F., Agüeros, M., Collantes, M., Peñuelas, I., Allemandi, D., & Irache, J. M., (2020). *In vivo* efficacy of bevacizumab-loaded albumin nanoparticles in the treatment of colorectal cancer. *Drug Delivery and Translational Research,* 1–11.
21. Dong, E., Yang, Z., Zhou, C., Wang, C., Li, S., Ouyang, Q., & Yang, P., (2019). pH-Responsive ultrasonic self-assembly Spinosad-loaded nanomicelles and their antifungal activity to *Fusarium oxysporum. Reactive and Functional Polymers, 141*, 123–132.
22. Donsì, F., Sessa, M., & Ferrari, G., (2012). Effect of emulsifier type and disruption chamber geometry on the fabrication of food nanoemulsions by high pressure homogenization. *Industrial & Engineering Chemistry Research, 51*(22), 7606–7618.
23. Dos Santos, P. D. F., Francisco, C. R. L., Coqueiro, A., Leimann, F. V., Pinela, J., Calhelha, R. C., et al., (2019). The nanoencapsulation of curcuminoids extracted from *curcuma longa* L. and an evaluation of their cytotoxic, enzymatic, antioxidant and anti-inflammatory activities. *Food & Function, 10*(2), 573–582.
24. Dufresne, A., (2010). Processing of polymer nanocomposites reinforced with polysaccharide nanocrystals. *Molecules, 15*(6), 4111–4128.
25. Dukhin, S. S., & Labib, M. E., (2013). Convective diffusion of nanoparticles from the epithelial barrier toward regional lymph nodes. *Advances in Colloid and Interface Science, 199*, 23–43.
26. Esfandiarpour-Boroujeni, S., Bagheri-Khoulenjani, S., Mirzadeh, H., & Amanpour, S., (2017). Fabrication and study of curcumin loaded nanoparticles based on folate-chitosan for breast cancer therapy application. *Carbohydrate Polymers, 168*, 14–21.
27. Esmaeili, H., Cheraghi, N., Khanjari, A., Rezaeigolestani, M., Basti, A. A., Kamkar, A., & Aghaee, E. M., (2020). Incorporation of nanoencapsulated garlic essential oil into edible films: A novel approach for extending shelf life of vacuum-packed sausages. *Meat Science,* 108135.
28. European Food Safety Authority (EFSA), (2009). The potential risks arising from nanoscience and nanotechnologies on food and feed safety. *EFSA Journal, 7*(3), 958.

29. Fadel, M., Kassab, K., Youssef, T., & El-Kholy, A. I., (2019). One-step synthesis of phyto-polymer-coated gold nanospheres as a delivery system to enhance resveratrol cytotoxicity. *Drug Development and Industrial Pharmacy, 45*(6), 937–945.
30. Fan, Y., Liu, Y., Gao, L., Zhang, Y., & Yi, J., (2018). Improved chemical stability and cellular antioxidant activity of resveratrol in zein nanoparticle with bovine serum albumin-caffeic acid conjugate. *Food Chemistry, 261*, 283–291.
31. FDA, US (US Food and Drug Administration), (2015). *Guidance for Industry: Use of Nanomaterials in Food for Animals*. Food and Drug Administration, US Department of Health and Human Services. https://www.fda.gov/media/88828/download (accessed on 11 February 2022).
32. FDA, US (US Food and Drug Administration). (2018). *FDA's Approach to Regulation of Nanotechnology Products.* https://www.fda.gov/science-research/nanotechnology-programs-fda/fdas-approach-regulation-nanotechnology-products#:~:text=FDA%20will%20regulate%20nanotechnology%20products,of%20product%20under%20its%20jurisdiction.&text=FDA%20intends%20to%20ensure%20transparent,in%20the%20best%20available%20science (accessed on 11 February 2022).
33. Foujdar, R., Bera, M. B., & Chopra, H. K., (2020). Phenolic nanoconjugates and its application in food. In: Pal, K., Banerjee, I., Sarkar, P., Kim, D., Deng, W. P., Dubey, N. K., & Mjumder, K., (eds.), *Biopolymer-Based Formulations* (pp. 751–780). Cambridge – USA: Elsevier Inc.
34. Foujdar, R., Chopra, H. K., Bera, M. B., Chauhan, A. K., & Mahajan, P., (2020). Effect of probe ultrasonication, microwave and sunlight on biosynthesis, bioactivity and structural morphology of *Punica granatum* peel's polyphenols-based silver nanoconjugates. *Waste and Biomass Valorization,* 1–20.
35. Gallocchio, F., Belluco, S., & Ricci, A., (2020). Nanotechnology and food: Brief overview of the current scenario. *Procedia Food Science, 5*, 85–88.
36. Ganta, S., Devalapally, H., Shahiwala, A., & Amiji, M., (2008). A review of stimuli-responsive nanocarriers for drug and gene delivery. *Journal of Controlled Release, 126*(3), 187–204.
37. García-Pinilla, S., Villalobos-Espinosa, J. C., Cornejo-Mazón, M., & Gutiérrez-López, G. F., (2019). 12 nanotechnologies in food processing. *Advances in Processing Technologies for Bio-Based Nanosystems in Food, 1*, 259.
38. Gaucher, G., Dufresne, M. H., Sant, V. P., Kang, N., Maysinger, D., & Leroux, J. C., (2005). Block copolymer micelles: preparation, characterization and application in drug delivery. *Journal of Controlled Release, 109*(1–3), 169–188.
39. Global Food Nanotechnology Market, (2019–2023). *Growth Analysis and Forecast.* https://www.businesswire.com/news/home/20190117005434/en/Global-Food-Nanotechnology-Market-2019-2023-Growth-Analysis-and-Forecast-Technavio (accessed on 11 February 2022).
40. Gómez-Estaca, J., Balaguer, M. P., Gavara, R., & Hernandez-Munoz, P., (2012). Formation of zein nanoparticles by electrohydrodynamic atomization: Effect of the main processing variables and suitability for encapsulating the food coloring and active ingredient curcumin. *Food Hydrocolloids, 28*(1), 82–91.
41. Gupta, A., (2020). Nanoemulsions. In: Chung, E. J., Leon, L., & Rinaldi, C., (eds.), *Nanoparticles for Biomedical Applications: Fundamental Concepts, Biological Interactions and Clinical Applications* (pp. 371–384). Cambridge – USA: Elsevier Inc.

42. Gupta, A., Eral, H. B., Hatton, T. A., & Doyle, P. S., (2016). Nanoemulsions: Formation, properties and applications. *Soft Matter, 12*(11), 2826–2841.
43. Gupta, A., Eral, H. B., Hatton, T. A., & Doyle, P. S., (2016b). Controlling and predicting droplet size of nanoemulsions: Scaling relations with experimental validation. *Soft Matter, 12*(5), 1452–1458.
44. Gupta, P., Singh, M., Kumar, R., Belz, J., Shanker, R., Dwivedi, P. D., & Singh, S. P., (2018). Synthesis and *In Vitro* studies Of PLGA-DTX nanoconjugate as potential drug delivery vehicle for oral cancer. *International Journal of Nanomedicine, 13*(T-NANO 2014 Abstracts), 67.
45. Hannon, J. C., Kerry, J., Cruz-Romero, M., Morris, M., & Cummins, E., (2015). Advances and challenges for the use of engineered nanoparticles in food contact materials. *Trends in Food Science & Technology, 43*(1), 43–62.
46. He, X., Deng, H., Aker, W. G., & Hwang, H. M., (2019). Regulation and safety of nanotechnology in the food and agriculture industry. In: *Food Applications of Nanotechnology* (pp. 1–12). Boca Raton – FL: CRC Press.
47. Hernández-Nava, R., López-Malo, A., Palou, E., Ramírez-Corona, N., & Jiménez-Munguía, M. T., (2020). Encapsulation of oregano essential oil (*Origanum vulgare*) by complex coacervation between gelatin and chia mucilage and its properties after spray drying. *Food Hydrocolloids,* 106077.
48. Holsapple, M. P., Farland, W. H., Landry, T. D., Monteiro-Riviere, N. A., Carter, J. M., Walker, N. J., & Thomas, K. V., (2005). Research strategies for safety evaluation of nanomaterials, part II: Toxicological and safety evaluation of nanomaterials, current challenges and data needs. *Toxicological Sciences, 88*(1), 12–17.
49. Huang, Y., Gu, C., He, S., Zhu, D., Liu, X., & Chen, Z., (2020). Development and characterization of an edible chitosan–whey protein nanocomposite film for chestnut (*Castanea mollissima* Bl.) preservation. *Journal of Food Science,* 75–86.
50. Husseini, G. A., Myrup, G. D., Pitt, W. G., Christensen, D. A., & Rapoport, N. Y., (2000). Factors affecting acoustically triggered release of drugs from polymeric micelles. *Journal of Controlled Release, 69*(1), 43–52.
51. Ilyas, R. A., Sapuan, S. M., Ishak, M. R., & Zainudin, E. S., (2018). Development and characterization of sugar palm nanocrystalline cellulose reinforced sugar palm starch bionanocomposites. *Carbohydrate Polymers, 202*, 186–202.
52. Israelachvili, J. N., (1992). Intermolecular and surface forces. In: *Intermolecular and Surface Forces* (p. 567). London – UK: Academic Press.
53. Joyner, J. J., & Kumar, D. V., (2015). Nanosensors and their applications in food analysis: A review. *The International Journal of Science and Technology, 3*(4), 80.
54. Kanikireddy, V., Varaprasad, K., Rani, M. S., Venkataswamy, P., Reddy, B. J. M., & Vithal, M., (2020). Biosynthesis Of CMC-guar gum-Ag0 nanocomposites for inactivation of food pathogenic microbes and its effect on the shelf life of strawberries. *Carbohydrate Polymers, 236*, 116053.
55. Kaput, J., Kussmann, M., Mendoza, Y., Le Coutre, R., Cooper, K., & Roulin, A., (2015). Enabling nutrient security and sustainability through systems research. *Genes & Nutrition, 10*(3), 12.
56. Khan, A. U., Wei, Y., Ahmad, A., Khan, Z. U. H., Tahir, K., Khan, S. U., & Yuan, Q., (2016). Enzymatic browning reduction in white cabbage, potent antibacterial and antioxidant activities of biogenic silver nanoparticles. *Journal of Molecular Liquids, 215*, 39–46.

57. Kirdar, S. S., (2012). A survey on chemical, biochemical and microbiological characteristics of a traditional dairy product in Mediterranean region: Kes. *Journal of Animal and Veterinary Advances, 11*(3), 330–334.
58. Komaiko, J. S., & Mcclements, D. J., (2016). Formation of food-grade nanoemulsions using low-energy preparation methods: A review of available methods. *Comprehensive Reviews in Food Science and Food Safety, 15*(2), 331–352.
59. Kumar, D. D., Mann, B., Pothuraju, R., Sharma, R., & Bajaj, R., (2016). Formulation and characterization of nanoencapsulated curcumin using sodium caseinate and its incorporation in ice cream. *Food & Function, 7*(1), 417–424.
60. Langer, R., & Peppas, N. A., (2003). Advances in biomaterials, drug delivery, and bionanotechnology. *Aiche Journal, 49*(12), 2990–3006.
61. Li, Q., Gao, R., Wang, L., Xu, M., Yuan, Y., Ma, L., & Yang, X., (2020). Nanocomposites of bacterial cellulose nanofibrils and zein nanoparticles for food packaging. *ACS Applied Nano Materials, 3*(3), 2899–2910.
62. Lucchini, R. G., Dorman, D. C., Elder, A., & Veronesi, B., (2012). neurological impacts from inhalation of pollutants and the nose–brain connection. *Neurotoxicology, 33*(4), 838–841.
63. Luechinger, N. A., Loher, S., Athanassiou, E. K., Grass, R. N., & Stark, W. J., (2007). Highly sensitive optical detection of humidity on polymer/metal nanoparticle hybrid films. *Langmuir, 23*(6), 3473–3477.
64. Madl, A. K., Plummer, L. E., Carosino, C., & Pinkerton, K. E., (2014). Nanoparticles, lung injury, and the role of oxidant stress. *Annual Review of Physiology, 76*, 447–465.
65. Mcclements, D. J., & Rao, J., (2011). Food-grade nanoemulsions: Formulation, fabrication, properties, performance, biological fate, and potential toxicity. *Critical Reviews in Food Science and Nutrition, 51*(4), 285–330.
66. Mesquita, L. D. R., Rahal, S. C., De Faria, L. G., Miceli, Y., Govoni, V. M., Gomide, P., De Souza, J. A. L., & Siqueira, E. G. M., (2013). Nanoskin®-nanotechnology applied to skin healing in a dog. *Acta Veterinaria Brasilica, 7*(Suppl. 1), 125–127.
67. Mishra, B. B. T. S., Patel, B. B., & Tiwari, S., (2010). Colloidal nanocarriers: A review on formulation technology, types and applications toward targeted drug delivery. *Nanomedicine: Nanotechnology, Biology and Medicine, 6*(1), 9–24.
68. Mohammadi, M., Jafari, S. M., Hamishehkar, H., & Ghanbarzadeh, B., (2020). Phytosterols as the core or stabilizing agent in different nanocarriers. *Trends in Food Science & Technology*, 25–32.
69. Muhammad, D. R. A., Gonzalez, C. G., Doost, A. S., Van De, W. D., Van, D. M. P., & Dewettinck, K., (2019). Improvement of antioxidant activity and physical stability of chocolate beverage using colloidal cinnamon nanoparticles. *Food and Bioprocess Technology, 12*(6), 976–989.
70. Muñoz-Pina, S., Ros-Lis, J. V., Argüelles, Á., & Andrés, A., (2020). Use of nanomaterials as alternative for controlling enzymatic browning in fruit juices. In: *Nanoengineering in the Beverage Industry* (pp. 163–196). London – UK: Academic Press.
71. Muzzarelli, R. A., (2011). Chitin nanostructures in living organisms. In: *Chitin* (pp. 1–34). Dordrecht – Netherlands: Springer International Publishing.
72. NANO (National Nanotechnology Initiative), (2014). *What's so Special about the Nanoscale?*. https://www.nano.gov/nanotech-101/special (accessed on 11 February 2022).

73. Narsaiah, K., Sharma, M., Sridhar, K., & Dikkala, P., (2019). Garlic oil nanoemulsions hybridized in calcium alginate microcapsules for functional bread. *Agricultural Research, 8*(3), 356–363.
74. Nassir, A. M., Shahzad, N., Ibrahim, I. A., Ahmad, I., Md, S., & Ain, M. R., (2018). Resveratrol-loaded PLGA nanoparticles mediated programmed cell death in prostate cancer cells. *Saudi Pharm. J., 26*, 876–885.
75. Navaeishoorvarzi, S., Shahraki, F., Shafaei, N., Karimi, E., & Oskoueian, E., (2020). *Citrus aurantium* L. bloom essential oil nanoemulsion: Synthesis, characterization, cytotoxicity, and its potential health impacts on mice. *Journal of Food Biochemistry, 44*(5), E13181.
76. Nazir, S., & Azad, Z. A. A., (2019). Food nanotechnology: An emerging technology in food processing and preservation. In: Malik, A., Erginkaya, Z., & Erten, H., (eds.), *Health and Safety Aspects of Food Processing Technologies* (pp. 567–576). Basel – Switzerland: Springer International Publishing.
77. Nazrin, A., Sapuan, S. M., Zuhri, M. Y. M., Ilyas, R. A., Syafiq, R., & Sherwani, S. F. K., (2020). Nanocellulose reinforced thermoplastic starch (TPS), polylactic acid (PLA), and polybutylene succinate (PBS) for food packaging applications. *Frontiers in Chemistry, 8*, 5–79.
78. Oberdörster, G., Maynard, A., Donaldson, K., Castranova, V., Fitzpatrick, J., Ausman, K., & Olin, S., (2005). Principles for characterizing the potential human health effects from exposure to nanomaterials: Elements of a screening strategy. *Particle and Fibre Toxicology, 2*(1), 8.
79. Omanović-Mikličanina, E., & Maksimović, M., (2016). Nanosensors applications in agriculture and food industry. *Bull. Chem. Technol. Bosnia Herzegovina, 47*, 59–70.
80. Pande, V. V., & Sanklecha, V. M., (2017). Bionanocomposite: A review. *Austin J. Nanomednanotechnol., 5*(1), 1045.
81. Park, C. Y., Jang, C. H., Lee, D. Y., Cho, H. T., Kim, Y. J., Park, Y. H., & Imm, J. Y., (2017). Changes in hepatic gene expression and serum metabolites after oral administration of overdosed vitamin-E-loaded nanoemulsion in rats. *Food and Chemical Toxicology, 109*, 421–427.
82. Perotti, G. F., Tronto, J., Bizeto, M. A., Izumi, C., Temperini, M. L., Lugão, A. B., & Constantino, V. R., (2014). Biopolymer-clay nanocomposites: Cassava starch and synthetic clay cast films. *Journal of the Brazilian Chemical Society, 25*(2), 320–330.
83. Qin, Y., Liu, Y., Yuan, L., Yong, H., & Liu, J., (2019). Preparation and characterization of antioxidant, antimicrobial and pH-sensitive films based on chitosan, silver nanoparticles and purple corn extract. *Food Hydrocolloids, 96*, 102–111.
84. Radomski, A., Jurasz, P., Alonso-Escolano, D., Drews, M., Morandi, M., Malinski, T., & Radomski, M. W., (2005). Nanoparticle-induced platelet aggregation and vascular thrombosis. *British Journal of Pharmacology, 146*(6), 882–893.
85. Rai, M., Ingle, A. P., Gupta, I., Pandit, R., Paralikar, P., Gade, A., & Dos, S. C. A., (2019). Smart nanopackaging for the enhancement of food shelf life. *Environmental Chemistry Letters, 17*(1), 277–290.
86. Raviadaran, R., Ng, M. H., Manickam, S., & Chandran, D., (2020). Ultrasound-assisted production of palm oil-based isotonic W/O/W multiple nanoemulsion encapsulating both hydrophobic tocotrienols and hydrophilic caffeic acid with enhanced stability using oil-based sucragel. *Ultrasonics Sonochemistry, 64*, 104995.

87. Ravichandran, R., (2010). Nanotechnology applications in food and food processing: Innovative green approaches, opportunities and uncertainties for global market. *International Journal of Green Nanotechnology: Physics and Chemistry, 1*(2), P72–P96.
88. Ravichandran, R., Venugopal, J. R., Sundarrajan, S., Mukherjee, S., Sridhar, R., & Ramakrishna, S., (2012). Minimally invasive injectable short nanofibers of poly (glycerol sebacate) for cardiac tissue engineering. *Nanotechnology, 23*(38), 385102.
89. Rehan, F., Ahemad, N., & Gupta, M., (2019). Casein nanomicelle as an emerging biomaterial—A comprehensive review. *Colloids and Surfaces B: Biointerfaces, 179*, 280–292.
90. Renton, A., (2006). *Welcome to the World of Nanofoods*. Guardian Unlimited, UK, http://observer.guardian.co.uk/foodmonthly/futureoffood/story/0,1971266,00.html (accessed on 11 February 2022).
91. Santeramo, F. G., Carlucci, D., De Devitiis, B., Seccia, A., Stasi, A., Viscecchia, R., & Nardone, G., (2018). Emerging trends in european food, diets and food industry. *Food Research International, 104*, 39–47.
92. Schubert, H., Ax, K., & Behrend, O., (2003). Product engineering of dispersed systems. *Trends in Food Science & Technology, 14*(1, 2), 9–16.
93. Sessa, M., Balestrieri, M. L., Ferrari, G., Servillo, L., Castaldo, D., D'Onofrio, N., & Tsao, R., (2014). Bioavailability of encapsulated resveratrol into nanoemulsion-based delivery systems. *Food Chemistry, 147*, 42–50.
94. Setälä, H., Alakomi, H. L., Paananen, A., Szilvay, G. R., Kellock, M., Lievonen, M., & Kostiainen, M., (2020). Lignin nanoparticles modified with tall oil fatty acid for cellulose functionalization. *Cellulose, 27*(1), 273–284.
95. Sheth, T., Seshadri, S., Prileszky, T., & Helgeson, M. E., (2020). Multiple nanoemulsions. *Nature Reviews Materials,* 1–15.
96. Singh, T., Shukla, S., Kumar, P., Wahla, V., Bajpai, V. K., & Rather, I. A., (2017). Application of nanotechnology in food science: Perception and overview. *Frontiers in Microbiology, 8*, 1501.
97. Sukhanova, A., Bozrova, S., Sokolov, P., Berestovoy, M., Karaulov, A., & Nabiev, I., (2018). Dependence of nanoparticle toxicity on their physical and chemical properties. *Nanoscale Research Letters, 13*(1), 44.
98. Sukirtha, R., Priyanka, K. M., Antony, J. J., Kamalakkannan, S., Thangam, R., Gunasekaran, P., & Achiraman, S., (2012). Cytotoxic effect of green synthesized silver nanoparticles using Melia azedarach against *in vitro* Hela cell lines and lymphoma mice model. *Process Biochemistry, 47*(2), 273–279.
99. Sun, X., Jia, P., Zhe, T., Bu, T., Liu, Y., Wang, Q., & Wang, L., (2019). Construction and multifunctionalization of chitosan-based three-phase nano-delivery system. *Food Hydrocolloids, 96*, 402–411.
100. Taccola, S., Greco, F., Zucca, A., Innocenti, C., De JuliánFernández, C., Campo, G., & Mattoli, V., (2013). Characterization of free-standing PEDOT: PSS/iron oxide nanoparticle composite thin films and application as conformable humidity sensors. *ACS Applied Materials & Interfaces, 5*(13), 6324–6332.
101. Tran, T. T., & Tran, P. H., (2019). Nanoconjugation and encapsulation strategies for improving drug delivery and therapeutic efficacy of poorly water-soluble drugs. *Pharmaceutics, 11*(7), 325.
102. Walia, N., & Chen, L., (2020). Pea protein-based vitamin d nanoemulsions: Fabrication, stability and *in vitro* study using Caco-2 cells. *Food Chemistry, 305*, 125475.

103. Wang, H., Yang, Z., He, Z., Zhou, C., Wang, C., Chen, Y., & Li, P., (2019). Self-assembled amphiphilic chitosan nanomicelles to enhance the solubility of quercetin for efficient delivery. *Colloids and Surfaces B: Biointerfaces, 179*, 519–526.
104. Wang, Y., & Duncan, T. V., (2017). Nanoscale sensors for assuring the safety of food products. *Current Opinion in Biotechnology, 44*, 74–86.
105. Wang, Y., Zhang, R., Qin, W., Dai, J., Zhang, Q., Lee, K., & Liu, Y., (2020). Physicochemical properties of gelatin films containing tea polyphenol-loaded chitosan nanoparticles generated by electrospray. *Materials & Design, 185*, 108277.
106. WHO (World Health Organization), (2013). *State of the Art on the Initiatives and Activities Relevant to Risk Assessment and Risk Management of Nanotechnologies in the Food and Agriculture Sectors: FAO/WHO Technical Paper*. World Health Organization.
107. Xu, W., Zhu, D., Xu, B., Huang, L., Xiong, Y., Ge, P., & Wang, Z., (2019). Preparation, antibacterial and antioxidant properties of green tea seed oil nanoemulsions by self-emulsification method. *Micro & Nano Letters, 14*(12), 1219–1222.
108. Zardini, A. A., Mohebbi, M., Farhoosh, R., & Bolurian, S., (2018). Production and characterization of nanostructured lipid carriers and solid lipid nanoparticles containing lycopene for food fortification. *Journal of Food Science and Technology, 55*(1), 287–298.
109. Zhai, X., Li, Z., Shi, J., Huang, X., Sun, Z., Zhang, D., & Gong, Y., (2019). A colorimetric hydrogen sulfide sensor based on gellan gum-silver nanoparticles bionanocomposite for monitoring of meat spoilage in intelligent packaging. *Food Chemistry, 290*, 135–143.
110. Zhang, L., Han, C., Liu, M., Yang, H., Zhang, F., Liu, B., & Meng, X., (2020). The formation, stability of DHA/EPA nanoemulsion prepared by emulsion phase inversion method and its application in apple juice. *Food Research International, 133*, 109132.

# CHAPTER 4

# POTENTIAL OF POLYSACCHARIDE NANOPARTICLES IN FOODS

HIMJYOTI DUTTA

## ABSTRACT

With the advent and advancement of nanotechnology in the past decades, biological nanomaterials (NMs) have gained tremendous importance due to an abundance of raw materials, biodegradability, and hazard-free nature. Top-down extraction and bottom-up synthesis procedures are mostly adopted to develop polysaccharide-based nanostructures. Most of their uses in the food industry have been in designing biodegradable packaging materials, edible nanoemulsions for retaining storage quality of fresh horticultural produce, as encapsulating material of functional compounds for site-specific delivery, and for stabilization of food emulsions. Attempts have been carried out to develop new, safe, economic, and low-cost technologies for their production to expand their application in the food industry. Few lesser-known polysaccharides (e.g., xylan, mannan, minor gums, etc.), have also been identified to have important nanoparticulate characteristics suitable for specific food uses.

## 4.1 INTRODUCTION

Nanotechnology is one of the greatest scientific advancements in exploring structural attributes of materials in its miniature forms. Nanotechnology studies generally encompass extraction, synthesis, or development of materials in nanometer (typically 1–100 nm) ranges, followed by their characterization for possible applications in different utility sectors. Nanosized

Nanotechnology Horizons in Food Process Engineering, Volume 1. Megh R. Goyal, PhD, PE, Junaid Ahmad Malik, PhD, Satish Kumar, PhD & Ritesh B. Watharkar, PhD (Eds.)

materials have several novel properties that lead to their intended use in industries that are designing and producing innovative products. Studies on inorganic nanomaterials (NMs) like those from heavy metals paved the way for nanotechnology in different industries, including electronics, telecommunication, gadgets, cosmetics, detergents, and medicine. However, non-renewability, non-biodegradability, toxicity, and lower availability compelled scientists to look for biological organic alternatives, raw materials that are readily available, biodegradable, and most importantly, biocompatible, and low cost. Such NMs derived from biological sources were collectively termed as 'bionanomaterials' and have been extensively researched during the recent decades for potential industrial applications.

Bionanomaterials that are mostly designed from proteins and carbohydrates have been used for the development of drugs and their delivery, fabrication of biosensors, tissue engineering, and diagnostic probes. Here, by the term nanomaterials, we are not considering the nanosized entities that are naturally present in a biological system, but referring to those which are intentionally engineered and developed, stabilized, and applied for their specific functional roles. They are often synthesized in conjugation with inorganic molecules like different metal and metal oxides for imparting better stability and target-based application. The utilization of NMs and the principles of nanotechnology in the food industry is relatively a newer field. Nanotechnology has seen very recent progress in different areas of nanotextured foods (as nanosized ingredients), nanocarrier systems for effective and improved delivery of nutrients and supplements (nanoencapsulation), additives (nanosized colors, preservatives, flavorings, antimicrobials), packaging (nanoparticle impregnated polymers for improved mechanical, functional, barrier, and antimicrobial properties), coatings, and contact surfaces and for water decontamination. Broadly, the role of nanotechnology has been recognized for improved quality of food, extended shelf life, and adding to nutritional benefits in a very short period. A larger segment of bionanomaterials encompasses the application of active inorganic molecules along with their organic counterparts. Toxicity risks have suppressed their utility in edible products. Available literature suggests that studies that have been carried out on nanotechnology use in foods are mostly limited to packaging, barrier, and coating materials. Bionanomaterials with only biological components in them are more feasible for incorporation in food systems and require more extensive research and development to meet the current demand for innovative and safe food alternatives. However, few of the metal and metal oxide nanoparticles have been used for their prospective role as preservative, and for binding and intra-colon antimicrobial activities. Safe and efficient extraction and synthesis procedures for obtaining appropriate bionanomaterials for food use have been developed and standardized over the period as potential improvements in the field of food nanotechnology.

Carbohydrates, comprising more than 80% of the earth's biomass are the most abundant class of biopolymers with a strong tendency to form supramolecular networks. The polysaccharide is the most complex class of carbohydrates formed by repeated binding of (more than 10) monosaccharide units by glycosidic bonds. They have been again categorized into storage polysaccharides (e.g., starch and glycogen) and structural polysaccharides (e.g., cellulose and chitin). Based on the type of repeating units, they can also be categorized as homopolysaccharides (only one repeating monosaccharide, e.g., starch and cellulose) and heteropolysaccharides (repeating disaccharide units, e.g., pectin and heparin). In each repeating disaccharide unit of a heteropolysaccharide, one of the two monosaccharide units is N-acetylglucosamine or N-acetyl galactosamine and the other is a uronic acid. The glycosidic chains in the polysaccharide molecules are so arranged that many of them develop portions of crystalline domains due to regular arrangements. The semicrystalline nature of the polysaccharide relates to their versatile mechanical, structural, and thermal properties, enabling them towards applications as raw materials and ingredients for various food and non-food applications.

Biopolymeric properties of polysaccharides can be suitably modified by altering the glycosidic chain lengths, hydrogen, and hydrophobic bonding characteristics and chemically binding with other molecules. Their sensitivity to changing environment can be well utilized as per physiological changes within the human gastrointestinal tract. Most of the polysaccharide-based NMs used in biomedical fields are in nanoparticulate and nanogel forms. Polysaccharides extracted from various natural sources have been identified as substrates for material designing and may behave as alternatives to synthetic compounds for designing engineered NMs. Considering the context here, mention of the widely reported synthesis on inorganic nanoparticles and their growth on macro-structured polysaccharide substrates for nonfood uses have been avoided. The general tendency of glycosidic chains to form supramolecular networks makes them suitable for designing nano-clusters or nano-gels for food application. Various physical, chemical, and enzymatic modifications have been successfully applied to polysaccharides to improve their material behavior, solubility, dispersibility, stability, and chemical responsiveness. Such functional changes have been suitably applied for nanoengineering of polysaccharide-based standalone or conjugated nanoparticles and target them for food applications.

In this chapter, innovations in the production of different polysaccharide nanoparticles of plant, animal, and microbial origins and their nature have been thoroughly discussed. This chapter also deliberates applications of the individual class of polysaccharide nanoparticles in the food industry.

## 4.2 MAJOR POLYSACCHARIDES

### *4.2.1 CELLULOSE*

Cellulose is the most abundant biopolymer on earth. It has low density, low thermal expansion, good mechanical properties, and strength. It is a structural polysaccharide, primarily found in plant cell walls and also in animals like tunicates and some algae (Genus *Cladophorales* and *Siphonocladales*) and bacteria. It is a homopolymer of D-glucose monomeric units linked by β-1,4 glycosidic linkages, forming straight long chains. Multiple parallel chains stack together to form elementary fibrils, which again cluster together by van der Waals forces and H-bonding interactions to form larger cellulose microfibrils. The crystallinity of semicrystalline cellulose is due to the highly ordered regions of the microfibrils. Non-ordered portions and non-cellulosic materials like lignin and hemicellulose adhering to the microfibrils contribute to the amorphousness. In nature, cellulose fibrils are embedded within matrices of variable masses of these amorphous biomaterials.

To extract and obtain the pure form of the micron-sized functional cellulose, these components have to be removed. Several treatments and their combinations have been proposed for different plant substrates for effective removal of non-cellulosic components before hydrolytically defragmenting the pure cellulose to nanosized structures. As cellulose is not digested in the human gastrointestinal tract, it is a major fiber which provides non-energy liberating bulkiness to foods and may add to colon health by exerting a prebiotic effect. The cellulose source, cellulose extraction, and nanocellulose preparation method and process conditions decide the morphology, dimensions, yield, surface chemistry, and crystallinity of the nanoparticles. Use of chemicals for purification of the extracted cellulose and breaking the cellulose down to nanocellulose often restrict the use of these methods for preparation of their products aimed for food usage. Therefore, cellulose produced by bacteria can be a potential alternative.

### 4.2.2 STARCH

Starch is the most abundant edible polysaccharide on earth. Physically, starch exists as granules of varying sizes and shapes. It occurs within amyloplasts of plant cells and is the form of stored energy in plants. Chemically, starch is a polymer of D-glucose units linked by α-glycosidic linkages. The macromolecule is composed of two structurally different fractions, namely amylose, which is the linear fraction with only α-1,4 bonds, and primarily responsible for starch's amorphousness and amylopectin, the branched fraction formed by additional α-1,6 branchpoint linkages. Inter-glycosidic side-chain supercoiling of amylopectin majorly contributes to the ordered regions of the polysaccharide, resulting in crystallinity.

Glycosidic chains, under specific conditions, tend to coil lipid molecules to form starch-lipid crystalline complexes. Although long chains of amylose are known to form amylose-lipid complexes, the formation of amylopectin-lipid complexes have also been evidenced by starch researchers. In an early review of researches on NMs extracted from starch, Rajeswari et al. [118] identified two kinds of starch nanostructures based on extraction protocols, namely starch nanocrystals which are crystallite portions obtained after selective acid hydrolysis of the amorphous portions and starch nanoparticles which are produced from gelatinized starch and are largely amorphous in nature.

### 4.2.3 GLYCOGEN

Glycogen is a glucose polysaccharide occurring in most animal tissues and some plants. It is the readily available source of stored glucose in animal liver and muscle tissues. Therefore, in a way, it corresponds to starch in plants. Similar to starch, the glycosidic bonding pattern in glycogen is also an admixture of α-1,4 and α-1,6 bonds. Each granule of glycogen looks like a spherical hyperbranched tree with several branches originating from a core located at the center. The core consists of glycogenin, an enzyme which acts as a primer for the initial construction of the glycosyl monomers and linkages.

Glycogen is highly hydrophilic and can hold water up to 3–4 times its own dry weight. The hygroscopicity of the biomaterial can be explored for its use in suitable micro- and nanoengineered systems. Phytoglycogen is a naturally occurring nanoparticulate glycogen that has gained recent popularity. Phytoglycogen can be extracted from its sources like sweet corn kernels. Very few reports on NMs extracted from this biopolymer and their application are available in the current scientific literature.

### 4.2.4 PECTIN

Pectin is a group of high molecular weight heteropolysaccharide found in the cells walls and intercellular tissues of certain plants. The primary backbone is mainly composed of repeating galacturonic acid monomers linked by α-1,4 glycosidic bonds, with substitution of α-1,2 rhamnopyranose units at certain points. From these points, side chains of galactose, mannose, glucose, and xylose appear. Methyl esterification of galacturonic acid residues occurs naturally. Pectin has been classified as high methoxyl (>50%) and low methoxyl types. The precursor protopectin develops to pectin with maturity. Pectin present in the peel and pulp of fruits helps in maintaining the shape and texture of the fruits. With ripening, pectin degrades into simple sugars, thereby softening the tissue texture. Traditionally, pectin has been used as a popular gelling agent due to its inherent property to form a gel on addition of water. In the modern food industry, pectin has been used in beverages and semisolid food systems to give the desired thickness and mouthfeel. In sugar-free beverages, where the desired mouth feel is lost due to the removal of sugars, a low concentration of pectin is used to regain the body.

Generally, high methyl-esterified pectin is used in beverages. Low methoxyl pectin requires the addition of extraneous calcium to form a strong gel. Generally, pectin is extracted from apple pomace and the peel of citrus fruits. Pectin gel does not get digested in the human intestine and is a fiber that is degraded by the colon microflora. In the gastrointestinal tract, pectin gel can bind food components, ions, and bile acid and hence regulate their digestion, absorption, and utilization. This also relates to its cholesterol and lipid regulating and gastric emptying properties. Such properties of the polysaccharide have been well-utilized for its application in encapsulating and targeted delivery of drugs and functional molecules in the colon. Its other popular food applications are as adsorbent, emulsion stabilizer, and bulking agent. Rather than synthesizing or extracting as a nanomaterial, the specific structure of pectin nanoparticle/nanomaterial has been usually related to its synthesized and applied nanodimensional forms.

### 4.2.5 CHITIN AND CHITOSAN (CS)

Chitin is a primary component in the cell walls of fungi, the exoskeleton of insects and crustaceans, arthropods, scales of fish, and some amphibians. It is a major waste of the seafood processing industries and is the second most abundant biopolymer, after cellulose. The macromolecule is composed of linear repeating β-1,4 N-acetyl-D-glucosamine units, resembling the structure of cellulose except for the second carbon of cellulose structure is replaced by acetamido group in chitin. Its white color, inert, non-toxic nature, strength, biodegradability, and biocompatibility make it an ideal biomaterial for designing eco-friendly standalone and composite structures. In nature, chitin exists in tightly bound microfibrillar arrangements like in cellulose. Each microfibril is composed of alternate crystalline and amorphous domains and is a bundle of nanofibers. CS is the major functional derivative of chitin. CS is prepared by deacetylation of chitin. The most important factor in chitin and CS from food processing and packaging perspective is their antimicrobial properties. The inhibition of microbial cells is based on two possible mechanisms:

- The polycationic property may lead to electrostatic stacking at the bacterial cell surface, thereby interfering with cellular metabolism; and
- Intracellular penetration into the DNA of the microorganisms and blocking transcription.

Both chitin and CS are not hydrolyzed by intestinal enzymes and hence considered as part of dietary fiber and may act as prebiotic. Chitin and chitosan nanoparticles (CNP) have been prepared by several researchers.

Tran et al. [148] produced five distinct chitin NMs from shrimp shell: chitin nanofiber, chitin whisker, CS whisker, 2,2,6,6-tetramethylpiperidin-1-oxyl (TEMPO)-mediated chitin whisker, and zwitterionic chitin whisker. The source material, the extent of interfibrillar interactions, the final size, and applicability of the nanoparticles are the primary factors to be considered for choosing the appropriate isolation method, process combinations, and process conditions. Various chemical modifications of the surface layers of chitin nanoparticles to bring about necessary changes in their dissolution and binding properties have been reported by scientists.

### 4.2.6 GUMS

Polysaccharide gums principally occur in plants and microbial sources as storage materials, components of the cell wall, and as exudates or extracellular substances. They are complex heteropolysaccharides constituted by a large range of monomeric units. Some of the gums like gum Arabic also contain a protein component. Due to wide versatility in chemical structures and monomeric arrangements, the functionality and properties of gums cannot be generalized. The primary functional chemical groups associated with most gums are hydroxyl (-OH), acetyl ($-CH_3CO-$), carboxyl (-COO-), ether (C-O-C), aliphatic (-CH) and carbonyl (-C=O). A large number of hydroxyl groups along the backbone of any gum molecule facilitates in the formation of metal ions. This is an important characteristic of the use of gums for use in organo-metal ion composites. The (-OH) groups also help them to come in ready contact with water, solubilize, and form the characteristic colloidal structure and 'gummy' texture.

Popular gums like guar gum, locust bean gum, carrageenan, xanthan, and gum Arabic are commercial entities. Some other gums like ghatti gum, karaya gum, and tragacanth gum are also attaining gradual popularity for their application in foods. Based on the source, structural, and functional properties and targeted application, gums have been classified into bacterial polysaccharides (gellan gum dextran and xanthan), botanicals (guar gum locust bean gum and pectin), tree gums (Arabic, karaya, tragacanth, ghatti, and kondagogu) and seaweed polysaccharides (alginate, carrageenan, and agar). Gums can reduce the interfacial tension between different phases of matter and confer stability by steric, electrostatic, and hydration forces. Amongst the gums used in the food industry, galactomannans are the most frequently used. They are linear polysaccharides consisting of a mannose backbone formed by β-1,4 linkages and with galactose units arranged in side chains by β-1,6 linkages. Gums extracted from locust bean, tara, cassia, and guar are galactomannans. Amongst these sources, guar gum is the cheapest and most widely used galactomannan. Nanoparticles of gum have been mostly synthesized by self-assembly; ionic desolvation, electrospraying, and nanospray drying techniques. The gel properties of the particles are governed by the source material, type of processing, processing conditions and whether the material has been used singly or as a mixed system in conjugation with other gum or non-gum components.

### 4.2.7 HEMICELLULOSE

Hemicellulose is a class of non-cellulose polysaccharides present in the plant cell wall. It is the second most abundant polysaccharide in plants. Chemically, they contain monomers, namely mannose, galactose, glucose, glucuronic acid, xylose, and arabinose with a low degree of polymerization in their chains. Some of the hemicelluloses like arabinogalactan from larch, β-glucan, and xyloglucans are sometimes also considered under the category of 'gum' due to their hot water solubility and gummy texture. Several important structural types of functional hemicelluloses based on backbone chain constructions, namely xylans, mannans, mixed linkage β-glucans, and xyloglucans have been identified. Xylans are composed of β-1,4 linked D-xylopyranose monomers with short side chains comprising of D-glucuronic acid and its methyl ester, L-arabinose, and oligosaccharides composed of D-xylose, L-arabinose, D- or L-galactose, and D-glucose.

The backbone of mannans is either made up of β-1,4 linked D-mannopyranose (galactomannans) or both β-1,4 linked D-mannopyranose and β-1,4 linked D-glucopyranose. In both cases, D-galactopyranose form β-1,6 links for side-chain generation at a different ratio. The mixed-linkage (1 → 3,1 → 4)-β-d-glucans (β-glucans) are unique in their structures and have obtained tremendous popularity in the field of food and nutritional sciences. Xyloglucans have a β-1,4 glucopyranose backbone similar to cellulose, which relates to its intimate binding with the later in plant cell walls. The backbone is also added with sixth carbon-linked α-D-xyloparanose residues.

Xyloglucan is also a primary component of tamarind seed polysaccharide. Hemicellulose is generally considered as a byproduct of cellulose extraction and is removed from the adhering cellulose fibers by the application of alkali, pressure blast steam and high-pressure hot water treatments coupled with alcoholic precipitation. Hemicellulose obtained from either of these methods is often not pure and gets degraded from its original form. Process-induced depolymerization of the hemicellulose results in a further reduction in chain length and chain orders (if any), often leading to complete amorphousness of the obtained material. Techniques like graded ethanol and ammonium sulfate precipitation and ion-selective chromatographic separation have been used to obtain purer fractions of the different hemicelluloses. The commercially isolated polysaccharides have been used to fabricate different nanostructured materials. Being mostly amorphous, hemicelluloses cannot easily yield stable nanoparticles of themselves. However, nanoformulations have been made in conjugation or reinforcement with other compounds and polymers.

### 4.2.8 MICROBIAL POLYSACCHARIDES

Some microorganisms form exopolysaccharides. These protective polysaccharides either remain adherent to the microbe's body or are released into the growing medium as visible slimy substances which often form biofilm for adherence of the producing cell as well as other microorganisms to form colonies. A large range of bacterial species of the genus *Acetobacter, Agrobacterium, Achromobacter, Aerobacter, Azotobacter, Sarcina ventriculi, Salmonella, Escherichia, and Rhizobium* have been found to produce extracellular cellulose which further self-assemble and crystallize to cellulose fibrils and form a network membrane. A typical example of bacterial cellulose is the network frame of a symbiotic colony of acetic acid bacteria and yeast (SCOBY), which is characteristic of kombucha drink fermentation [35].

The cellulose network in the SCOBY holds the bacteria and yeast population in it and protect them from UV light, external harmful physical agents and help them float at the surface of the fermenting liquor to ensure sufficient oxygen availability for the aerobic species. Unlike cellulose from plant sources, bacterial cellulose is free from any adhering non-cellulosic materials suggesting its remarkably high natural purity and crystallinity as high as 85%. The higher specific area due to network-type arrangement relates to its higher water holding capacity and longer (up to seven times its weight) drying time as compared to plant cellulose. Moreover, if the generating bacterial culture is grown in liquid foods taken as meda (kombucha, fruit syrup, tea infusion, etc.), it can impart health benefit by secreting useful metabolites into the food and also take up characteristics like flavor and pigment from it. *Nata de coco* and *Nata de pina* are bacterial celluloses produced by growing *Komagataeibacter xylinus* in coconut water and pineapple juice respectively and consumed as a dessert in the Philippines, Vietnam, Indonesia, and Thailand.

The cellulosic substance is cut into pieces, dried, and soaked in suitable syrup before consuming. Once sufficiently explored, biosynthesized bacterial cellulose and nanocellulose can be a safer alternative to chemically extracted currently popular plant-based products. Bacterial cellulose as a dietary fiber has been approved as a "generally recognized as safe" (GRAS) food by the Food and Drug Administration (FDA). Bacterial species of *Agrobacterium* and *Rhizobacterium* can produce curdlan, a $\beta$(1–3) linked glucan of low molecular weight. Curdlan forms a weak gel upon mild heating and cooling, which can be further thickened by heating at high temperature (100°C and above). At very high temperatures, a strong gel is formed with triple-helical complex formation between the glycosidic chains. Curdlan strong gels are resistant to glucanase enzyme. It has been used and can be further explored as a food texturizer in different processed food systems. Curdlan and its derivatives conjugated with several organic and inorganic molecules have a high tendency towards self-assembling and form nanoparticles to be used as nanoscale drug carriers. Scleroglucans is a class of other β(1–3) linked glucans with β(1–6) glucosyl residues attached to the main β(1–3) backbone chain. They are produced by fungal species such as *Sclerotium rolfsii* and *Schizophyllum commune*. They form temperature resistant pseudoplastic gels and have antitumor properties which have been explored for drug delivery purposes. The polysaccharide forms a triple helix in its gel which has been utilized as a substrate for other functional molecules to function as drug and protein carriers. There is ample scope for top-down extraction of scleroglucan nano-gel for use in low glycemic foods. *Aureobasidium pullulans*, a fungus, produces pullulan by metabolizing starch. Pullulan is composed of maltotriose (3 G-glucose units linked by α-1,4 bonds) chain units linked to each other by α-1,6 linkages. Although pullulan does not form a viscous solution by itself, it has very efficient binding properties and has been used to design different conjugated products. Gellan is produced by the *elodea* species of bacteria. It is composed of repeated tetrasaccharide (two residues of D-glucose and one of each residue of L-rhamnose and D-glucuronic acid) units and two acyl substituents, namely L-glyceryl and acetyl.

Based on the degree of acyl substitution, gellan has been classified into high and low acyl gellans. Gellans form a weak gel by double-helical arrangement of their glycosidic chains in aqueous phase. However, in the presence of cationic ions, the double helices become more rigid due to cation-mediated aggregation, forming a strong gel. Xanthan is secreted by *Xanthomonas campestris. The chemical structure of xanthan primarily contains a cellulose-like backbone of D-glucose with added side chains of a trisaccharide (acetylated* α-D-mannose + β-D-glucuronic acid + β-D-mannose) linked to a pyruvic acid group. *It is soluble in both cold and hot water and forms pseudoplastic gels stable throughout a wide range of temperatures and pH. It has been used in food for its stabilizing and emulsifying actions. Another important food-grade microbial polysaccharide product is alginate. Alginates are actually sodium and calcium salts of alginic acid, the actual polysaccharide secreted by brown algae such as Ascophyllum, Durvillaea, Ecklonia, Laminaria, Lessonia, Macrocystis, Sargassum, Turbinaria and a few bacteria. They are linear structures formed by α-1,4 linked* β-D-mannuronic acid and α-L-guluronic acid residues. Alginate gels are highly hygroscopic and can take up about 20 times water per unit dry weight of the material. It is one of the most prominent microbial polysaccharides which has been effectively synthesized in nano-forms for a number of applications.

Agar is the most popular of microbial polysaccharides due to its rapid thickening property, extensive use in food product development, and prevalence in microbial solid media preparation. Agar is extracted from red algae (*Rhodophyta*) species of *agarophytes*. It occurs as a mixture of two components, namely agarose and agaropectin. Agarose is the predominant linear fraction formed by repeating units of agarobiose (a disaccharide of D-galactose and 3,6-anhydro-L-galactopyranose). Agaropectin is composed of a mixture of D- and L-galactose units with acidic side-groups. Agar can form a gel at room temperature. It melts at 85°C and forms the characteristic strong gel when cooled to 32–40°C.

Lentinan is isolated from the mushroom *Lentinus edodes*. It has a backbone of β-1,3 linked glucose units with β-1,6 linked pentameric glucose branches. Some other microbial polysaccharides, namely zymosan, grifolan, ganoderan are extracted from specific fungus and yeast strains.

## 4.3 PRODUCTION TECHNIQUES

Both 'top-down' and 'bottom-up' approaches are applied for obtaining nanosized polysaccharide structures. In the 'top-down' approaches, generally, complex semicrystalline polysaccharide molecules are gradually separated from adhering unnecessary organic components, cleaned, isolated, and reduced to nanoparticulate sizes by different size reduction techniques. Therefore, 'extraction' is a suitable term used for top-down approaches.

In the 'bottom-up' approaches, gels or hydrolyzed masses of the polysaccharide material are agglomerated under suitably standardized physical and chemical conditions to obtain the nanostructures. Hence, 'synthesis' is the right term used for such approaches. Solvent-compatibility, pH stability and specificity, thermal stability, strength, and desired specificity for targeted uses often limit the formation and applicability of nanostructures from a single polymer. Polysaccharides have been conjugated with other polysaccharides, non-polysaccharide organic and inorganic polymers, and crosslinking agents to develop necessary properties. This section discusses different extraction and synthesis techniques for the production of polysaccharide nanoparticles. The techniques can be classified into the major categories – hydrolysis, mechanical disintegration, reaction, molecular assembly, electro-spraying, and electrospinning and their sub-types based on the operating principles.

### 4.3.1 HYDROLYSIS

#### 4.3.1.1 ACID HYDROLYSIS

Nanoparticles and other nanostructures like nanofibers and nanowhiskers extracted from isolated cellulose fibrils have been widely reported for application in food and non-food sectors. In most studies, the crystalline domains of the fibrillar cellulose structures have been isolated by selectively hydrolyzing the amorphous domains using acids and enzymes. Such isolated nanoparticles have also been popularly termed as 'cellulose nanocrystals (CNC)' for their high crystallinity and related functional properties. This popular approach and its variations are generally used for CNC, nanofiber, and nanowhisker extraction from their parent matrices. In nature, cellulose exists in intimate complexation with non-cellulosic components, namely hemicellulose, and lignin. A series of alkali and chlorite bleaching treatments are generally used for delignification. More than hemicellulose, which gets solubilized and removed during dilute alkali treatment, lignin adheres inherently to the cellulose bundles and serves as a more intimate protective layer. Therefore, the removal of lignin is more challenging.

Alkali treatment with sodium or potassium hydroxide can be carried out along with steam blasting or explosion of the material to loosen up the compact fiber structures, exposing the non-cellulosic components to the alkaline environment for effective reaction and removal. In steam explosion, the lignocellulosic pulp is exposed to steam under pressure followed by – rapid release of the steam pressure. This sudden pressure change causes loosening of the bonds between the lignin and cellulose, thereby helping the removal of the former. Cellulose being stable to the delignification conditions remains largely intact. However, in some reports, change in polymorphic features of crystalline cellulose has been reported. It is necessary to remove the alkali traces adhering to the cellulosic matter and attain neutrality before further processing. Repeated washing with distilled water is done for this. Bleaching solution oxidizes the glycosidic chains, inducing inter-fibril repulsion for effective separation. It also gives the necessary whiteness to the extracted cellulose material.

Chemical bleaching typically uses chlorine and chlorite-based compounds. However, the liberation of $Cl^*$ radical and formation of toxic organochlorine necessitated intervention and the use of chlorine-free bleaching agents. Hydrogen peroxide ($H_2O_2$) has gained recent popularity due to its effective penetration, bleaching action, and reduced risks. A strong oxidant 2,2,6,6-Tetramethylpiperidin-1-oxyl, popularly called as TEMPO and ammonium sulfate have been effectively used to defibrillate cellulose to obtain cellulose nanofibers for further processing to nanocrystals. Such oxidation methods can be used with the physical or chemical methods to obtain better yield and uniform quality of the cellulose nanoparticles.

Cellulose-dissolution in several solutions like alkali/urea and NaOH/thiourea aqueous solution, ionic, N-methylmorpholine-N-oxide (NMMO), LiCl/N,N-dimethylacetamide (LiCl/DMAc), metal complex solutions, and molten inorganic salt hydrates have been successfully used to dissociate the cellular microfibrils into desired nano-ranges and can be used as pretreatments before mechanical disintegration for efficient and uniform results. To extract CNCs, isolated cellulose fibers are hydrolyzed with variably diluted acid which selectively removes the irregular amorphous fractions from the polymeric structure, sparing the crystalline particles or fibers (Figure 4.1(a)). Mostly, the use of HCl and $H_2SO_4$ for hydrolysis has been reported by researchers. Although considered as more efficient, hydrolysis using $H_2SO_4$ results in negatively charged nanoparticles with superior dispersibility in water as compared to HCl hydrolyzed particles and also differ in properties owing to this difference in surface charges. After repeated washing and dialysis, neutral particles can be obtained.

In ‘top-down’ extraction approaches involving acid hydrolysis of starch, type of starch, type of acid used, concentration, and time play the most vital roles. Sufficiently diluted hydrochloric acid (HCL) and sulfuric acid have been popularly used for the two-to-three phase hydrolysis of the macromolecule. In the initial phase, the amorphous portions get readily and rapidly hydrolyzed. Intercrystalline domain amorphous linkages get slowly hydrolyzed in the later phase(s) of the prolonged treatments generally lasting for more than 30 days. Acid hydrolysis is still the highest explored method for the isolation of starch nanoparticles with higher crystallinity and thermal stability compared to other methods [6]. Extraction of high crystalline nanoplatelets from starch by using weaker concentrations of HCl and $H_2SO_4$ for prolonged durations has been tried. However, very high crystallinity as in CNCs is often not achieved in the case of starch nanoparticles and ready agglomeration occurs between the particles and could be seen as micron-level clusters of nano-entities under the transmission electron microscope.

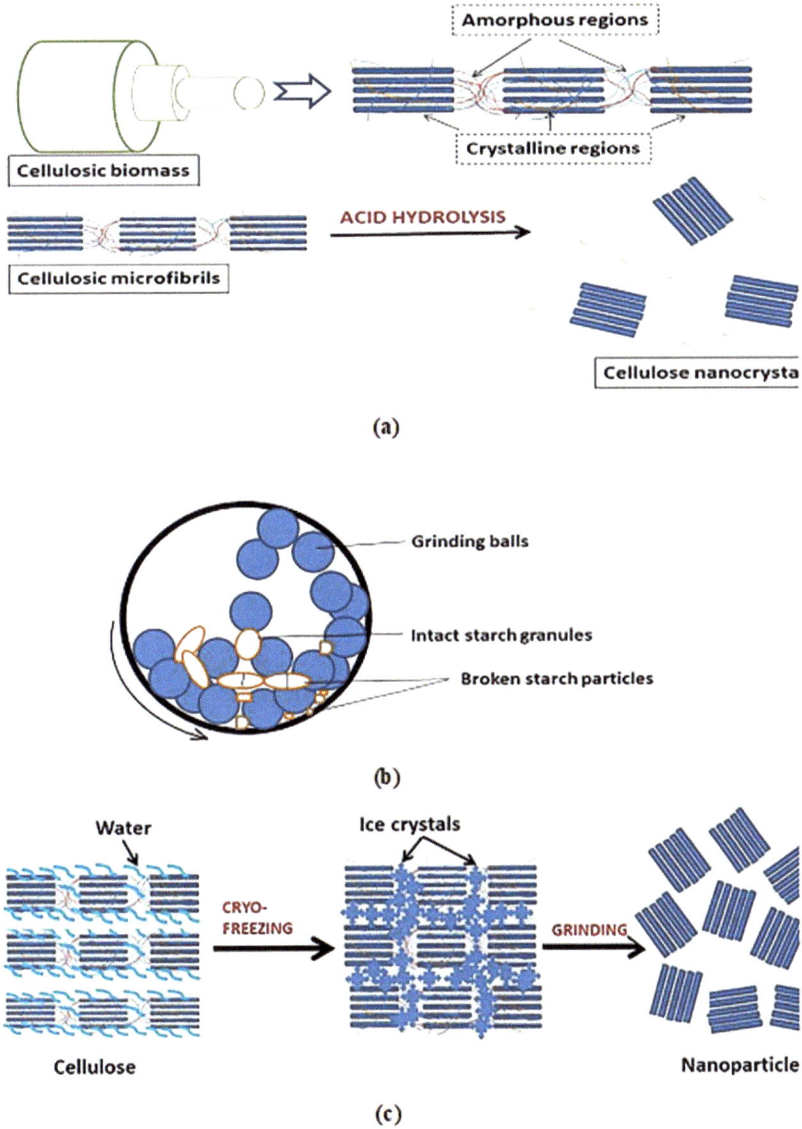


**FIGURE 4.1** Extraction of cellulose nanoparticles from cellulosic biomass by: (a) acid hydrolysis; (b) ball milling; and (c) cryocrushing.

Sulfate esterification of the nanoparticle surfaces, when hydrolyzed with sulfuric acid, can enhance their particulate stability. However, this also reduces the thermal stability of the particles. The addition of a small amount of ammonia in the suspension may significantly enhance the thermal stability of the dispersed particles. In a study on taro and arrowroot starches, it was found that HCl treatment resulted in higher particle yield than $H_2SO_4$ hydrolysis. The crystallinity of amorphous particles obtained on extraction from different methods can also be enhanced by storing the suspension in a cold environment, allowing recrystallization of the material to occur [6]. Approaches involving grafting and crosslinking of the nanoparticles with a number of chemicals to induce separation and dispersibility have been proposed. However, for food use, potential toxicity imparted by such chemicals must be avoided. TEMPO-oxidation has been also used for starch degradation. Oxidation by TEMPO can enhance the surface porosity of starch granules which can assist in the effective application of other synthesis methods for the extraction of nanoparticles from the granules. Freeze-drying, sonication coupled with freeze-drying, and ethanol treatment can be effective for obtaining dried stable starch nanoparticles.

Similar to cellulose, chitin nanoparticle extraction also involves preliminary purification steps of alkali (dilute NaOH or KOH solution) treatment for removal of adhering protein and lipids on the raw material, demineralization followed by bleaching and discoloration. These steps are repeated multiple times to ensure high purity of raw chitin followed by acid hydrolysis using HCl, $H_2SO_{4,}$ or other acids. Neutralized chitin nanoparticles, nanowhiskers, or nanorods can be obtained after sufficient dialysis against appropriate solvent. Upon strong acid hydrolysis, the amorphous regions of chitin are dissoluted and the bundles open longitudinally to release the nanofibers in the form of nanowhiskers with high crystallinity. The nanoparticles in the acidic colloidal solution experience repulsive forces between them owing to positive charges on their surfaces imparted by protonation of the amino groups. However, at above a critical concentration, the particles can undergo isotropic-anisotropic nematic transition. Ultrasonication has been used for effective individualization and dispersion of the chitin nanowhiskers with a high aspect ratio. TEMPO-mediated oxidation has been identified to be used for obtaining high chitin nanowhisker yield when coupled with simple mechanical treatment like sonication, which may even replace the acid hydrolysis step due to the more eco-friendly nature. However, the extraction technique bears a direct impact upon the chitin nanoparticle characteristics and their application.

Very few reports on NMs extracted from glycogen and their application are available. In one report, glycogen was oxidized with periodate. It was found that oxidation and subsequent nanoaggregation with other functional groups were more prominent at the surface layers of the glycogen granules [14]. Partial acid hydrolysis of sodium alginate by HCl was followed by crosslinking with $CaCl_2$ to yield calcium alginate nanoparticles for probable use in drug delivery. Guar gum nanoparticles can be extracted after acid hydrolysis followed by repeated centrifugation and neutralization of the precipitate. Konjac glucomannan extracted from the roots of the plant konjac is a popular food emulsifier and thickener whose nanocrystals can be obtained by acid hydrolysis.

#### *4.3.1.2 ENZYMATIC HYDROLYSIS*

Enzymes have been used for extraction as well as synthesis of polysaccharide nanoparticles owing to their reaction specificity and non-toxicity to human health. In a research, spherical cellulose nanoparticles were prepared from cotton. Cellulase enzyme at 5 g/l substrate concentration was applied for 175 hours. The reaction was terminated by heat denaturation of the enzyme. After washing under centrifugation, the hydrolysate was sonicated and the supernatant containing nanoparticles was centrifuged and freeze-dried. Particles lower than 100 nm in size were obtained [97]. Enzymatic hydrolysis of bacterial cellulose results in rough-surfaced nanoparticles with an axial groove resulting in a C-shaped cross-section as has been also reported for cellulose nanoparticles from eucalyptus bark [32]. In a bottom-up method, bacterial cellulose synthase enzyme from *Rhodobacter sphaeroides* immobilized on nickel film was used to synthesize nanofibers. Analyzes indicated that both amorphous and characteristic crystalline domains were formed during synthesis [13]. Highly pure cellulose nanostructures can be obtained by further insight and development of this method. Selective termination of naturally synthesized microbial cellulose followed by requisite technique for nanoparticulate aggregation and precipitation can be another option.

Pullulanase enzyme debranched waxy retrograded starch paste, can form nanoparticles by self-assembly of the short-chain amylose. Glucan chain length and retrogradation (temperature and time) conditions are particularly decisive of the physico-chemistry of the generated nanoparticles. More recently, enzymatic hydrolysis using amylolytic α-amylase, β-amylase, glucoamylase, and pullulanase have been applied singly and in combination with other treatments targeting safer and rapid extraction of starch nanoparticles and nanocrystals. Gelatinized proso millet starch treated with pullulanase followed by retrogradation to induce self-assembly amongst the glucan chains, formed nanoparticles of 20 nm–100 nm in size, as obtained after freeze-drying of the self-assembled structures [141]. Xyloglucan nanoparticles can be prepared by glycoside-hydrolase enzyme hydrolysis.

### 4.3.2 MECHANICAL DISINTEGRATION

The risks associated with the use of corrosive chemicals limit the suitability of using acid hydrolyzed polysaccharide NMs for food uses. Mechanical means of disintegrating micron-sized fibrils to nanosized structures have been employed. Treatments like high-pressure homogenization, microfluidization, extrusion, high-intensity ultrasonication, cryocrushing, and milling have been used for the preparation of polysaccharide nanoparticles.

#### 4.3.2.1 MILLING

Milling is a mechanism of reducing the size of an object by using some other device. Milling may encompass either impact frictional or compressive forces, or a mixed application of two or all of these at a time on the material. Different types of milling machineries are developed based on the physical and chemical characteristics of the object to be processed and those of the size-reduced products. Frictional grinding has been effectively used for the extraction of cellulose nanofibers. A grinding machinery specifically designed for cellulose nanofiber production was developed by Masuko Sangyo Co., Ltd. The equipment consists of a static and a rotating grindstone that revolves at 1,500 rpm, shearing the fibers in between, causing individualization of the nanofibers. For pine fibers, 5 passes in this machinery are sufficient to obtain the optimum separation of nanofibrils. Different structures of nanofibrils can be observed from cellulose extracted from different sources. Prolonged friction can result in the reduction of the nanofibrils to nanowhiskers.

For starch, wet milling is preferred over dry milling to avoid unwanted temperature rise and the corresponding change in crystallinity and molecular arrangements. Ball milling of starch suspension (Figure 4.1(b)) with its modern nomenclature as 'stirred media mills' using differently sized grinding media for the breaking down of the fed starch granules particles is the most popular technique used for the reduction of starch size. Size reduction from 3–20 μm to 245 nm by stirred media milling of starch suspension has been reported [113]. It can be opined that while the milling media reduces the size of the particles, the solvent provides enough friction to the particles to attain regularity in shape. Addition of selected surfactants during stirred media milling can further reduce the sizes of starch nanoparticles. Although not intended, wet milling using this technology may result in a significant loss in crystallinity of the formed particles. Cellulose nanofibers can be also obtained using ball milling. Ball size and ball-to-cellulose mass ration should be precisely controlled in order to control the extracted nanofibers' morphology.

Wet media milling has also been used for the size reduction of CS to obtain nanoparticles. The treatment can decrease the degree of polymerization, indicating glycosidic degeneration and enhanced viscosity when made into a solution. Ultrafine ball milling of CS can yield CS nanopowder.

#### *4.3.2.2 CRYOCRUSHING*

Cryogenic crushing or cryocrushing method is applied for the mechanical size reduction of flexible materials. It involves dipping the material in liquid refrigerants like liquid nitrogen, liquid carbon dioxide which have very low boiling points. At such low temperature, the interdimensional sliding inside the solid material is reduced to almost zero. This makes the material brittle and easily crushable by impact or friction. However, abrupt cracking of the microstructures may occur, destroying the desired crystalline entities. Cryocrushed fine organic materials can be further treated by suitable gentle mechanical treatments like homogenization and ultrasonication or mild chemical or enzymatic treatments to obtain suitable nanosized particles.

Cryocrushing has been used for mechanical defibrillation of cellulose followed by crushing with mortar and pestle. The treatment results in the formation of ice crystals which further help in the size refinement to submicron level sizes of the resultant fibers (Figure 4.1(c)). Cryocrushing can be followed by high pressure homogenization to obtain more efficient size reduction and separation of the particles. The extent of wetting (moistening) also plays a decisive role in particle size obtained by cryo crushing treatments.

#### 4.3.2.3 HIGH PRESSURE HOMOGENIZATION AND MICROFLUIDIZATION

Microfluidization technique involving high-pressure homogenization is based on the application of high shear force within a product stream through micro-channels causing defragmentation and splitting of suspended particles from micron to nano-sizes (Figure 4.2(a)). Cellulose pulp is partially hydrolyzed with endoglucanase enzyme in a buffer (pH 7) before shearing using multipass treatment in a fluidizer (a type of high-pressure homogenizer) device. As cellulose bears strong intermolecular bonds resulting in hard texture, acid or enzymatic hydrolysis is necessary to loosen the structure before processing through mechanical disintegration. Chlorine oxidation can be used instead of or in combination with hydrolysis before high pressure homogenization. High pressure homogenization resulted in size reduction of bacterial cellulose to nanofibrils which were used further for emulsion formation and stability [112]. Starch nanoparticles of size 10 nm can be obtained by passing slurry multiple times in a microfluidizer. However, crystallinity is often lost, leading to the formation of amorphous particles. The technique can be used for reducing the sizes of resistant starch particles and linking it to water-in-oil emulsion to form resistant starch-rich emulsion products [7]. Similarly, other nanoemulsions like pectin-oil and guar gum-oil can be also prepared by high-pressure homogenization. Crosslinking agents like Tween 80 are used to obtain stability of the nanoemulsions and nanogels.

#### 4.3.2.4 ULTRASONICATION

Ultrasonication carried out in sonicating devices uses high-frequency ultrasonic waves which cause acoustic cavitation in a dispersion phase. The microbubbles grow and collapse in the liquid which produces high energy transformed to high pressure and temperature which disintegrates the suspended particles by breaking down the bonds (Figure 4.2(b)). Ultrasonication has been used to obtain nanofibers and other nanoparticles like nanorods and nanowhiskers from a wide range of materials like chitin fibers, cotton, bamboo, ramee, hemp with sizes below 500 nm. Ultrasonication can act on both crystalline and amorphous zones of a polysaccharide, however mostly, the native polymorphism of starch and cellulose are retained after the isolation. It can be applied in combination with mechanical grinding and acid hydrolysis to obtain nanoparticles with improved optical properties, transparency, and superior crystalline packing. Sono-chemical assisted maleic acid hydrolysis of wood cellulose, Avicel, and recycled pulp of wood pulp resulted in nanoparticles of variable shapes [38]. It has been applied to bacterial microfibrillar cellulose, CS, and starch. In starch, crystalline lamellae of the glycosidic chains may get disrupted causing the resultant particles to be largely amorphous. Platelet-shaped starch nanoparticles with high crystallinity can be obtained by acid hydrolysis followed by ultra-sonication. However, a perfect combination of the two treatments must be optimized to avoid excessive impact of the acid and the applied sonic energy.

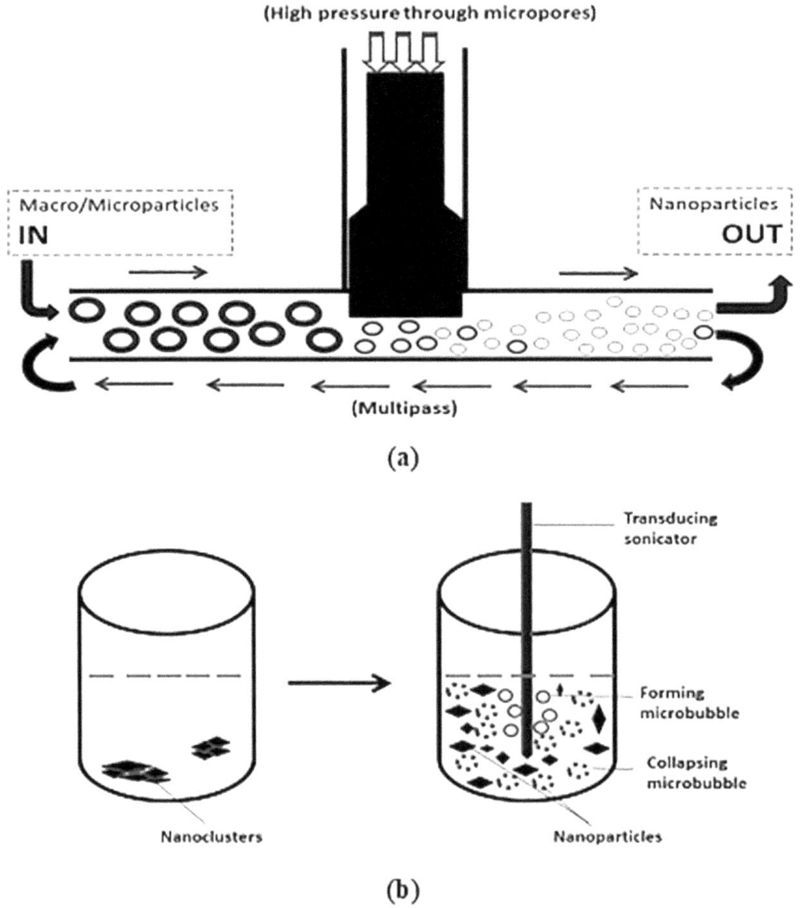


**FIGURE 4.2** Schematic diagram showing: (a) high pressure homogenization for size reduction of particles; and (b) ultrasonication for declustering of clumped nanoparticles (nanowhiskers).

Acid hydrolysis has a higher impact on the chemistry of starch than ultrasonication and a higher resemblance of the physicochemical properties of the nanoparticles to the native starch can be obtained in the later. Oxidation modification of the native starch followed by ultrasonication would also result in lower crystallinity than native starch. Oxidation by TEMPO can result in surface porosity of starch granules, enhancing the effect of ultrasonication to disintegrate into nanoparticles. For sonication induced size reduction efficiency, wave frequency, time, and suspension concentration play critical roles. High-intensity ultrasonication can cause damage to the nanoparticle structure affecting its functionality. Most organic nanoparticles (ONPs) produced by different extraction or synthesis techniques have a tendency for agglomeration during storage. Ultrasonication can deagglomerate such structures in suspension, thereby improving separation and dispersion. Deagglomeration of freeze-dried or cryo-immersion dried nanoparticles can be achieved using the right combination of the cryoprotectant and ultrasonication time. This can reduce the nanoparticles to still smaller sizes.

In a novel approach proposed by Chang et al. [20] in which commercial chitin flakes were treated with 3 N HCl for three times followed by mechanical disruption using ultrasonication and centrifugation. Heavier non-degraded particles were retreated with ultrasonication, and the process was repeated 6 times till clear suspension was obtained. After severe centrifugation at 12,000 *g* for 20 minutes, the nanoparticulate residue was collected and dried [20]. The inclusion of dialysis to the process may enhance the collection volume of the obtained nanoparticles. A combination of acid hydrolysis of deacetylated CS and sonication can effectively yield chitin nanoparticles. Such nanoparticles have a self-clustering tendency, which can be tackled by sufficient sonication followed by freeze-drying. Zein–pectin composite nanoparticles with a hydrophilic pectin shell covering a hydrophobic zein core were developed by the ultrasonication method [30]. An alcoholic solution of Zein (0.01%) was first sonicated to obtain a single phase. To this, pectin solution (1.04 mg/ml) was added dropwise (2.5 ml/min) under sonication and continued for 5 minutes. Nanoparticles were obtained after lyophilizing. Whey protein concentrate along with maltodextrin was coupled with pectin to form nanoparticles using ultrasonication treatment (350 W for 10 min) [43].

#### *4.3.2.5 NANO SPRAY DRYING*

Spray drying involves high pressure spraying of a solution through a nozzle accompanied by instantaneous drying using heat flow on the path of the sprayed particles (Figure 4.3(a)). Depending on the size of the spraying nozzle (atomizer), droplets ranging from micrometer (μm) to nanometer sizes can be obtained. Resultant dried solid particle sizes also vary accordingly. For an evenly dissolute solution, particles of uniform size can be obtained. Ultrasound or high pressure-assisted homogenization can be effectively used to obtain such uniform dispersions. In a nanospray dryer, the atomization occurs through an actuator with micron-sized release pores operating at a specific ultrasonic frequency. This produces an ultrafine spray of particles that dry very rapidly. With the surface charge, they accumulate at an electrode surface. Surface tension amongst the phases and agglomeration tendency of the solute molecules are well utilized in nanospray drying for organic nanomaterial production. For enhancing nano-scale agglomeration, crosslinking agents can also be added to the solutions. Nanoparticles obtained from different polysaccharides by spray drying approaches are listed in Table 4.1.

**TABLE 4.1** Polysaccharides, Crosslinking Agents, Conjugating Compounds, Sizes, and Potential Applications of Resultant Nanoparticle Obtained by Spray Drying

| Poly-Saccharide | Conjugating Compounds/ Crosslinking Agents | Particle Size | Application | References |
|---|---|---|---|---|
| Cellulose | Polypropylene | 150 nm | Dispersible mechanically strong nanoparticles | [69] |
| Starch | Tween-80 and Span-80 | 1,000–2,000 nm | As dispersed particles in o/w emulsions | [134] |
| | Epichlorohydrin, $POCl_3$ | > 800 nm | Trans-nasal insulin delivery | [59] |
| Chitosan | Iron oxide | 200–450 nm | Super magnetic properties for drug delivery, cell separation and cellular labeling, enzyme immobilization, magnetic resonance imaging | [53] |
| | Tripolyphosphate (TPP) ionic gelation pretreated chitosan | 166–1,230 nm | Antibacterial | [106] |
| Pectin | Sodium chloride, calcium acetate | 200–500 nm | Nanocarrier of cancer therapeutics | [167] |

**TABLE 4.1** *(Continued)*

| Poly-Saccharide | Conjugating Compounds/ Crosslinking Agents | Particle Size | Application | References |
|---|---|---|---|---|
| | Zein, caseinate | < 200 nm | Eugenol encapsulation | [153] |
| | Zein, sodium alginate | 100–200 nm | Curcumin encapsulation and enhanced dispersibility | [56] |
| Alginate | Maghemite, sodium silicate | 50–200 nm | Magnetic properties related applications | [15] |
| | Cashew gum | 200–400 nm | Encapsulation of essential oils | [28] |
| Gum Arabic | Cashew nut gum, sodium alginate and carboxymethyl cellulose | 100–500 nm | Encapsulation of bioactive compounds | [28, 85] |
| | Lecithin | < 500 nm | Eugenol entrapment in nanoemulsion | [52] |

Microwave irradiated pectin solution can be nanospray dried in the presence of inorganic salts such as NaCl as a superheating promoter. However, the added salt would also affect the physicochemical properties of the obtained nanoparticles. Similarly, alginate, cashew gum, or their conjugate nanoparticles can be prepared using spray drying and used for entrapment of bioactive substances and essential oils. Büchi Nano Spray Dryer B-90 is a spray drying device that has been used for the preparation of a number of polysaccharide nanoparticles including those from gum Arabic, modified starch, and maltodextrin. Gums, alginate, and carboxymethyl cellulose (CMC) are feasible choices for preparing spray-dried nanoparticles. Nano spray drying has been extensively used for the preparation of coating materials and simultaneous nanocoating of functional compounds like drugs and nutraceuticals.

## 4.3.3 REACTION AND MOLECULAR ASSEMBLING

### 4.3.3.1 REACTIVE EXTRUSION

Extrusion technology involves simultaneous mixing, shearing, cooking, and pressing of masses to a semi-viscous dough consistency. The technique has been most popularly applied for the synthesis of starch nanoparticles. It is because the starch macromolecule undergoes easy plasticizing and pre-gelatinization, which are required for the use of this technology. When applied with suitable crosslinker and plasticizer compounds, plasticized starch granules undergo simultaneous nano-splitting and crosslinking, forming nanoparticles (Figure 4.3(b)).

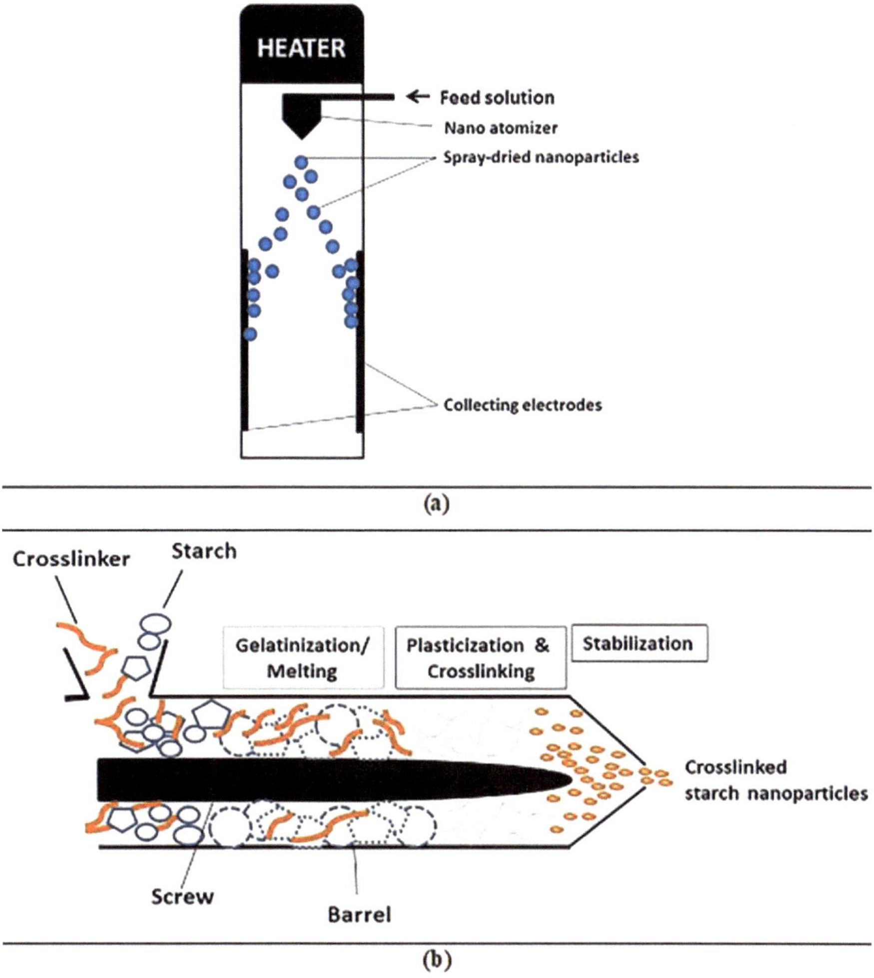


**FIGURE 4.3** Schematic diagram showing the mechanisms of: (a) nanospray drying; and (b) reactive extrusion for crosslinked starch nanoparticle synthesis.

This approach has been termed as 'reactive extrusion' as a reaction between the defragmented starch fractions and the crosslinking chemicals take place inside the extrusion barrel. Glyoxal, glycerol, and other reaction crosslinkers in recommended amounts can be used for obtaining the lowest possible sizes, higher shear, and stability of the nanoparticles. Reactive extrusion can be also used to prepare nanosized starch-lipid complex spherulites.

#### *4.3.3.2 WET HEAT PROCESSING*

Based on the mechanism of starch-lipid complex formation, steam jet cooking can be used. Starch chains can helically bind lipid molecules to form nanosized starch-lipid complex, also popularly known as amylose-lipid complex (Figure 4.4(a)). Slurries containing high amylose starch and fatty acid were subjected to cooking followed by cooling with stirring to obtain submicron to nanosized spherulites of amylose-lipid complexes. Suitable starch and fatty acid molecules can be pasted by steam jet cooked to allow the formation of the complexes. Instead of starch, dextrins obtained by acid or alcohol-acid hydrolysis of high amylose starches can be used for nanocrystal complex formation with fatty acid and n-butanol. Particles with <100 nm sizes can be formed under standardized process conditions and recovered after selective acid hydrolysis of the amorphous portions.

#### *4.3.3.3 IRRADIATION*

Gamma irradiation can rapidly and conveniently cleave glycosidic linkages by forming reactive free radicals, forming minor structures. Gamma rays dosage of 15 to 25 kGy and dosage rate of 10–15 kGy/h can form amorphous nanoparticles in starch suspension and other polysaccharide colloids. Stability of irradiation-produced starch nanoparticles can be enhanced by carboxymethylation or other chemical modification treatments. Such clustered formation of native and modified starch nanoparticles is popularly called starch nanogels. Similar approaches can also be used to obtain nanostructured CS and gums. Upon gamma irradiation, a significant loss in the crystallinity of cellulose occurs, liquefying the material. However, the application of gamma irradiation for the production of cellulose nanoparticles or nanowhiskers could not be found in the reviewed literature.

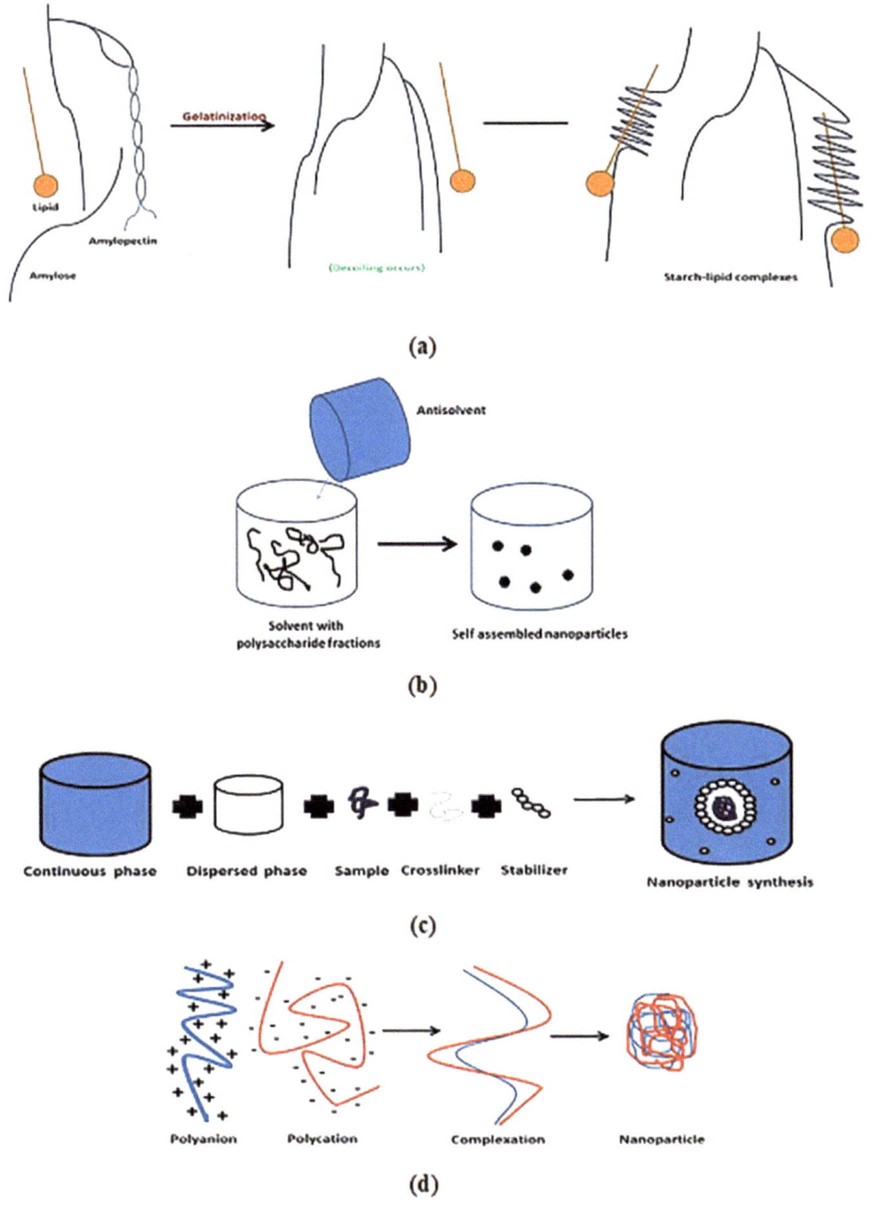


**FIGURE 4.4** Mechanisms of formation of: (a) starch-lipid complexes; (b) self-assembled dispersed polysaccharide fractions by antisolvent treatment; (c) nanoparticles within dispersed phased droplet in an emulsion system; and (d) polyelectrolyte complexation and subsequent stabilized nanoparticle.

It is most likely that irradiation coupled with mechanical and chemical treatments (acid degradation, crosslinking) under controlled process conditions can be used for extracting nanostructured cellulose. Similarly, gamma-ray depolymerizes and degrades CS. Acetylated CS can reduce to nanoparticles when dispersed in the form of colloid and treated with gamma radiation. Radiation may increase the antibacterial activity of CS and thus of chitosan nanoparticles (CNPs) produced by this technique. Gamma irradiation-induced crosslinking of gums has been used to form hydrogels that act as substrate for several metals for induced nucleation and formation of metal nanoparticles. A green approach can be exploration towards the synthesis of gum nanoparticles for food use by crosslinking with edible non-toxic compounds.

#### *4.3.3.4 ANTISOLVENT NANOPRECIPITATION*

It is one of the simple, fast, and reproducible bottom-up techniques used to fabricate nanoparticles using self-agglomeration of fragments under induced supersaturated conditions. A non-solvent is added to the dispersion containing the polysaccharide fragments. With an increasing non-solvent concentration in the liquid mixer, supersaturation occurs, causing nucleation, particle growth and finally precipitation of the formed nanoparticles (Figure 4.4(b)). A super-critical antisolvent nanoprecipitation technique has been developed in which supercritical fluid is taken as the antisolvent. On the application of high pressure, a phase change occurs and the solute particles get separated.

Acid hydrolyzed glycosidic portions of starch could be crystallized using alcohol as antisolvent. Starch nanoparticles with high crystallinity could be attained by this technique. Spherical nanoparticles can be obtained by nano-precipitation of high-amylose corn starch and n-butanol followed by purification using enzymatic hydrolysis. In a study, waxy maize starch acetate nanospheres were prepared. The acetylated starch was dispersed in acetone and added drop-wise to distil water. Nanospheres were formed, which could be purified by evaporating the acetone [142].

Upon application of acetic acid and ethanol precipitation, starch nanoparticles with V-type crystallinity (typical to starch-lipid complexation) can be obtained. Morphology of starch nanoparticles may change with change in the ratio of starch solution and ethanol. Spherical to worm-like structures of amylopectin can be obtained by changing the ethanol to starch ratio [116].

The use of surfactants and coupling the reduction method with gamma-irradiation, ultrasonication during nanoprecipitation can also further reduce particle sizes. Although a good number of research have been carried out on precipitation techniques, industrial-scale applicability of the method for high yield is the major challenge. In a solvent-antisolvent (sodium hydroxide-ethanol) precipitation method, corn starch nanoparticles were prepared in a spinning disc reactor (SDR). The rotational speed of the disc and the disc texture (grooved) affected the size and quality of the obtained nanoparticles [121].

Cellulose nanofibers do not have very strong natural agglomeration tendency. However, functionally derivatized cellulose compounds have been used for nanoprecipitation. Cellulose acetate nanoparticles have been prepared by nanoprecipitation commercial cellulose acetate in a mixture of water and tetrahydrofuran taken as antisolvent. Cellulose acetate dissolved in tetrahydrofuran was added drop-wise till precipitation occurred. Letting the reaction occur in an ultrasonic bath resulted in delayed dispersion, inhibiting coagulation and formation of smaller and uniform nanoparticles. The particles were filtered and freeze-dried [75]. CMC is another possible alternative for nanoparticle synthesis. Acetylated CS in aqueous solution can be precipitated by diffusing into methanol taken as antisolvent. Magnesium chloride can act as an ionic crosslinker for pectin which can be nanoprecipitated using suitable antisolvent alcohol. The technique can be also used to obtain nanoparticles of maltodextrin, guar gum and almond gum with crosslinkers (Tween 20, Tritin X-100, etc.), and surfactants (borate, polysorbate 80, saponin, etc.), aiding the respective reactions and precipitation. Conjugated nanoparticles can also be precipitated using this technique and targeted for specific applications like drug loading and delivery. Self-assembling hydrogel nanoparticles of gum and pullulan derivatives can be formed by dialyzing in a suitable antisolvent.

#### *4.3.3.5 EMULSION*

Any emulsion contains two immiscible liquids and can be categorized as oil-in-water or water-in-oil emulsion, depending on the dispersed and dispersion phases. Emulsions are stabilized either by high-pressure homogenization or application of stabilizing agents or, more practically, a combination of both. Crosslinking agents are used for rapid inter-glycosidic bonding of the dispersed glucan chains within the water droplets in the emulsion. Ionic liquid with immiscibility in both water and oil phases can be used to form emulsions with both the phases and synthesize starch nanoparticles. A schematic diagram showing the use of emulsion technique for the formation of polysaccharide nanoparticles is presented in Figure 4.4(c).

A low-energy emulsification method was developed by Spernarth and Magdassi [139] for obtaining nanoparticles of ethyl cellulose. A measured combination of non-ionic surfactants decaglycerol monolaurate and Span 20 was added to the toluene medium. Ethylcellulose dispersed in an aqueous phase containing dissolved NaCl was added dropwise at a constant rate under constant stirring of the emulsion. Nanoparticles of 50–120 nm were obtained after evaporation of toluene [139].

The technique can be assisted by ultrasound to obtain higher separation and dispersibility followed by solvent evaporation. Emulsification technique results in more uniform-sized nanoparticles and hence opined to be superior to the other solvent-based techniques for polysaccharide nanoparticles synthesis. High methoxyl pectin nanoparticles synthesized using oil-in-water nanoemulsion technique has been used for entrapment of drugs and a range of essential oils. Tween 80 is an effective crosslinker for the assembling reaction to occur. Similarly, dextrin, guar gum, curdlan, pullulan, alginate, and agar nanoparticles can be formed by using appropriate crosslinkers, stabilizers, and detergents in the forming solution. The formed nanoparticles, after sufficient stabilization in the solution, can be isolated using centrifugation, filtration, dialysis, and/or freeze-drying.

#### *4.3.3.6 POLYELECTROLYTE COMPLEXATION*

Organic molecules can be naturally or artificially induced with charges. Under aqueous conditions, such charged particles can bind with one another depending on the surface charges to form polyelectrolyte complexes (Figure 4.4(d)). This phenomenon is also often termed as ionic gelation. Polyelectrolyte complexed nanoparticles from a number of polysaccharides like cellulose, starch, dextran, CS, and alginate have been fabricated. In cellulose, artificially induced cationic thiol-group to cellulose can enable it to form a complex with anionic tripolyphosphates to be fabricated into spherical nanoparticles. The ionic strength of the materials, temperature, and the concentration and pH of the medium play important roles in the feasibility of the complex formation and stability.

Using this principle, nanocrystalline cellulose-alginate complexed nanoparticles were synthesized, and the sizes of the ion-complexed nanoparticles were reduced to 100 nm using sonication [147]. Nanoparticulate aggregation of anionic CMC and cationic quaternary cellulose can encapsulate proteins. Some other ideal polyelectrolytic complexations between polysaccharides are carboxymethyl starch-chitosan, carboxymethyl kondagogu gum-chitosan, and angico gum-chitosan. CS being a known charged entity has been most abundantly used for polyelectrolytic complexation with a range of other compounds, including polysaccharides.

Pectin, carboxymethyl cashew gum, gum Arabic, carboxymethylated moringa gum, sulfated locust bean gum, carboxymethyl gum katira, gum ghatti, and hyaluronic acid are a few examples. Generally, with increasing concentration of the polymeric solution, reduction in particle size is observed. Not only double but even triple or higher levels of conjugations are possible by this technique. Starch can be chemically modified to induce polycationic or zwitterionic properties to make it suitable to undergo electrolytic complexation reactions.

Amphiphilic molecules have a tendency to form self-assembled micelles at stable pH. Self-assembly is the most safe, cost-effective bottom-up approach for the synthesis of polysaccharide nanoparticles. During aggregation under polyelectrolyte solutions of salt, alcohol, and acetone, the gum molecules can undergo different conformational changes which also decides the final arrangement of molecules in the nanoparticles, and hence their properties. Disassembly of the aggregated molecules can be obtained by nano-shearing using suitable physical tools.

### *4.3.4 ELECTROSPRAYING AND ELECTROSPINNING*

Electrospraying is another upcoming technology for the processing of polysaccharide nanoparticles. It is based on the principle that when an electric field is applied to a liquid droplet released from the needle tip of a syringe, Coulomb force is generated inside the droplet which acts against the cohesive force to deform the interface, releasing the surface tension and form nanodroplets (Figure 4.5). Electrospraying has been extensively used for nanoencapsulation of functional biomolecules with biopolymeric materials.

Electrospraying forms smaller and more uniformly sized nanocapsules and nanoparticles than most of the other synthesizing techniques. For obtaining cellulose nanoparticles, a sufficiently diluted homogeneous solution of cellulose prepared in solvents like N, N-dimethylacetamide, and lithium chloride can be electrosprayed at room temperature using high voltage (20 kV or more). The charged nanoparticles will be sprayed over an aluminum sheet placed at a distance of about 15 cm in the setup. Nozzle diameter, solution concentration, feed rate, voltage, and collection distance are critical in deciding the size of the obtained nanoparticles. Nanoparticles from other polysaccharides and derivatives, namely ethylcellulose, native, and modified CS (in acetic acid), gums, alginate, and their conjugates can be also prepared using this technique. To prepare conjugated nano encapsulating matrices, a coaxial electrospraying technique involving coaxial separate streams of the two polymers is more effective.

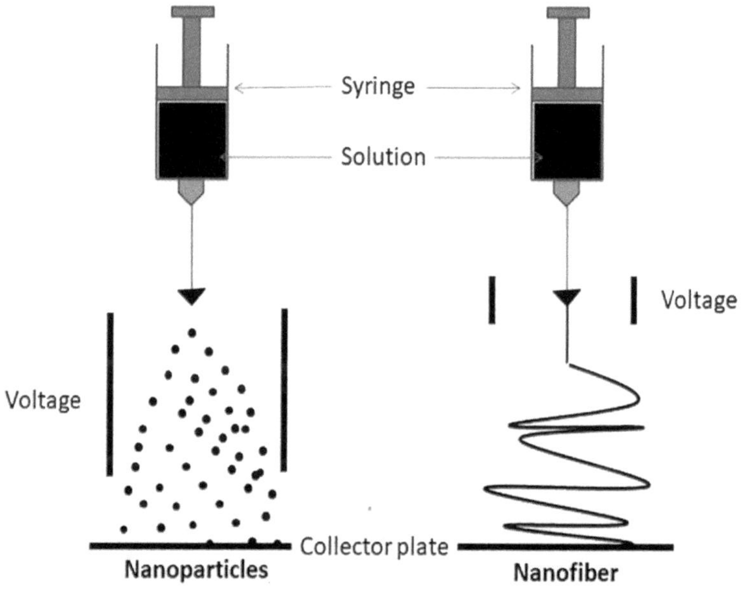


**FIGURE 4.5** Nanospraying (left) and nanospinning (right) techniques for the formation of nanoparticles and nanofibers, respectively.

Electrospinning can be considered as a modification to electrospraying, however with some notable differences in the working principle (Figure 4.5). Electrospinning involved the use of high voltage to a thicker concentration of the homogeneous polymeric solution. Due to Coulomb force and repulsion between the like forces within the released droplet, the droplet turns into a cone from its original hemispherical shape. This cone is called as Taylor cone. A jet of the polymer solution gets ejected from the tip of the Taylor cone and dries up before reaching the collector with appropriate drying conditions provided and adjustment of the nozzle to collector distance. Charge differences results in 'whipping' of the jet causing its elongation before reaching the collector. Still wet nanofibers can be dried with appropriate rapid drying techniques like alcohol treatment, freeze-drying or *in vacuo* evaporation. Nanometric fibers are hence collected.

Almost all major polysaccharides that have been processed to form nanoparticles using appropriate conditions of electrospraying have also been subjected to electrospinning for specifically targeted usage. Gums have been also used for the necessary thickening of electrospraying and electrospinning solutions meant to produce nanoparticles and nanocapsules with other primary polymers like resistant starch and meant for enhanced uptake of nutraceuticals like folic acid. Non-uniform mixing and bonding between the components could form beaded fibers instead of uniform fibers, suggesting the importance of standardizing the components in the solution with electrospinning conditions.

Electrospun nanofibers have possible applications in 3D designing of packaging films and encapsulating materials for bioactive compounds. Combination of the two 'sister techniques' has been recently developed for designing smart and more effective nanoengineered systems. Such a hybrid application of the two mechanisms can be used to prepare membranes made of nanothreads obtained by electrospinning and nanobeads impregnated in them giving a nanodimensional rough surface. This can be done by bushing in different concentrations of the polysaccharide material solutions repeatedly to obtain layer by layer formation of nanofibers/nanomats and nanoparticles of the polysaccharide. This would result in strong adherence materials onto it, suggesting its multiple applications such as substrate for cell proliferation, adsorbent surface, etc. Besides, difference in surface charges of the formed surfaces can enhance the intimacy of the formed hybrid NMs.

## 4.4 APPLICATIONS OF POLYSACCHARIDE NANOPARTICLES

### *4.4.1 CELLULOSE*

#### *4.4.1.1 DIETARY FIBER*

As not digestible by human digestive enzymes, cellulose, and its nanoparticles, once present in food, behaves like dietary fiber. They add bulk to the food without imparting nutrition. Therefore, these can be used in fat-free, low-cholesterol, and low-calorie food commodities. Bacterial cellulose has been identified to have lipid and cholesterol-lowering activities, efficient as a fat replacer in emulsified meat products, can be developed into a 'vegetarian meat' in combination with a red *Monascus purpureus* mold extract, texture enhancer of tofu and as humidifying agents in wet foods like ice-cream. Applications of cellulose as dietary fiber capable of enhancing the sensory quality of various foods have been reported. Nanosized cellulose particles may further improve sensory properties due to finer dispersion with a minimum characteristic mouth feel of cellulose which is imparted by added macrocellulose particles and considered as a drawback.

#### 4.4.1.2 PACKAGING MATERIAL

In the food industry, the use of cellulose nanoparticles and their chemical derivatives have been mostly limited to the packaging sector. Cellulose and nanocellulose have been identified as excellent reinforcing agents and have been used to design a number of nanocomposite materials for food packaging. Besides, above a critical concentration in aqueous solutions, CNCs obtained by acid hydrolysis approach tend to self-assemble and form films with high mechanical integrity. Packaging materials of organic composites made from nanocellulose with other biopolymers like polylactic acid (PLA) and polyhydroxyalkanoate have been commercially developed. Incorporation of sulfuric acid hydrolysis extracted cellulose nanoparticles could reduce both solubility and moisture migration through alginate films and increase tensile strength (TS) [2]. However, cloudiness appeared at higher concentrations of the nanoparticles, suggesting their intermingling in the film casting solution. Interestingly, an opposite trend of opacity was observed when olive oil containing CS film was reinforced with increasing concentrations of CNCs. Improved strength and moisture barrier properties of composite food packaging films on the incorporation of nanocellulose have been also reported. Nanocrystalline cellulose has been developed as a porous aerogel with high water holding capacity for use as food packaging material. The edible film made from mango puree showed improvement in TS and water vapor barrier properties when reinforced with cellulose nanofibers [10].

A similar result was also observed on reinforcing bacterial cellulose nanofibers in silver carp myofibrillar protein to form edible films [125]. Cellulose-based biodegradable films and packaging materials reinforced with metal and metal oxide nanoparticles of silver, gold, cadmium, zinc, and copper have been used for fresh food packaging, active food packaging, vegetable packaging, light barrier films, controlled atmosphere packaging, antimicrobial packaging, and photocatalytic surfaces. Metal ion-induced cellulose nanofiber-based material has been designed as sensors for detecting pesticides on food surfaces. Overall, researches on the incorporation of cellulose nanoparticles in edible food products have been fewer as compared to its use in packaging materials.

#### *4.4.1.3 EMULSION STABILIZER AND THICKENER*

Nanocrystalline cellulose has a substantial ability to be used as a stabilizer in Pickering emulsions. Pickering emulsions are characterized by solid particles being used to stabilize the emulsion. To be Pickering stabilizer, the particles should have partial wettability by both oil and water phases of the emulsion. Cellulose nanoparticles, especially those from microbial origins are hydrophilic due to the high concentration of hydroxyl groups on the surface. Particles with high crystallinity and high concentration of inter-glycosidic chain hydrogen bonds also tend to induce partial hydrophobicity. This peculiar amphiphilic property of cellulose nanoparticles makes them ideal as nanostabilizers of emulsions. High thermal stability of the nanosized cellulose particles and fibers adds to its suitability for application in processed food emulsions. The incorporated nanoparticles also serve as a thickener for such products as the density of the emulsions markedly increases. The effect of nanocellulose on rheological properties of protein-rich ingested foods inside the digestive tract was reported [93].

It was found that nanofibers from bacterial cellulose could stabilize an oil-in-water Pickering emulsion with peanut oil concentration of 15% (v/v). The emulsion with only 0.05% (w/v) of the nanofibers could stabilize the emulsion at pH$\geq$7 till 4$^{th}$ week of its storage [169]. A complex of ethyl lauroyl arginate and CNCs was successfully used to stabilize edible oil-in-water emulsions [12]. Phenyl-trimethoxysilane covered cellulose nanofibers were also effective as Pickering emulsion stabilizer and showed good redispersibility in water. Redispersibility is essential as cellulose nanofibers show a strong tendency to agglomerate during dehydration. It is therefore necessary to increase the surface hydrophobicity of the particles.

A 'green' surface-modification approach to coat cellulose nanofibrils (CNF) with soya protein was reported [170]. This improved the interfacial rheological properties with smaller droplets size, higher viscosity, and better physical stability. In addition, the emulsion showed resistance to oxidative changes and hydrolysis of lipids. The emulsion stability is dependent upon the size of the nanoparticles and the corresponding time of acid hydrolysis to obtain the same. A positive correlation can be observed between particle size and particle-stabilized emulsion size. Nanoparticles obtained after the 6$^{th}$ and 10$^{th}$ hour of hydrolysis time formed the best stable emulsions for storage time, pH, and salt concentration of the emulsions. Bread dough containing cellulose nanofibers reportedly resulted in softer bread with and higher and even volume rise which may be also associated with its emulsion stabilization capacity [140]. The stability of the cellulose nanomaterial's emulsifying properties can be altered by adjusting surface charges.

#### *4.4.1.4 NANOENCAPSULATION*

In recent decades, consumer awareness of the bioactivity of functional food compounds has increased the demand for encapsulated products with shelf-stability and acceptable sensory attributes. Nanoencapsulation is one of the most researched and evolved domains of nanotechnology being used in the food industry. In a nanocapsule, the functional compound is entrapped within a nanosized outer material, which is often called a shell. Nanoparticles of cellulose have been used to design shell of nanocapsules. Cellulose nanospheres have been mostly used to carry functional biomolecules to be released in specific sites of the body. In many of these fabrications, the encapsulation of the material is carried out simultaneously during the formation of the nanospheres. Spherical nanospheres of different size, charge, and functionality could be used as carrier units for hydrophobic substances in aqueous media. The targeted site-specific delivery of the loaded nutraceutical biomolecule is dependent on the integrity of the capsule materials and its degradation in the gastrointestinal tract. For this, *in vitro* studies on release under simulated gastric conditions are carried out.

CNC encapsulated peppermint oil system showed high encapsulation and loading capacities when applied at 50% concentration of the former and released in simulated saliva for 160 minutes [66]. CNCs crosslinked with sodium tripolyphosphate was successfully used as a carrier of bovine serum albumin taken as a model protein drug. High loading and slow *in vitro* release rates were observed. Nanofibers extracted from the cellulose of turmeric spent were developed as an encapsulating material for asafetida, curcumin, curcuminoids, carvacrol, and gingerol. The encapsulation was obtained using a spray drying technique. Controlled release in the gut suggesting positive treatment against amelioration of ulcerative colitis and enhanced growth of broilers were recorded. Further, nanofiber weaving technology has been developed and used for better encapsulation of nanoparticulate nutraceuticals in powdered form. CNC grafted with CS oligosaccharide by ionic complexation was used as a carrier of vitamin-C. A 20 day long sustained in vitro release was recorded from a calorimetric thermogram. The released compound had remarkably high antioxidant potential. A potential grafting of cellulose nanoparticles with alginate to form stable tablets was proposed [57].

#### *4.4.1.5 STARCH-BOUND EDIBLE PRODUCTS*

Starchy foods are the most abundant products in the food market. Recent studies have revealed that hydrogen bonding between cellulose nanoparticles and processed starches can result in desirably changed edible food matrices. Hydrothermally processed gelatinized starch undergoes retrogradation upon cooling and storage due to the inter-coiling of the glycosidic chains and the simultaneous release of entrapped water. The addition of cellulose nanoparticles could inhibit retrogradation in gelatinized starches. This is due to crosslinking between the two polysaccharides, which can be further enhanced by the application of a suitable crosslinking agent. Ready crosslinking occurs with amorphous amylose fractions resulting in a rapid anti-staling effect. This could also result in the formation of chemically induced resistant starch fractions within the modified starch matrix. With altered physicochemical properties, crystallinity, and digestibility, cellulose nanoparticle-crosslinked starches can be used as novel ingredients in different food systems.

#### *4.4.1.6 CALORIE INHIBITOR*

Certain compounds in foods inhibit the digestion of carbohydrates and fats and absorption of hydrolyzed monosaccharides and fatty acids, thereby serving as a calorie inhibitor. When used as a stabilizer in Pickering solution, oxidized CNF inhibited the digestion of oils in the food emulsion, as seen when compared with a control conventional protein-stabilized emulsion [91].

The particles were entrapped in the intestinal mucosal layer, failing to reach the underlying epithelial layers. These indicated both safety and calorie reducing effects of the cellulose nanoparticles. Nanocellulose particles of different shapes and sizes and those prepared using different extraction techniques show different adhesion properties to the intestinal mucosa of different gastrointestinal regions [90]. This might also alter the digestion and absorption of the primary food components. Such calorie inhibiting effect of cellulose nanoparticles could assist in weight and obesity management. Such site-specific absorption characteristic of the particles can also be explored for further application for the delivery of drugs and nutraceuticals. Cellulose nanoparticles could also delay protein digestion, increase the viscosity of digesting food, altering the bowel movement of starchy and proteinous foods, and promote satiety [92].

#### *4.4.1.7 FILTRATION SUBSTRATE*

Filtration serves important roles in the processing of beverages for clarification and effluent treatments in the food industry. Due to non-toxicity and safe handling, suitable organic filtering devices are being designed to replace traditional inorganic filtering materials and adsorption substrates. Nano- to microporous cellulose composite material can be used as a filtering material to effectively screen out yeast and bacterial species from juices. This creates potential for research on designing three-dimensional crosslinked cellulose nanoparticles, nanowhiskers, and nanorod substrates for use as similar but uniformly porous beverage filtering devices.

#### *4.4.1.8 METAL-COMPLEXED NANOSYSTEMS*

One important property of cellulose and cellulose nanoparticles is that they can easily form bonds with active metal ions and other cations. This property has been well-utilized for the synthesis of a large number of cellulose-based engineered hybrid NMs with metals and metal oxides, with tremendous functionality for application in various sectors. Nanostructured celluloses and also other polysaccharides have been effectively used as capping agents for such highly reactive, thermo-sensitive inorganic nanoparticles. The different approaches to attach nanocellulose to inorganic nanoparticles have been categorized as follows [111]:

- Simple blending of cellulose and inorganic nanoparticle components;
- Synthesis using an external reducing agent;

- Synthesis by surface modification of nanocellulose;
- Synthesis using cellulose materials without any external reducing or capping agent.

Most of the synthesis protocols for such inorganic nanoparticle use hazardous chemicals that limit their application in foods and food packaging systems. Green synthesis methods for metal and metal oxide nanoparticles are those where environment-friendly non-toxic biological systems like fruit juice and spice extract or biological pathways of microorganisms can be alternatively used. Such green approaches have gained popularity and have been explored for use in nanocellulose-metal nanoparticle hybridized systems also. While polysaccharide-metal complexes have been abundantly used in drugs, their food usage and associated safety are currently being researched by several groups of food scientists. In the recent years, metal nanoparticles of silver, zinc, titanium, and silica and their oxides have been applied for both extraneous and intravenous functionality in food packaging and food matrix modification, owing to their preservative and material properties.

The use of cellulose nanoparticles within food systems as an ingredient is still a major challenge due to potential hazards that may be related to their sizes, shapes, and toxicity associated with the chemicals used in their extraction and synthesis. Although a number of researches have proven the non-toxicity of cellulose nanoparticles to human cells, life-cycle analysis of the material and its fate in the food system must be thoroughly carried out. If used in food, the behavior of any nanomaterial inside the gastrointestinal tract and its effect on general digestion and assimilation must be analyzed.

Mucoadhesion property of any newly developed material designed to be ingested encompassing the pathways of absorption onto the mucosal layer of the gastrointestinal tract and the longevity of the material in this layer requires crucial analyzes. Mucoadhesive nanoparticles could alter the structure of the mucosal gel. A detailed idea of the probable fate of CNCs with the human gastrointestinal tract was proposed [72] primarily based on the size and surface charges of the particles; and it was opined by the authors that the cellulose nanoparticles can be used as feasible nanocarriers of drugs and nutraceuticals. Oxidized particles would tend to degrade rapidly in human blood plasma. Spherical particles would result in higher cellular uptake.

Negatively charged particles will not adhere to mucous and can penetrate into the buccal and intestinal mucosa. The alkaline atmosphere in the duodenum can alter the surface charges of the particles, causing aggregation and thereby influence their interaction with the gut microflora.

While available literature on food uses of CNCs are mostly limited to uses in food packaging development, microcrystalline cellulose with larger structural dimensions than nanocrystalline cellulose has found use in a number of food matrices as a functional ingredient. It has been used as a texture enhancer, fat replacer, emulsifier, thickener, stabilizer, anti-caking agent, and encapsulating material in different food systems. Similar insight into cellulose nanoparticle induced food systems would be highly beneficial for innovations in the food industry and consumer market.

## 4.4.2 STARCH

### 4.4.2.1 PACKAGING MATERIAL

Nanoparticulate starches have been widely explored as nano-filler or reinforcing agents in composites for developing packaging films and other biodegradable materials with added functionality due to their viscoelastic, hygroscopic, enzyme, and bioactive components adhering and variable gas and moisture barrier properties. To enhance their particulate nature, thermal properties, crystallinity, dispersibility, and resulting functionalities, a range of chemical, physical, and enzymatic modification methods have been proposed by different authors. Mostly, the high availability and ready-reactivity of the polysaccharide have been utilized for chemical modification and binding to develop composite biomaterials having important properties. However, for food applications, keeping a check on the toxicity of the chemical used for modification of starch is necessary.

Starch has been extensively used as the base material for the development of edible films and coating materials. Researches on starch nanoparticles incorporated starch-based films as the most compatible mode of using starch nanoparticles in edible films and coatings are abundant. Corn starch films reinforced with taro starch nanoparticles were produced by enzymatic and gelatinization-recrystallization methods [25]. Starch nanoparticle incorporation resulted in increased TS, significant reduction in water vapor permeability (WVP), and induced thermostability in the films. However, the surface of the films became uneven which created a possibility for easier physical addition of other particulate components or layers on the surface. In another study, a similar increase in roughness was observed upon incorporation of starch nanoparticles in starch film. However, reduction in crystallinity and film strength were recorded with increased density and cloudiness of the nanoparticle added films [133].

Pea starch films incorporated with varying levels of potato starch nanoparticles resulted in improved mechanical properties, reduced WVP, and solubility and increased melting temperature [61]. Citric acid modification of starch followed by nanoparticle synthesis could highly increase the thermal resistance of the nanoparticles [95]. Addition of such nanoparticles in the starch film can result in even distribution and much improved mechanical and thermal stability with decrease in water vapor transmittance. A 1:1 mixture of montmorillonite (MMT) nanoclay and banana starch nanoparticles at 5% (w/w) proportion in banana flour film resulted in improvement in barrier properties [110]. Similar results were obtained when curcumin-loaded starch nanoparticles were loaded in a composite film matrix of aloe vera and banana starch [107]. Starch nanoparticles also improved the surface hydrophobicity and TS of gelatin films [145]. Besides, nanoparticulate starch has been researched for application as a food additive with a range of potential functions like stabilizing emulsions, fat replacing, modifying food texture, being enzyme resistant prebiotic entity and as nanoencapsulating material. They have been also used for wastewater treatment [77].

#### 4.4.2.2 EMULSION STABILIZER

As a stabilizer, starch nanoparticles have been used for stabilizing oil-in-water emulsions. A paraffin-water emulsion stabilized by starch nanoparticles showed stability till 2 months. However, the thermal instability of the nanoparticles led to coalescence of the dispersed oil droplets when temperature was raised to above 80°C. A much longer stability of polymer dispersion could be obtained when nanoparticle concentration was applied above 4% (w/w) [47].

Esterified starch nanospheres prepared by nanoprecipitation were used to stabilize glyceryl trioctanoate-water emulsion with long-term stability [143]. Up to an extent, increasing concentration of the emulsifying component increased the emulsion stability. In a study on the influence of type and size of starch nanoparticles on the stability of Pickering emulsions, particles within the range of 100–220 nm showed better stability than particles below or above this range [41]. When taro starch nanoparticles were used as stabilizer in a Pickering emulsion, thick stable layers of the nanoparticles were formed, which could encapsulate tea polyphenols dissolved in the oil phase [128].

A particle concentration of 7% gave the most stable emulsion. Both octenylsuccinylation treated soluble maize starch nanoparticle and insoluble starch nanoparticle extracted by acid hydrolysis could form stabilizing layers between oil and water interfaces [165]. Chemically modified starch nanoparticles with improved wettability for imparting stability to food-grade Pickering emulsions and protecting the entrapped molecules within the oil droplets have been also proposed. Both hydrophobic and hydrophilic starch nanoparticles can be developed, and their effectiveness as emulsion stabilizers under varying concentrations for specific ratio of water-in-oil or oil-in-water emulsions be targeted.

#### *4.4.2.3 FAT MIMETIC*

Both raw and modified starch nanoparticles have been suitably used as fat replacer or fat mimetic in a number of solid to semisolid food systems. Replacing fats with starch can result in reduced calorie intake with retention of taste, mouthfeel and volume of the food. Starch nanoparticles, being small in size can be easily applied as fat mimetic. Cereal starch nanoparticles have been used as fat replacer in salad dressing. Nano-scale complexation of maize starch with stearic acid and monoglyceride under wet-heat processing conditions produced the corresponding starch-lipid complexes with smooth creamy texture like fat [4]. In a previously reported work, wet-heat processing followed by annealing resulted in the formation of starch-lipid complexes and the system matrix could be used to partially replace fat in whipping cream [58]. Use of starch nanocrystals for emulsion stabilization as discussed above can also reduce the fat content in the emulsion, owing to its own proportion in the emulsion [60].

#### *4.4.2.4 TEXTURIZER*

As starch shows variable viscosity patterns with respect to process condition, heating (gelatinization) and cooling (retrogradation), the same is also reflected in many of their nano-scale variants. When added into food matrices, the rheological behavior of starch nanoparticles exerts a marked effect on the texture of the products as a function of shear, frequency, and temperature. Factors like particle size, nature of fluid and aggregation or disaggregation tendency would play a vital role in deciding the rheological behavior of such food systems. Starch nanocrystals prepared by acid hydrolysis with higher crystallinity than starch and nanoparticles are more resistant to temperature rise. Once starch nanoparticles are added to starchy systems, they can result in increased strength of the matrix and resistant to melting, as have been observed in starch solutions intended to form packaging films. The same would be also applicable when added to starchy food systems.

#### 4.4.2.5 NANOENCAPSULATION

Starch nanoparticles are effective in entrapping several bioactive compounds, have a large surface area, and show low viscosity at high concentration, which is suitable for its migration in a physiological medium. Having a neutral taste, starch nanoparticle-encapsulated materials can efficiently preserve the taste and aroma of a natural compound, while retaining its chemical nature. Curcumin is the bioactive compound with multifacet positive health effects. It is extracted from turmeric (*Curcuma longa*). It has low water solubility, high degradation tendency, and low bioavailability. Curcumin has been extensively studied for encapsulation by starch nanoparticle-based encapsulating materials to minimize the loss of desirable properties in it. Nanoparticles obtained from acid-hydrolyzed cassava starch were effectively loaded with curcumin under acetone environment [118].

Similarly, curcumin was loaded in a sago starch nanoparticle carrier to find sustained-release under *in vitro* buffered condition [22]. It was observed that enhanced solubility of the encapsulated compound in aqueous media. Curcumin loaded in native and acetylated banana starch nanoparticles showed control release in gastric fluid. Encapsulation with the acetylated nanoparticles showed better efficiency. It was observed that intimate bonding between the compound and the carrier nanomaterial occurred, as revealed by crystallinity and infrared spectrum analyses [3]. Similar prompt encapsulation was also found when catechin was encapsulated by nanoparticles obtained from horse chestnut, water chestnut, and lotus stem starches. The bioactive properties of the encapsulated catechin and the *in vitro* digestibility of the encapsulation system were retained [5].

Hydroxyethyl starch-deoxycholic acid nanoparticles were stabilized by a self-assembling technique and used to encapsulate morin, a flavonoid extracted from Osage orange (*Maclura pomifera*) used as a treatment for high level of uric acid in blood and several other disorders [82]. *In vivo* pharmacokinetics and *in vitro* studies suggested controlled release, suggesting effectiveness. Similarly, controlled release of salvianolic acid improved after encapsulation with a nanocomposite of gum and starch nanoparticles [86]. The fact that amylose has a tendency to bind lipid molecule can be well-explored for nanoencapsulation of fat-soluble components. Vitamin D3 was successfully loaded in high amylose corn starch and potato starch nanoparticles-based encapsulating matrices [51]. Arrowroot starch nanoparticle was used for spherically encapsulating hydrophilic andrographolide with controlled-release [158].

#### 4.4.2.6 RESISTANT STARCH

Resistant starch has been identified as that part of raw or processed starch which does not undergo hydrolysis within the gastrointestinal tract by the human digestive enzymes and behaves as dietary fiber. They also serve as prebiotic in the colon. Thereby, their consumption does not contribute to calorie gain. Several methods have been developed for the synthesis of resistant starch and their isolation from food matrix to be used as individual functional ingredient. Both resistant starch and resistant starch-rich foods are now available in the market. It has been found that starch nanoparticles synthesized under certain conditions may behave as resistant starch showing prolonged rates of *in vitro* hydrolysis. Acid hydrolyzed starch chains can recoil and recrystallize under alcohol precipitation and other treatments to form enzyme-resistant nanocrystalline forms [102].

Enzyme resistance can also be attained after crosslinking of starch nanoparticles with suitable agents inducing hydrophobic functional groups. Enhance crystallinity of starch nanocrystals also relate to prolonged enzymatic hydrolysis and the same can be used in food and beverage systems to lower glycemic indices. Contrary to this, naturally enzyme resistant high amylose maize starch and nanoparticles obtained from it have low crystallinity. Nanoparticles formed by high pressure homogenization treatments of high amylose maize starch had lower crystallinity but were resistant to enzymatic hydrolysis [7].

#### 4.4.2.7 WASTE WATER TREATMENT

The easy availability, low cost, and non-toxicity of starch nanoparticles can be well utilized for developing into substrates for wastewater treatment. Stearic acid grafted waxy maize starch nanoparticles were used for adsorption and removal of aromatic organic compounds, namely trichlorobenzene, xylene, quinoline, nitrobenzene, 2-naphthol, dichlorobenzene, and chlorobenzene present in water as pollutant. After 6 hours of exposure, variable rates of adsorption depending on affinity level and concentration of the compounds were observed [77].

### 4.4.3 GLYCOGEN

#### 4.4.3.1 SMART PACKAGING

Glycogen has been used for binding and growth of silver NPs for antimicrobial activity against *Staphylococcus aureus*, *Escherichia coli*, and *Candida albicans* pathogens [16]. Hence, this nanoencapsulated system can be suitably fabricated for designing antimicrobial food packages. A nanostructured hybrid Glycogen-*graft*-poly(2-alkyl-2-oxazolines) thermo-responsive material was designed, which may have application as a thermal sensing component in food packages [115].

#### 4.4.3.2 ENCAPSULATION

Lactosylated glycogen NPs containing terminal β-galactoside moieties were developed [80]. It was found that the particles could bind lectin and have a strong tendency to cause aggregation of prostate cancer cells. A glycogen-lectin co-polymeric nanoparticulate coating system was designed through self-assembly of the two components for incorporating drug molecules, like transferrin. This suggests an opportunity for researchers to work on the food use of glycogen nanoparticles as bioactive carriers.

#### 4.4.3.3 TEXTURE ENHANCER

Monodisperse glycogen and phytoglycogen nanoparticle with claimed applicability in food formulations as rheology modifier has been patented [71]. The nanoparticles can be obtained and used in both native as well as chemically-modified forms for the intended application.

### 4.4.4 PECTIN

#### 4.4.4.1 ENCAPSULATION

Stable dispersion of pectin-caseinate nanocomplex was developed and used to encapsulate docosahexaenoic acid [11]. β-lactoglobulin conjugated with low methoxyl pectin produced a nanocomplex which could protect docosahexaenoic acid when subjected to accelerated shelf-life test with minimal degradation over the period of 100 hours [171]. A composite nanoparticle of sodium caseinate and zein coated over by pectin using electrodeposition technique could increase the efficiency of the complex to encapsulate curcumin and its sustained release under gastrointestinal conditions [19]. Curcumin has also been encapsulated by a heat-denatured lactoferrin-pectin complex [162]. Nanoencapsulation matrix of whey protein isolate and high methoxyl pectin may form the smallest particle size at pH 3.5 and can entrap lactoferrin, a bioactive compound generally used in infant formulae [117].

Spray drying technique has been used in developing nanocapsules from a complex of pectin and sodium caseinate for encapsulating curcumin. The encapsulated curcumin showed better bioactive properties than native curcumin and released slowly under simulated gastrointestinal conditions [156]. Whey protein-pectin nanocomposite has also been used for encapsulation of d-limonene [42]. Eugenol oil extracted from clove with strong antioxidant and antimicrobial activity was encapsulated in nanoparticles made from zein-sodium caseinate-pectin complex [153]. Zein-pectin conjugate was also used for effective encapsulation of resveratrol, a polyphenolic phytoalexin with antioxidant and potential anticarcinogenic activities [55].

Nanohydrogel systems have been developed by crosslinking pectin with proteins due to their differences in surface charges and applied for entrapment and targeted release of flavor oil, vitamin, and bioactive compounds. Phenolic extract from olive leaf was encapsulated in whey protein concentrate and pectin using a water-in-oil-in-water multiple emulsion [99]. Double layer emulsion with spray drying formed nanocapsules with better-controlled release of bioactive folic acid [8]. Using similar techniques, other important bioactive compounds like oleuropein, gallic acid, and saffron could be also encapsulated. Nisin loaded pectin nanoparticles showing significant antimicrobial properties meant for use in food systems were developed [73]. Electrospinning technique was used to develop high and low methoxyl pectin nanofibers and pectin and pullulan conjugated ultrafine fibers and mats for potential food application [24]. The later could be loaded with probiotic culture of *Lactobacillus rhamnosus*. A composite of bacterial nanocellulose and pectin was developed for enhancing the survivability of probiotic *Bacillus coagulans* in the human digestive system and its targeted release in the gut. The encapsulation was further modified by the addition of starch and carboxymethyl cellulose (CMC) [67]. Pectin based nanoencapsulation can be suitably used for targeted delivery of the encapsulated materials like citrus flavonoids, curcumin, rutin, Vitamin E and Vitamin B12 in the colon as pectin is hydrolyzed by the colon microflora but not by the enzymes in the digestive tract.

#### 4.4.4.2 EMULSION STABILIZER

Pectin alone or in conjugation with other polysaccharides and proteins have been used as nanoemulsion stabilizers. Food-grade oil-in-water nanoemulsions containing essential oils from sources like oregano, thyme, and lemongrass have been prepared by using high methoxyl pectin as emulsifying agent. Nanoemulsion containing lemongrass oil showed the highest stability. Nanoemulsions prepared using lemongrass and mandarin oil showed higher negative potential than those fabricated using oil from oregano and thyme when stabilized using pectin. This was also related to their higher stability and lower droplet size in the respective emulsions with stronger antimicrobial activity [46]. Electrostatic deposition technique was used for spontaneous oil-in-water emulsion stabilization by pectin. Dissolution of lauric arginate generated positive charge in the oil droplets. At high pectin concentration, the anionic pectin molecules got absorbed to the cationic oil droplets, forming a stable emulsion with a net negative charge on the droplets [168]. Pectin was also used for stabilizing emulsion created by rice glutelin. It was found that a better stability could be imparted by pectin to the emulsion at lower pH (3.5) than neutral [161].

In another study, pectin was found to impart increased emulsion stability than CMC used as stabilizer in an orange beverage stored for two months at 25°C [98]. Temperature, pH, and component ratio play an important role in the stability of an emulsion. Clove oil nanoemulsion prepared by high-speed homogenization and stabilized by a mixture of sodium caseinate and pectin was found to be stable under different process temperature conditions and salt concentrations but was not stable within the pH range of 3–5 [130]. In an apricot gum-apple pectin complex emulsion stabilizer, a weight ratio of 21.4 (gum to pectin, w/w) of the mixture was found to exert the highest stability to the sunflower oil-in-water nanoemulsion at acidic pH of 2 and temperature of 60°C [126].

#### 4.4.4.3 COATING

Pectin can act as a carrier of various functional compounds like antimicrobials and flavors and serve as a suitable layer in designing edible packaging materials. Pectin has oxygen barrier properties. When applied in nano-scale multilayer coating, coated mangoes showed longer shelf lives than control [96]. This also inhibited browning in fresh-cut persimmon [123]. A novel bio-nano hybrid active coating based on pectin incorporated with layered double hydroxide-sodium salicylate was designed. On application over fresh apricots, the coating enhanced the shelf life by showing anti-fungal activity [80]. Reinforcement with nanocrystalline cellulose resulted in increased TS and reduced water vapor transmission rate (WVTR) of pectin-based film [18].

#### 4.4.4.4 WATER PURIFICATION

Pectin can form bond with several functional groups and produce gels with different structural characteristics. Modified pectin hydrogels have been used to adsorb and separate pollutants from water. An interpenetrating polymer network superabsorbent substrate of pectin-g-(sodium acrylate-coN-isopropylacylamide) (PC-g-(SA-co-NIPA)) hydrogel with sodium acrylate as monomer and methylene-bis-acrylamide as crosslinking agent was developed [135]. This was used for adsorption of dyes and metal ions from aqueous solutions. Polyacrylic acid grafted pectin hydrogel could adsorb $Cd^{2+}$ ions [37]. Pectin-g-(sodium acrylate-co-N-iso-propylacyl-amide) hydrogel was synthesized by free radical polymerization protocol and was found to adsorb $Pb^{2+}$, $Zn^{2+}$, $Cu^{2+}$ and $Co^{2+}$ ions to different extents [131].

The pH-sensitive adsorption substrate of pectin-poly(acrylamide-co-acrylamidoglycolic acid) has been developed which can sufficiently adsorb $Cu^{2+}$ and $Ni^{2+}$ with a higher adsorption affinity for $Cu^{2+}$ [136]. Pectin-cellulose microfiber beads also showed efficient adsorption of $Cd^{2+}$, $Cu^{2+}$ and $Fe^{2+}$ [78]. On applying at nano-scale, the adsorption capacity of such pectin-based hydrogels would significantly improve.

#### 4.4.4.5 CHOLESTEROL METABOLISM

Although never applied in its nano-form, the proven cholesterol and blood glucose-lowering properties of pectin can be explored for application in special nanotechnologically designed food matrices meant for therapeutic consumption.

### 4.4.5 CHITIN AND CHITOSAN (CS)

#### 4.4.5.1 ANTI-OBESITY

Chitin and CS are not digested by the intestinal enzymes and hence behave as dietary fibers and can be used as prebiotics. They have been found to stimulate beneficial colon bacteria such as *Bifidobacterium* and *Lactobacillus* species [49]. Carboxymethyl derivatized chitin has the important property of being soluble in neutral and alkaline media and can be synthesized into nanoparticles. Carboxymethyl-chitin was found to have an anti-obesity effect on Mouse 3T3-L1 preadipocyte cells. The analyzes showed that triglyceride content in blood decreased with simultaneous increase in glycerol release [70]. Chitin glucan also exerted a similar effect on obese mice showing significant reduction in weight [105]. These derivatized compounds of chitin could show higher efficiency once reduced to nano sizes and it has scope for research.

#### 4.4.5.2 ANTIOXIDANT

Chitin shows antioxidant activity and can be used as a safe, non-toxic alternative to synthetic antioxidants for stabilizing unsaturated food lipids. The residual free amino group in the structure can react with free radicals. A 70% decrease in thiobarbituric acid reactive substances in meat stored at 4°C for 3 days was observed after addition of 1% CS [146]. Addition of CS lowered peroxide value in thiobarbituric acid reactive substances and volatile aldehydes in herring (*Clupea harengus*) flesh [64]. Once applied in nanoparticulate form in food and packaging systems, chitin, and CS can impart vital antioxidant and anti-inflammatory effects.

#### 4.4.5.3 EMULSION STABILIZER

As mentioned before, chitin nanowhiskers can undergo isotropic-anisotropic nematic transition above a critical acidic solution concentration. The nematic gel can be further strengthened by altering pH and temperature. This gel-forming property of the nanoparticles has been used to stabilize oil-in-water emulsions by forming interdroplet network of the nano-gel [150, 151].

#### 4.4.5.4 THICKENER

Acetylated derivatives of chitin and CS with the ability to form gels can be used as thickeners in vegetable oils instead of conventionally used metallic soaps and polyuria derivatives [122]. Chitin nanoparticles and whey protein mixed dispersion showed gel-like behavior with increasing concentration of either of the solutes. The gel strength irreversibly increased on further heating, suggesting synthesis of a novel gel with potential applicability as a food thickener [150, 151].

#### 4.4.5.5 ENCAPSULATION

Carboxymethyl CS was used in combination with CS and alginate to encapsulate probiotic *Lactobacillus casei* by an extrusion method and observed efficient multiplication under *in vitro* gastric conditions [84]. The same approach would be feasible and could be more effective if CS nanowhiskers were used instead of macromolecular CS. With polyvinyl alcohol, chitin nanowhiskers could form a composite hydrogel for encapsulation and release of bovine serum albumin taken as model drug [104]. The same can be also used for nutraceutical encapsulations.

#### 4.4.5.6 ANTIMICROBIAL PACKAGING

Although both chitin and CS in their macro and nanoforms can form highly thermostable standalone films, the films lack mechanical strength. Therefore, they have been often used to form composite with nanocrystalline cellulose and other biopolymers to attain the desired mechanical properties. Both chitin and CS and their chemically derived forms have proven bactericidal and fungicidal properties and are used widely as biopesticides for treating stored seeds before sowing. Ternary film made of chitin, CS, and rectorite had a smooth texture and showed moisture resistance, high thermostability beside anti-bacterial properties [87]. Chitin nanowhisker reinforced with hyaluronic acid film showed strong activity against the growth of two each of gram-positive and gram-negative pathogenic bacterial strains. Biocompatibility of this edible film was supported by cytotoxicity analysis [1].

CS-cellulose nanowhisker composite showed bacterial inhibition with 3.1 log cycle reduction in the population of lactic acid bacteria in ground meat stored at 25°C for 6 days [29]. In another study, oligochitosan and CS nanowhiskers were used as preservative in minced pork meat. Apart from inhibitory affect against *Staphylococcus aureus*, *Listeria monocytogenes*, *Bacillus cereus, Salmonella enteritidis*, and *Escherichia coli* O157:H7, the two forms of CS also helped in extension of the shelf-life of the meat by retarding lipid oxidation myosin heavy chain protein degradation and formation of biogenic amines [21]. CS coated cellulose film also showed antimicrobial action against *Staphylococcus aureus*. Antimicrobial films using a blend of polyvinyl alcohol, cellulose nanowhiskers, and CS was also developed [79]. Carrageenan-chitin nanofibril composite film showed strong antibacterial activity against *Listeria monocytogenes* [127].

Kelp grouper fish species *Epinephelus bruneus* showed resistance to *Vibrio alginolyticus* infection when fed with a diet rich in chitin and CS. It has been proven that chitin supplements in diet can modulate growth of gut microflora and inhibit pathogens in fishes [50]. Similar trials in human may be necessarily carried out. Based on the antimicrobial properties, chitin, and CNPs can be designed as food preservative. Several metal nanoparticles have been embedded, synthesized, and grown on chitin and CS substrates for use as antimicrobial surface. A composite material formed by immobilization of copper nanoparticles in chitin nanoparticles showed higher antimicrobial activity when compared to the individual materials [138]. Such cumulative effect of both the agents on selected microbial species was also reported [89]. In a study encompassing five different chitin nanoparticle variants concluded that the difference in antimicrobial property is governed by the extent of positive surface charge on the particles [148].

#### 4.4.5.7 WATER TREATMENT

Nanosized chitin and CS are active adsorbent NMs capable of cleaning water streams from metal ions and dyes. CS nanorods were used for absorption of hexavalent chromium and observed an uptake capacity of up to 323.6 mg/g of the substrate [137]. In another study, absorption of 358 mg/g was attained for bivalent cadmium [124]. Nanofibers prepared by electrospinning could absorb 263.2 mg/g and 485.4 mg/g of $Pb^{2+}$ and $Cu^{2+}$, respectively [48].

A nanodispersion of CS was used to adsorb Refractive Red 120 dye. The extent of adsorption by the dispersion was found to be much higher (910 mg/g of dye) than that of the dissolved CS (51 mg/g) [100]. A number of other ions and dyes like arsenate ion, methylene blue (MB), Bromophenol Blue, and Coomassie Brilliant Blue, Acid Green 27, Acid Red 73, Acid Orange

10, Acid Orange 12, and Acid Red 18 van be also significantly adsorbed by chitin nanoparticles. Chitin nanowhisker-based membranes have been used for purification of water. A membrane with chitin nanowhiskers having high virus absorption property suggested its potential use for drinking water filtration and purification [94]. Chitin nanowhiskers prepared from shell of shrimp could chemically as well as electrostatically adsorb melanoidin in sugar syrup produced in the sugar industry [31]. A composite membrane of nanocrystalline cellulose in CS was observed to be nanoporous and could adsorb cationic dyes, namely Victoria Blue 2B, Methyl Violet 2B and Rhodamine 6G dissolved in water [65].

Water filtration mats were prepared by electrospinning of cellulose acetate and coated with chitin nanocrystals wherein a stable webbed film structure of the nanoparticles was formed by self-assembly. This strengthened the mats and also increased stiffness. Significant increase in hydrophilicity was attained and a notable reduction in biofouling and biofilm formation was observed [44]. Carbonized aerogel from isolated and freeze-dried prawn chitin nanowhiskers could retain the original fine network of the nanowhiskers providing large surface area as filtration/adsorption substrate [108].

#### *4.4.5.8 IMMOBILIZATION SUBSTRATE*

Chitin nanowhiskers and chitin-derived CNPs were successfully used for immobilization of bovine serum albumin taken as a model protein. CNPs showed high immobilizing efficiency at pH 6 and 25°C [103]. Lysozyme immobilized on chitin nanowhiskers showed enhanced antimicrobial activity due to the combined effect of both the components and change in the secondary structure of the former upon immobilization. Thereby, the enzyme and immobilization substrate formed a nature-based strong antimicrobial preservative system for food application and biosensor designing [62].

An electrochemical sensor for glucose oxidation was developed by immobilizing copper nanoparticles on chitin nanoparticles [138]. Electro-oxidation of glucose in 0.1 M NaOH solution at 0.45 V potential was observed with a lowest detection limit of 0.776 μM. The composite electrode matrix showed no cytotoxicity against brine shrimp (*Artemia salina*). TEMPO-oxidized chitin nanoparticles were induced with magnetism by amine-functionalizing and used as a substrate for chymotrypsin enzyme. Increased stability and activity of the enzyme upon immobilization was recorded [54]. A high proteinase activity (84.3%) was achieved upon immobilization over CNPs [144].

### 4.4.6 GUMS

#### 4.4.6.1 EMULSION STABILIZATION

Wheat gluten nanoparticle-xanthan gum complexes resulted in stable Pickering emulsion stabilization over a wide range of pH (4–8). β-carotene encapsulated in the Pickering solution showed high storage stability and thermal sterilization with increased bioaccessibility indicating its feasible delivery in this form [40]. High stability was exerted by a mixture of lecithin and gum Arabic encapsulating eugenol in an oil-in-water emulsion [52].

In another study, zein-gum Arabic nanoparticles were fabricated and used to stabilize the oil-in-water emulsion of a Pickering emulsion at a concentration of 6.25% [81]. When the nanoparticles were loaded with thymol, effective antimicrobial activity inhibiting the growth of *E. coli* was observed with a controlled release of the encapsulated thymol. Complexation of ovotransferrin with gum Arabic was attained by turbidimetric titration to obtain nanoparticles which could stabilize oil-in-water Pickering emulsion [157]. Zein-gum Arabic electrostatically assembled colloidal nanoparticles were prepared [26]. A core-shell structure of gum Arabic over zein nanoparticles was formed. On application in Pickering emulsion, the nanoparticles formed densely packed layer over lipid droplets, resulting in stability of the emulsion.

#### 4.4.6.2 TEXTURE ENHANCER

Gum Arabic is used in the beverage, confectionery, food emulsion, flavor modularity, baking, and beer industries to resist sugar crystallization. This creates the possibility for the use of synthesized gum and gum-based nanoparticles as texture enhancer in such food ranges.

#### 4.4.6.3 NANOENCAPSULATION

While designing an encapsulating material to be used for food application, it should be ensured that the polymers used are natural, edible, biocompatible, and free of any toxic solvents. Most of the reported researches on gum nanoparticles application in food are concentrated on their uses as encapsulating nanomaterial for the delivery of functional compounds. Alginate/cashew gum nanoparticles prepared by spray drying for encapsulation of

*Lippia sidoides* essential oil showed encapsulation efficiency up to 55% with encapsulated oil levels ranging from 1.9 to 4.4% [28]. Maximum release time of 50 h was recorded in distilled water taken in dialysis bag of 14 kDa pore size. A nano-spray drying vibrating mesh technology was used to form nanoencapsulating polymer matrices containing different gums like gum Arabic and cashew nut gum [109].

The matrix was used to encapsulate vitamin B12. An encapsulating material matrix of gum Arabic and lecithin was prepared and eugenol oil was encapsulated by nano-spray drying of their homogeneous suspension. Nanospheres of size below 500 nm were obtained which showed excellent redispersibility in water and rehydration without adversely affecting the antimicrobial activity of the encapsulated eugenol. Negatively charged xanthan gum and lysozyme were conjugated to form nanoparticles under an alkaline atmosphere (pH = 11) and at high temperature, which caused denaturation of the natural structures of both the components. However, the internal hydrophobic regions remained unaltered and could be utilized for holding suitable functional molecules [52]. Gliadin, a wheat protein has been used as a drug carrier in its nanoparticulate form. Gum Arabic coating over gliadin nanoparticles improved its stability to temperature and pH [160].

Cress seed mucilage-poly vinyl alcohol (PVA) conjugated nanofibers were used for entrapment of vitamin A [36]. Entrapment resulted in crystallization of the fibers and increase thermal stability of the entrapped vitamin. A low release rate was recorded under simulated gastric conditions. D-limonene was encapsulated in electrosprayed nanocapsules and nanofibers containing *Alyssum homolocarpum* gum cross-linked with Tween 20 [68]. A high encapsulation efficiency of approximately 90% was attained by this method. Using coacervation technique, kaffir lime oil was encapsulated in a matrix of gum Arabic and maltodextrin [149]. Xanthan gum-chitosan mix prepared in formic acid solution were molded into nanofibers using electrospinning and used to encapsulate curcumin. Lower release rate at acidic pH was recorded, suggesting feasibility of the nanofibers to be used as carrier of hydrophobic bioactive compounds with extended-release properties [132].

Tea polyphenol epigallocatechin-3-gallate was loaded into a matrix of gum Arabic and maltodextrin [119]. An encapsulation efficiency of 85% was attained. Tragacanth gum-sodium caesinate complex nanoparticles of the size 70 nm were formed in aqueous medium at pH 4 in a designed experiment [45]. This critical pH shifted to higher values when the protein to polysaccharide ratio was progressively increased. Use of polysorbate 80 and Quillaja saponin surfactants to almond gum and shellac composite nanoparticles precipitated at pH 1.2 resulted in a marked decrease in particle size. Quercetin loaded to the nanoparticles was stable at the forming pH with loading efficiency of up to 98% [33].

The nanoencapsulating material was further used for encapsulating polyphenol-rich cinnamon bark extract [101]. Increase in concentration of extract reduced the loading efficiency. Encapsulation resulted in significant thermal stability of the compounds and retention of antioxidant activity. More than 90% extract got released at gastric pH. Acetylated cashew gum self-assembled nanoparticles were used as a carrier for a hydrophobic model drug Indomethacin to observe controlled release [114]. There occurs an opportunity for the material to be used for multiple nutraceutical encapsulation.

#### 4.4.6.4 WATER PURIFICATION

Inorganic substrates used for the elimination of metallic wastes from water streams are based on the principles of adsorption, chemical precipitation, ion exchange and reverse osmosis. Drawbacks of these techniques are elimination of metal, deposition of toxic slurry. Besides, use of expensive chemicals and methods makes the set up costly. Renewable, naturally occurring engineered biosorbent materials are inexpensive alternatives. Guar gum grafted onto multiwalled carbon nanotube and iron oxide nanoparticles could adsorb neutral red and MB dyes form aqueous systems [163]. Guar gum was cross-linked with soy lecithin to form a nanohydrogel sheet which could remove thiophanate methyl, a fungicide from its diluted aqueous solution [129]. Synthesized guar gum-cinnamate conjugate self-dispersing nanoparticles were effective in killing *E. coli* and *Staphylococcus aureus to decontaminate water* [27].

#### 4.4.6.5 METAL-COMPLEXED NANOSYSTEMS

Different gums have been extensively used as green synthesis substrates for a number of metals and their oxides, as was discussed for cellulose. Such approaches are supposed to find specific applicability as antimicrobial substrates, packaging, and mode food fortification in coming times.

## 4.4.7 HEMICELLULOSE

### 4.4.7.1 ANTIMICROBIAL

Chagas disease (American Trypanosomiasis) is caused by a protozoan *Trypanosoma cruzi and its treatment requires oral medication. Xylan from corn cob when was conjugated with silver nanoparticles (*AgNPs*) as an effective delivery system for efficient destruction of the protozoa [17]. A thermally conjugated Maillard complex of xylan and* CS *was found to show higher antimicrobial activity than* CS *alone [*88*].* The material also showed good antioxidant capacity. If obtained in nano-sized form using physical size reduction, better effectiveness of the material can be expected.

### 4.4.7.2 PACKAGING

Xylan can form films in the presence of suitable plasticizing agents. Films and coatings of xylan show low oxygen and gas permeability and in certain cases, good TS. Mechanical properties of these films can be further enhanced by the addition of fillers like cellulose nanofibers.

### 4.4.7.3 NANOENCAPSULATION

Xylan-curcumin nanoconjugate particles were found to be highly pH-responsive and most of the curcumin was released at low pH [76]. The nanoparticles showed significant cytotoxic effect against human colon cancer cells and hence could be effective in cancer therapy.

### 4.4.7.4 BIOSENSOR

A microwave-assisted one-step method was used to develop polyamine-functionalized carbon dots from xylan-polyethyleneimine precursor which can detect the presence of tannic acid in water and white wine samples [164]. In another study, polyvinyl alcohol, CS, xylan, and hydroxyapatite nanocomposite film was prepared which could be used as a packaging material as well as a biosensor substrate to detect real-time freshness of Indian oil sardine fish by sharp change in color [152].

### 4.4.8 MICROBIAL POLYSACCHARIDES

#### 4.4.8.1 NANOENCAPSULATION AND NANO-CARRIER

Alginates are one of the most common polysaccharide sources for synthesizing nanogels to be used as effective carrier of bioactive compounds. This is due to their ready reactivity and thermal transition properties to form gel. It has been used alone as well as in combination with other biopolymers to develop suitable nanocarrier systems for a number of bioactive compounds, vitamins, essential oils and polyphenols. However, the same binding property of the alginate nanogels can be also applied to other functionalities like carrier of antimicrobials, development of nano-biosensors, etc. (Table 4.2).

**TABLE 4.2** Applications of Alginate Nanoparticles, Conjugating Compounds, and Their Sizes

| Conjugating Compounds | Obtained Size | Applicability | References |
|---|---|---|---|
| Chitosan | 235 nm | Carrier of isoniazid, rifampicin, pyrazinamide, drugs | [166] |
| | 250–350 nm | Carrier of paclitaxel drug | [155] |
| | 50 nm | Activity against Propionibacterium acnes, *which causes skin acne* | [39] |
| | 150–300 nm | Protein delivery | [83] |
| | > 300 nm | Carrier of DNA | [34] |
| Calcium ion, Ag, and Au nanoparticles | < 10 nm | 4-Nitrophenol Reduction | [120] |
| Calcium ion | 200–600 nm | Delivery of antisense oligonucleotide | [9] |
| Magnetic nanoparticles | 10–60 μm sized microspheres holding 5–15 nm-sized metal nanoparticles | Biosensor designing | [63] |

Polyelectrolyte nanocomplex spheres of curdlan and pectin were loaded with curcumin. The nanoencapsulated system exhibited high solubility and excellent antioxidant activity [159]. Microbeads of gellan-whey protein isolate were produced by emulsion extrusion and used to encapsulate flax-seed oil [74]. The beads could surpass stomach and release the oil in the small intestine under simulated conditions. Further refining of the process variable may lead to nano-sized beads with better effectiveness and functionality. Cinnamon bark extract was encapsulated in a shellac-xanthan gum conjugate nanoparticles. Xanthan gum imparted stability to the particles [101].

#### *4.4.8.2 FOOD ADDITIVE*

Pullulan has been extensively used in foods as a low-calorie additive. It can partially replace starch in some foods and help in the development and retention of texture, give glazing effect, entrapping moisture. It can impart thaw stability to frozen products. Several other applications of pullulan are known. However, no research to date has dealt with or not mentioned the application of the nanostructured forms of the polysaccharide in any food or food packaging system. Pullulan-based nanoconjugated particles have been used as drug carriers and other non-food matrices only, which draws immediate attention.

#### *4.4.8.3 ANTIMICROBIAL*

Chitosan-gellan gum and lactoferrin-gellan gum nanoparticles formed by electrostatic complexation were found to inhibit the growth of *Staphylococcus aureus* [154].

#### *4.4.8.4 WATER PURIFICATION*

Chemically cross-linked gellan-pullulan nanogels may efficiently strip MB from alkaline phase. The outcome was superior to using only gellan nanogel [23].

## 4.5 SUMMARY

Polysaccharide and polysaccharide-based nanoparticles have been fabricated by a number of extraction and synthesis techniques. Deliberated combinations of two or more of these techniques are more effective in isolating individual nanoparticles targeted for specific usage. Synthesizing techniques are being upgraded for obtaining the intended size and morphology of the nanoparticles. Major polysaccharides whose nano-sized particulate forms have been designed and applied in food systems are cellulose, starch, chitin, and CS. This is due to their natural abundance and inherent semicrystalline built-up of matrices. Pectin, gums, hemicelluloses, and most of the microbial polysaccharides being largely amorphous do not easily form nanocrystals by themselves. Crystallinity is an important factor responsible for structural stability and thermo-tolerance of foods they are induced in. However, they have been widely used in beverages and soft foods due to their surface properties. Pectin and alginates show-ready reactivity and stable gel-forming properties, which has been widely explored for designing nano-dimensional gel systems. In most instances, polysaccharide nanoparticles have been modified or conjugated with other polymers to develop into appropriate nanoparticulate forms. Such modification sometimes involves the use of chemicals that are not recognized as safe for consumption, which is the major limitation in the use of several of the polysaccharide nanoparticles in foods. It becomes clear from the general observation that most of the scientific reports on the application of polysaccharide nanoparticles for food usage are on their use in food packaging systems and not on the actual use in the food matrix. A large volume of research has been done on their use in drug delivery, but similar use in nanoencapsulation and delivery of functional food ingredients and nutraceuticals is limited. Larger cost involvement and the current choice of consumers of natural bulky diets could be considered as the major reasons behind this difference. Still, the development of polysaccharide nanoparticles for food usage has accelerated in recent years. Scientists and technologists have started working on glycogen, gums, curdlan, pullulan, and other polysaccharides which have attained lesser nanotechnological importance till now. It is expected that with changing customer demand towards safer, smarter, and bioactive-dense alternative natural food options, the development and use of polysaccharide nanoparticles would rise exponentially in the food industry during the next decades.

## ACKNOWLEDGMENTS

The author is thankful to Ms. Aditi Duarah for providing support in literature collection and survey during preparation of this chapter.

## KEYWORDS

- **carbohydrate**
- **encapsulation**
- **nanoemulsion**
- **nanoprecipitation**
- **pickering**
- **polyelectrolyte**

## REFERENCES

1. Abdelrahman, R. M., Abdel-Mohsen, A. M., Zboncak, M., Frankova, J., Lepcio, P., Kobera, L., & Brus, J., (2020). Hyaluronan biofilms reinforced with partially deacetylated chitin nanowhiskers: Extraction, fabrication, *In-Vitro* and antibacterial properties of advanced nanocomposites. *Carbohydrate Polymers, 235*, 115951.
2. Abdollahi, M., Alboofetileh, M., Behrooz, R., Rezaei, M., & Miraki, R., (2013). Reducing water sensitivity of alginate bio-nanocomposite film using cellulose nanoparticles. *International Journal of Biological Macromolecules, 54*, 166–173.
3. Acevedo-Guevara, L., Nieto-Suaza, L., Sanchez, L. T., Pinzon, M. I., & Villa, C. C., (2018). Development of native and modified banana starch nanoparticles as vehicles for curcumin. *International Journal of Biological Macromolecules, 111*, 498–504.
4. Agyei-Amponsah, J., Macakova, L., Dekock, H. L., & Emmambux, M. N., (2019). Sensory, tribological, and rheological profiling of "clean label" starch–lipid complexes as fat replacers. *Starch-Stärke, 71*(9, 10), 1800340.
5. Ahmad, M., Mudgil, P., Gani, A., Hamed, F., Masoodi, F. A., & Maqsood, S., (2019). Nano-encapsulation of catechin in starch nanoparticles: Characterization, release behavior and bioactivity retention during simulated in-vitro digestion. *Food Chemistry, 270*, 95–104.
6. Aldao, D. C., Šárka, E., Ulbrich, P., & Menšíková, E., (2018). Starch nanoparticles–two ways of their preparation. *Czech Journal of Food Sciences, 36*(2), 133–138.
7. Apostolidis, E., & Mandala, I., (2020). Modification of resistant starch nanoparticles using high-pressure homogenization treatment. *Food Hydrocolloids, 103*, 105677.
8. Assadpour, E., Jafari, S. M., & Maghsoudlou, Y., (2017). Evaluation of folic acid release from spray-dried powder particles of pectin-whey protein nano-capsules. *International Journal of Biological Macromolecules, 95*, 238–247.
9. Aynie, I., Vauther, C., Chacun, H., Fattal, E., & Couvreur, P., (1999). Spongelike alginate nanoparticles as a new potential system for the delivery of antisense oligonucleotides. *Antisense and Nucleic Acid Drug Development, 9*(3), 301–312.
10. Azeredo, H. M., Mattoso, L. H. C., Wood, D., Williams, T. G., Avena-Bustillos, R. J., & Mchugh, T. H., (2009). Nanocomposite edible films from mango puree reinforced with cellulose nanofibers. *Journal of Food Science, 74*(5), N31–N35.

11. Bahrani, S., Ghanbarzadeh, B., Khiabani, M. S., Ghanbarzadeh, S., & Hamishehkar, H., (2018). Application of response surface methodology in the preparation of pectin-caseinate nanocomplexes for potential use as nutraceutical formulation: A statistical experimental design analysis. *Pharmaceutical Sciences, 24*(1), 52–59.
12. Bai, L., Huan, S., Xiang, W., & Rojas, O. J., (2018). Pickering emulsions by combining cellulose nanofibrils and nanocrystals: Phase behavior and depletion stabilization. *Green Chemistry, 20*(7), 1571–1582.
13. Basu, S., Omadjela, O., Gaddes, D., Tadigadapa, S., Zimmer, J., & Catchmark, J. M., (2016). Cellulose microfibril formation by surface-tethered cellulose synthase enzymes. *ACS Nano, 10*(2), 1896–1907.
14. Bertoldo, M., Zampano, G., Suffner, L., Liberati, E., & Ciardelli, F., (2013). Oxidation of glycogen —molecular nanoparticles‖ by periodate. *Polymer Chemistry, 4*(3), 653–661.
15. Boissiere, M., Allouche, J., Chanéac, C., Brayner, R., Devoisselle, J. M., Livage, J., & Coradin, T., (2007). Potentialities of silica/alginate nanoparticles as hybrid magnetic carriers. *International Journal of Pharmaceutics, 344*(1, 2), 128–134.
16. Božanić, D. K., Dimitrijević-Branković, S., Bibić, N., Luyt, A. S., & Djoković, V., (2011). Silver nanoparticles encapsulated in glycogen biopolymer: Morphology, optical and antimicrobial properties. *Carbohydrate Polymers, 83*(2), 883–890.
17. Brito, T. K., Viana, R. L. S., Moreno, C. J. G., Da Silva, B. J., De Sousa, Jr. F. L., De Medeiros, M. J. C., & Rocha, H. A. O., (2020). Synthesis of silver nanoparticle employing corn cob xylan as a reducing agent with anti-trypanosoma cruzi activity. *International Journal of Nanomedicine, 15*, 965.
18. Chaichi, M., Badii, F., Mohammadi, A., & Hashemi, M., (2019). Water resistance and mechanical properties of low methoxy-pectin nanocomposite film responses to interactions of $Ca^{2+}$ ions and glycerol concentrations as crosslinking agents. *Food Chemistry, 293*, 429–437.
19. Chang, C., Wang, T., Hu, Q., Zhou, M., Xue, J., & Luo, Y., (2017). Pectin coating improves physicochemical properties of caseinate/zein nanoparticles as oral delivery vehicles for curcumin. *Food Hydrocolloids, 70*, 143–151.
20. Chang, P. R., Jian, R., Yu, J., & Ma, X., (2010). Starch-based composites reinforced with novel chitin nanoparticles. *Carbohydrate Polymers, 80*(2), 420–425.
21. Chantarasataporn, P., Tepkasikul, P., Kingcha, Y., Yoksan, R., Pichyangkura, R., Visessanguan, W., & Chirachanchai, S., (2014). Water-Based oligochitosan and nanowhisker chitosan as potential food preservatives for shelf-life extension of minced pork. *Food Chemistry, 159*, 463–470.
22. Chin, S. F., Yazid, M., Akmar, S. N., & Pang, S. C., (2014). Preparation and characterization of starch nanoparticles for controlled release of curcumin. *International Journal of Polymer Science*, 1–8.
23. Choudhury, A. R., (2020). Synthesis of a novel gellan-pullulan nanogel and its application in adsorption of cationic dye from aqueous medium. *Carbohydrate Polymers, 227*, 115291.
24. Cui, S., Yao, B., Sun, X., Hu, J., Zhou, Y., & Liu, Y., (2016). Reducing the content of carrier polymer in pectin nanofibers by electrospinning at low loading followed with selective washing. *Materials Science and Engineering: C, 59*, 885–893.
25. Dai, L., Qiu, C., Xiong, L., & Sun, Q., (2015). Characterization of cornstarch-based films reinforced with taro starch nanoparticles. *Food Chemistry, 174*, 82–88.

26. Dai, L., Sun, C., Wei, Y., Mao, L., & Gao, Y., (2018). Characterization of Pickering emulsion gels stabilized by zein/gum Arabic complex colloidal nanoparticles. *Food Hydrocolloids, 74*, 239–248.
27. Das, A., Kundu, S., Ghosh, S. K., Basu, A., Gupta, M., & Mukherjee, A., (2020). Guar gum cinnamate ouzo nanoparticles for bacterial contact killing in water environment. *Carbohydrate Research,* 107983.
28. De Oliveira, E. F., Paula, H. C., & De Paula, R. C., (2014). Alginate/cashew gum nanoparticles for essential oil encapsulation. *Colloids and Surfaces B: Biointerfaces, 113*, 146–151.
29. Dehnad, D., Mirzaei, H., Emam-Djomeh, Z., Jafari, S. M., & Dadashi, S., (2014). Thermal and antimicrobial properties of chitosan–nanocellulose films for extending shelf life of ground meat. *Carbohydrate Polymers, 109*, 148–154.
30. Dhanya, A. T., Haridas, K. R., Divia, N., & Sudheesh, S., (2012). Development of zein-pectin nanoparticle as drug carrier. *International Journal of Drug Delivery, 4*(2), 147.
31. Dolphen, R., & Thiravetyan, P., (2011). Adsorption of melanoidins by chitin nanofibers. *Chemical Engineering Journal, 166*(3), 890–895.
32. Domingues, A. A., Pereira, F. V., Sierakowski, M. R., Rojas, O. J., & Petri, D. F., (2016). Interfacial properties of cellulose nanoparticles obtained from acid and enzymatic hydrolysis of cellulose. *Cellulose, 23*(4), 2421–2437.
33. Doost, A. S., Muhammad, D. R. A., Stevens, C. V., Dewettinck, K., & Van, D. M. P., (2018). Fabrication and characterization of quercetin loaded almond gum-shellac nanoparticles prepared by antisolvent precipitation. *Food Hydrocolloids, 83*, 190–201.
34. Douglas, K. L., & Tabrizian, M., (2005). Effect of experimental parameters on the formation of alginate–chitosan nanoparticles and evaluation of their potential application as DNA carrier. *Journal of Biomaterials Science, Polymer Edition, 16*(1), 43–56.
35. Dutta, H., & Paul, S. K., (2019). Kombucha drink: Production, quality, and safety aspects. In: *Production and Management of Beverages* (pp. 259–288). Cambridge – UK: Woodhead Publishing.
36. Fahami, A., & Fathi, M., (2018). Development of cress seed mucilage/PVA nanofibers as a novel carrier for vitamin A delivery. *Food Hydrocolloids, 81*, 31–38.
37. Fares, M. M., Tahboub, Y. R., Khatatbeh, S. T., & Abul-Haija, Y. M., (2011). Eco-friendly, vascular shape and interpenetrating poly (acrylic acid) grafted pectin hydrogels; biosorption and desorption investigations. *Journal of Polymers and the Environment, 19*(2), 431.
38. Filson, P. B., & Dawson-Andoh, B. E., (2009). Sono-chemical preparation of cellulose nanocrystals from lignocellulose derived materials. *Bioresource Technology, 100*(7), 2259–2264.
39. Friedman, A. J., Phan, J., Schairer, D. O., Champer, J., Qin, M., Pirouz, A., & Kim, J., (2013). Antimicrobial and anti-inflammatory activity of chitosan–alginate nanoparticles: A targeted therapy for cutaneous pathogens. *Journal of Investigative Dermatology, 133*(5), 1231–1239.
40. Fu, D., Deng, S., Mcclements, D. J., Zhou, L., Zou, L., Yi, J., & Liu, W., (2019). Encapsulation of B-carotene in wheat gluten nanoparticle-xanthan gum-stabilized Pickering emulsions: Enhancement of carotenoid stability and bioaccessibility. *Food Hydrocolloids, 89*, 80–89.

41. Ge, S., Xiong, L., Li, M., Liu, J., Yang, J., Chang, R., & Sun, Q., (2017). Characterizations of Pickering emulsions stabilized by starch nanoparticles: Influence of starch variety and particle size. *Food Chemistry, 234*, 339–347.
42. Ghasemi, S., Jafari, S. M., Assadpour, E., & Khomeiri, M., (2018). Nanoencapsulation of D-limonene within nanocarriers produced by pectin-whey protein complexes. *Food Hydrocolloids, 77*, 152–162.
43. Ghasemi, S., Jafari, S. M., Assadpour, E., & Khomeiri, M., (2017). Production of pectin-whey protein nano-complexes as carriers of orange peel oil. *Carbohydrate Polymers, 177*, 369–377.
44. Goetz, L. A., Jalvo, B., Rosal, R., & Mathew, A. P., (2016). Super hydrophilic anti-fouling electrospun cellulose acetate membranes coated with chitin nanocrystals for water filtration. *Journal of Membrane Science, 510*, 238–248.
45. Gorji, S. G., Gorji, E. G., & Mohammadifar, M. A., (2014). Characterization of gum tragacanth (*Astragalus gossypinus*)/sodium caseinate complex coacervation as a function of pH in an aqueous medium. *Food Hydrocolloids, 34*, 161–168.
46. Guerra-Rosas, M. I., Morales-Castro, J., Cubero-Márquez, M. A., Salvia-Trujillo, L., & Martín-Belloso, O., (2017). Antimicrobial activity of nanoemulsions containing essential oils and high methoxyl pectin during long-term storage. *Food Control, 77*, 131–138.
47. Haaj, S. B., Magnin, A., Pétrier, C., & Boufi, S., (2013). Starch nanoparticles formation via high power ultrasonication. *Carbohydrate Polymers, 92*(2), 1625–1632.
48. Haider, S., & Park, S. Y., (2009). Preparation of the electrospun chitosan nanofibers and their applications to the adsorption of Cu (II) and Pb (II) ions from an aqueous solution. *Journal of Membrane Science, 328*(1, 2), 90–96.
49. Hamed, I., Özogul, F., & Regenstein, J. M., (2016). Industrial applications of crustacean by-products (chitin, chitosan, and chitooligosaccharides): A review. *Trends in Food Science & Technology, 48*, 40–50.
50. Harikrishnan, R., Kim, J. S., Balasundaram, C., & Heo, M. S., (2012). Immunomodulatory effects of chitin and chitosan enriched diets in *Epinephelus bruneus* against vibrio alginolyticus infection. *Aquaculture, 326*, 46–52.
51. Hasanvand, E., Fathi, M., & Bassiri, A., (2018). Production and characterization of vitamin D3 loaded starch nanoparticles: Effect of amylose to amylopectin ratio and sonication parameters. *Journal of Food Science and Technology, 55*(4), 1314–1324.
52. Hu, Q., Gerhard, H., Upadhyaya, I., Venkitanarayanan, K., & Luo, Y., (2016). Antimicrobial eugenol nanoemulsion prepared by gum Arabic and lecithin and evaluation of drying technologies. *International Journal of Biological Macromolecules, 87*, 130–140.
53. Huang, H. Y., Shieh, Y. T., Shih, C. M., & Twu, Y. K., (2010). Magnetic chitosan/iron (II, III) oxide nanoparticles prepared by spray-drying. *Carbohydrate Polymers, 81*(4), 906–910.
54. Huang, W. C., Wang, W., Xue, C., & Mao, X., (2018). Effective enzyme immobilization onto a magnetic chitin nanofiber composite. *ACS Sustainable Chemistry & Engineering, 6*(7), 8118–8124.
55. Huang, X., Dai, Y., Cai, J., Zhong, N., Xiao, H., Mcclements, D. J., & Hu, K., (2017). Resveratrol encapsulation in core-shell biopolymer nanoparticles: Impact on antioxidant and anticancer activities. *Food Hydrocolloids, 64*, 157–165.

56. Huang, X., Huang, X., Gong, Y., Xiao, H., Mcclements, D. J., & Hu, K., (2016). Enhancement of curcumin water dispersibility and antioxidant activity using core-shell protein-polysaccharide nanoparticles. *Food Research International, 87*, 1–9.
57. Huq, T., Vu, K. D., Riedl, B., Bouchard, J., Han, J., & Lacroix, M., (2016). Development of probiotic tablet using alginate, pectin, and cellulose nanocrystals as excipients. *Cellulose, 23*(3), 1967–1978.
58. Iftikhar, S. A., & Dutta, H., (2020). Use of raw and physically modified rice starches as fat replacer in whipping cream. *Current Research in Nutrition and Food Science Journal, 8*(1), 122–130.
59. Jain, A. K., Khar, R. K., Ahmed, F. J., & Diwan, P. V., (2008). effective insulin delivery using starch nanoparticles as a potential trans-nasal mucoadhesive carrier. *European Journal of Pharmaceutics and Biopharmaceutics, 69*(2), 426–435.
60. Javidi, F., Razavi, S. M., & Mohammad, A. A., (2019). Response surface optimization of reduced-fat O/W emulsions formulated with cornstarch nanocrystal as a novel fat replacer. *Starch-Stärke, 71*(7, 8), 1800311.
61. Jiang, S., Liu, C., Wang, X., Xiong, L., & Sun, Q., (2016). Physicochemical properties of starch nanocomposite films enhanced by self-assembled potato starch nanoparticles. *LWT-Food Science and Technology, 69*, 251–257.
62. Jiang, S., Qin, Y., Yang, J., Li, M., Xiong, L., & Sun, Q., (2017). Enhanced antibacterial activity of lysozyme immobilized on chitin nanowhiskers. *Food Chemistry, 221*, 1507–1513.
63. Joshi, A., Solanki, S., Chaudhari, R., Bahadur, D., Aslam, M., & Srivastava, R., (2011). Multifunctional alginate microspheres for biosensing, drug delivery and magnetic resonance imaging. *Acta Biomaterialia, 7*(11), 3955–3963.
64. Kamil, J. Y., Jeon, Y. J., & Shahidi, F., (2002). Antioxidative activity of chitosans of different viscosity in cooked comminuted flesh of herring (*Clupea harengus*). *Food Chemistry, 79*(1), 69–77.
65. Karim, Z., Mathew, A. P., Grahn, M., Mouzon, J., & Oksman, K., (2014). Nanoporous membranes with cellulose nanocrystals as functional entity in chitosan: Removal of dyes from water. *Carbohydrate Polymers, 112*, 668–676.
66. Kasiri, N., & Fathi, M., (2018). Entrapment of peppermint oil using cellulose nanocrystals. *Cellulose, 25*(1), 319–329.
67. Khorasani, A. C., & Shojaosadati, S. A., (2017). Starch-and carboxymethylcellulose-coated bacterial nanocellulose-pectin bionanocomposite as novel protective prebiotic matrices. *Food Hydrocolloids, 63*, 273–285.
68. Khoshakhlagh, K., Koocheki, A., Mohebbi, M., & Allafchian, A., (2017). Development and characterization of electrosprayed alyssum homolocarpum seed gum nanoparticles for encapsulation of D-limonene. *Journal of Colloid and Interface Science, 490*, 562–575.
69. Khoshkava, V., & Kamal, M. R., (2014). Effect of drying conditions on cellulose nanocrystal (CNC) agglomerate porosity and dispersibility in polymer nanocomposites. *Powder Technology, 261*, 288–298.
70. Kong, C. S., Kim, J. A., Bak, S. S., Byun, H. G., & Kim, S. K., (2011). Anti-obesity effect of carboxymethyl chitin by AMPK and aquaporin-7 pathways in 3T3-L1 adipocytes. *The Journal of Nutritional Biochemistry, 22*(3), 276–281.
71. Korenevski, A., Erzsebet, P. S., Dutcher, J. R., & Stukalov, O. U. S., (2019). *Patent No. 10,172,946.* Washington, DC: U.S. Patent and Trademark Office.

72. Koshani, R., & Madadlou, A., (2018). A viewpoint on the gastrointestinal fate of cellulose nanocrystals. *Trends in Food Science & Technology, 71*, 268–273.
73. Krivorotova, T., Cirkovas, A., Maciulyte, S., Staneviciene, R., Budriene, S., Serviene, E., & Sereikaite, J., (2016). Nisin-loaded pectin nanoparticles for food preservation. *Food Hydrocolloids, 54*, 49–56.
74. Kuhn, K. R., E Silva, F. G. D., Netto, F. M., & Da Cunha, R. L., (2019). Production of whey protein isolate–gellan microbeads for encapsulation and release of flaxseed bioactive compounds. *Journal of Food Engineering, 247*, 104–114.
75. Kulterer, M. R., Reischl, M., Reichel, V. E., Hribernik, S., Wu, M., Köstler, S., & Ribitsch, V., (2011). Nanoprecipitation of cellulose acetate using solvent/nonsolvent mixtures as dispersive media. *Colloids and Surfaces A: Physicochemical and Engineering Aspects, 375*(1–3), 23–29.
76. Kumar, S. U., Kumar, V., Priyadarshi, R., Gopinath, P., & Negi, Y. S., (2018). pH-responsive prodrug nanoparticles based on xylan-curcumin conjugate for the efficient delivery of curcumin in cancer therapy. *Carbohydrate Polymers, 188*, 252–259.
77. Kumari, S., Yadav, B. S., & Yadav, R. B., (2020). Synthesis and modification approaches for starch nanoparticles for their emerging food industrial applications: A review. *Food Research International, 128*, 108765.
78. Lessa, E. F., Medina, A. L., Ribeiro, A. S., & Fajardo, A. R., (2020). Removal of multi-metals from water using reusable pectin/cellulose microfibers composite beads. *Arabian Journal of Chemistry, 13*(1), 709–720.
79. Li, H. Z., Chen, S. C., & Wang, Y. Z., (2015). Preparation and characterization of nanocomposites of polyvinyl alcohol/cellulose nanowhiskers/chitosan. *Composites Science and Technology, 115*, 60–65.
80. Li, J., Qu, X., Payne, G. F., Zhang, C., Zhang, Y., Li, J., & Liu, C., (2015). Biospecific self-assembly of a nanoparticle coating for targeted and stimuli-responsive drug delivery. *Advanced Functional Materials, 25*(9), 1404–1417.
81. Li, J., Xu, X., Chen, Z., Wang, T., Lu, Z., Hu, W., & Wang, L., (2018). Zein/gum Arabic nanoparticle-stabilized Pickering emulsion with thymol as an antibacterial delivery system. *Carbohydrate Polymers, 200*, 416–426.
82. Li, J., Yang, Y., Lu, L., Ma, Q., & Zhang, J., (2018). Preparation, characterization and systemic application of self-assembled hydroxyethyl starch nanoparticles-loaded flavonoid morin for hyperuricemia therapy. *International Journal of Nanomedicine, 13*, 2129.
83. Li, T., Shi, X. W., Du, Y. M., & Tang, Y. F., (2007). Quaternized chitosan/alginate nanoparticles for protein delivery. *Journal of Biomedical Materials Research Part A, 83*(2), 383–390.
84. Li, X. Y., Chen, X. G., Sun, Z. W., Park, H. J., & Cha, D. S., (2011). Preparation of alginate/chitosan/carboxymethyl chitosan complex microcapsules and application in *Lactobacillus casei* ATCC 393. *Carbohydrate Polymers, 83*(4), 1479–1485.
85. Li, X., Anton, N., Arpagaus, C., Belleteix, F., & Vandamme, T. F., (2010). Nanoparticles by spray drying using innovative new technology: The büchi nano spray dryer B-90. *Journal of Controlled Release, 147*(2), 304–310.
86. Li, X., Ge, S., Yang, J., Chang, R., Liang, C., Xiong, L., & Sun, Q., (2017). Synthesis and study the properties of steps/gum nanoparticles for salvianolic acid B-oral delivery system. *Food Chemistry, 229*, 111–119.

87. Li, X., Li, X., Ke, B., Shi, X., & Du, Y., (2011). cooperative performance of chitin whisker and rectorite fillers on chitosan films. *Carbohydrate Polymers, 85*(4), 747–752.
88. Li, X., Shi, X., Wang, M., & Du, Y., (2011). Xylan chitosan conjugate-A potential food preservative. *Food Chemistry, 126*(2), 520–525.
89. Li, Z., Zhang, M., Cheng, D., & Yang, R., (2016). Preparation of silver nanoparticles immobilized onto chitin nanocrystals and their application to cellulose paper for imparting antimicrobial activity. *Carbohydrate Polymers, 151*, 834–840.
90. Lin, Y. J., Shatkin, J. A., & Kong, F., (2019). Evaluating mucoadhesion properties of three types of nanocellulose in the gastrointestinal tract *in vitro* and ex vivo. *Carbohydrate Polymers, 210*, 157–166.
91. Liu, B., Zhu, Y., Tian, J., Guan, T., Li, D., Bao, C., & Li, Y., (2019). Inhibition of oil digestion in Pickering emulsions stabilized by oxidized cellulose nanofibrils for low-calorie food design. *RSC Advances, 9*(26), 14966–14973.
92. Liu, L., Kerr, W. L., Kong, F., Dee, D. R., & Lin, M., (2018). Influence of nano-fibrillated cellulose (NFC) on starch digestion and glucose absorption. *Carbohydrate Polymers, 196*, 146–153.
93. Liu, L., & Kong, F., (2019). Influence of nanocellulose on *In Vitro* digestion of whey protein isolate. *Carbohydrate Polymers, 210*, 399–411.
94. Ma, H., Burger, C., Hsiao, B. S., & Chu, B., (2011). Ultrafine polysaccharide nanofibrous membranes for water purification. *Biomacromolecules, 12*(4), 970–976.
95. Ma, X., Jian, R., Chang, P. R., & Yu, J., (2008). Fabrication and characterization of citric acid-modified starch nanoparticles/plasticized-starch composites. *Biomacromolecules, 9*(11), 3314–3320.
96. Medeiros, B. G. D. S., Pinheiro, A. C., Carneiro-Da-Cunha, M. G., & Vicente, A. A., (2012). Development and characterization of a nanomultilayer coating of pectin and chitosan–evaluation of its gas barrier properties and application on 'Tommy Atkins' mangoes. *Journal of Food Engineering, 110*(3), 457–464.
97. Meyabadi, T. F., Dadashian, F., Sadeghi, G. M. M., & Asl, H. E. Z., (2014). Spherical cellulose nanoparticles preparation from waste cotton using a green method. *Powder Technology, 261*, 232–240.
98. Mirhosseini, H., Tan, C. P., Aghlara, A., Hamid, N. S., Yusof, S., & Chern, B. H., (2008). Influence of pectin and CMC on physical stability, turbidity loss rate, cloudiness and flavor release of orange beverage emulsion during storage. *Carbohydrate Polymers, 73*(1), 83–91.
99. Mohammadi, A., Jafari, S. M., Assadpour, E., & Esfanjani, A. F., (2016). Nano-encapsulation of olive leaf phenolic compounds through WPC–pectin complexes and evaluating their release rate. *International Journal of Biological Macromolecules, 82*, 816–822.
100. Momenzadeh, H., Tehrani-Bagha, A. R., Khosravi, A., Gharanjig, K., & Holmberg, K., (2011). Reactive dye removal from wastewater using a chitosan nanodispersion. *Desalination, 271*(1–3), 225–230.
101. Muhammad, D. R. A., Doost, A. S., Gupta, V., Bin, S. M. D., Van De, W. D., Van, D. M. P., & Dewettinck, K., (2020). Stability and functionality of xanthan gum–shellac nanoparticles for the encapsulation of cinnamon bark extract. *Food Hydrocolloids, 100*, 105377.

102. Nagahata, Y., Kobayashi, I., Goto, M., Nakaura, Y., & Inouchi, N., (2013). The formation of resistant starch during acid hydrolysis of high-amylose cornstarch. *Journal of Applied Glycoscience*. Jag-JAG.
103. Nakorn, P. N., (2017). Chitin nanowhisker and chitosan nanoparticles in protein immobilization for biosensor applications. *Journal of Metals, Materials and Minerals, 18*(2), 40–49.
104. Narayanan, D., Jayakumar, R., & Chennazhi, K. P., (2014). Versatile carboxymethyl chitin and chitosan nanomaterials: A review. *Wiley Interdisciplinary Reviews: Nanomedicine and Nanobiotechnology, 6*(6), 574–598.
105. Neyrinck, A. M., Possemiers, S., Verstraete, W., De Backer, F., Cani, P. D., & Delzenne, N. M., (2012). Dietary modulation of clostridial cluster xiva gut bacteria (*Roseburia* Spp.) by chitin–glucan fiber improves host metabolic alterations induced by high-fat diet in mice. *The Journal of Nutritional Biochemistry, 23*(1), 51–59.
106. Nguyen, T. V., Nguyen, T. T. H., Wang, S. L., Vo, T. P. K., & Nguyen, A. D., (2017). Preparation of chitosan nanoparticles by TPP ionic gelation combined with spray drying, and the antibacterial activity of chitosan nanoparticles and a chitosan nanoparticle–amoxicillin complex. *Research on Chemical Intermediates, 43*(6), 3527–3537.
107. Nieto-Suaza, L., Acevedo-Guevara, L., Sánchez, L. T., Pinzón, M. I., & Villa, C. C., (2019). Characterization of aloe vera-banana starch composite films reinforced with curcumin-loaded starch nanoparticles. *Food Structure, 22*, 100131.
108. Nogi, M., Kurosaki, F., Yano, H., & Takano, M., (2010). Preparation of nanofibrillar carbon from chitin nanofibers. *Carbohydrate Polymers, 81*(4), 919–924.
109. Oliveira, A. M., Guimarães, K. L., Cerize, N. N., Tunussi, A. S., & Poço, J. G., (2013). Nano spray drying as an innovative technology for encapsulating hydrophilic active pharmaceutical ingredients (API). *Journal of Nanomedicine and Nanotechnology, 4*(2).
110. Orsuwan, A., & Sothornvit, R., (2017). Development and characterization of banana flour film incorporated with montmorillonite and banana starch nanoparticles. *Carbohydrate Polymers, 174*, 235–242.
111. Oun, A. A., Shankar, S., & Rhim, J. W., (2020). Multifunctional nanocellulose/metal and metal oxide nanoparticle hybrid nanomaterials. *Critical Reviews in Food Science and Nutrition, 60*(3), 435–460.
112. Pääkkö, M., Ankerfors, M., Kosonen, H., Nykänen, A., Ahola, S., Österberg, M., & Lindström, T., (2007). Enzymatic hydrolysis combined with mechanical shearing and high-pressure homogenization for nanoscale cellulose fibrils and strong gels. *Biomacromolecules, 8*(6), 1934–1941.
113. Patel, C. M., Chakraborty, M., & Murthy, Z. V. P., (2016). Fast and scalable preparation of starch nanoparticles by stirred media milling. *Advanced Powder Technology, 27*(4), 1287–1294.
114. Pitombeira, N. A., Neto, J. G. V., Silva, D. A., Feitosa, J. P., Paula, H. C., & De Paula, R. C., (2015). Self-assembled nanoparticles of acetylated cashew gum: Characterization and evaluation as potential drug carrier. *Carbohydrate Polymers, 117*, 610–615.
115. Pospisilova, A., Filippov, S. K., Bogomolova, A., Turner, S., Sedlacek, O., Matushkin, N., & Hruby, M., (2014). Glycogen-graft-poly (2-alkyl-2-oxazolines)–the new versatile biopolymer-based thermoresponsive macromolecular toolbox. *RSC Advances, 4*(106), 61580–61588.
116. Qiu, C., Qin, Y., Zhang, S., Xiong, L., & Sun, Q., (2016). A comparative study of size-controlled worm-like amylopectin nanoparticles and spherical amylose nanoparticles:

Their characteristics and the adsorption properties of polyphenols. *Food Chemistry, 213*, 579–587.
117. Raei, M., Shahidi, F., Farhoodi, M., Jafari, S. M., & Rafe, A., (2017). Application of whey protein-pectin nano-complex carriers for loading of lactoferrin. *International Journal of Biological Macromolecules, 105*, 281–291.
118. Rajeswari, L. S., Moorthy, S. N., & Rajasekharan, K. N., (2011). Preparation of cassava starch nanoparticles and their application as a carrier system for curcumin delivery. *International Journal of Nanotechnology and Application, 5*, 193–201.
119. Rocha, S., Generalov, R., Pereira, M. D. C., Peres, I., Juzenas, P., & Coelho, M. A., (2011). Epigallocatechin gallate-loaded polysaccharide nanoparticles for prostate cancer chemoprevention. *Nanomedicine, 6*(1), 79–87.
120. Saha, S., Pal, A., Kundu, S., Basu, S., & Pal, T., (2010). Photochemical green synthesis of calcium-alginate-stabilized ag and au nanoparticles and their catalytic application to 4-nitrophenol reduction. *Langmuir, 26*(4), 2885–2893.
121. Sana, S., Boodhoo, K., & Zivkovic, V., (2019). Production of starch nanoparticles through solvent-antisolvent precipitation in a spinning disc reactor. *Green Processing and Synthesis, 8*(1), 507–515.
122. Sánchez, R., Stringari, G. B., Franco, J. M., Valencia, C., & Gallegos, C., (2011). Use of chitin, chitosan and acylated derivatives as thickener agents of vegetable oils for bio-lubricant applications. *Carbohydrate Polymers, 85*(3), 705–714.
123. Sanchis, E., Gonzalez, S., Ghidelli, C., Sheth, C. C., Mateos, M., Palou, L., & Pérez-Gago, M. B., (2016). Browning inhibition and microbial control in fresh-cut persimmon (*Diospyros kaki* thunb. CV. *Rojo brillante*) by apple pectin-based edible coatings. *Postharvest Biology and Technology, 112*, 186–193.
124. Seyedi, S. M., Anvaripour, B., Motavassel, M., & Jadidi, N., (2013). Comparative cadmium adsorption from water by nanochitosan and chitosan. *International Journal of Engineering and Innovative Technology, 2*(9), 145–148.
125. Shabanpour, B., Kazemi, M., Ojagh, S. M., & Pourashouri, P., (2018). Bacterial cellulose nanofibers as reinforce in edible fish myofibrillar protein nanocomposite films. *International Journal of Biological Macromolecules, 117*, 742–751.
126. Shamsara, O., Jafari, S. M., & Muhidinov, Z. K., (2017). Fabrication, characterization and stability of oil in water nano-emulsions produced by apricot gum-pectin complexes. *International Journal of Biological Macromolecules, 103*, 1285–1293.
127. Shankar, S., Reddy, J. P., Rhim, J. W., & Kim, H. Y., (2015). Preparation, characterization, and antimicrobial activity of chitin nanofibrils reinforced carrageenan nanocomposite films. *Carbohydrate Polymers, 117*, 468–475.
128. Shao, P., Zhang, H., Niu, B., & Jin, W., (2018). Physical stabilities of taro starch nanoparticles stabilized Pickering emulsions and the potential application of encapsulated tea polyphenols. *International Journal of Biological Macromolecules, 118*, 2032–2039.
129. Sharma, G., Kumar, A., Devi, K., Sharma, S., Naushad, M., Ghfar, A. A., & Stadler, F. J., (2018). Guar gum-crosslinked-soya lecithin nanohydrogel sheets as effective adsorbent for the removal of thiophanate methyl fungicide. *International Journal of Biological Macromolecules, 114*, 295–305.
130. Sharma, M., Mann, B., Sharma, R., Bajaj, R., Athira, S., Sarkar, P., & Pothuraju, R., (2017). Sodium caseinate stabilized clove oil nanoemulsion: Physicochemical properties. *Journal of Food Engineering, 212*, 38–46.

131. Sharma, R., Ahuja, M., & Kaur, H., (2012). Thiolated pectin nanoparticles: Preparation, characterization and ex vivo corneal permeation study. *Carbohydrate Polymers, 87*(2), 1606–1610.
132. Shekarforoush, E., Ajalloueian, F., Zeng, G., Mendes, A. C., & Chronakis, I. S., (2018). Electrospun xanthan gum-chitosan nanofibers as delivery carrier of hydrophobic bioactives. *Materials Letters, 228*, 322–326.
133. Shi, A. M., Wang, L. J., Li, D., & Adhikari, B., (2013). Characterization of starch films containing starch nanoparticles. Part 2: Viscoelasticity and creep properties. *Carbohydrate Polymers, 96*(2), 602–610.
134. Shi, A. M., Wang, L. J., Li, D., & Adhikari, B., (2012). The effect of annealing and cryoprotectants on the properties of vacuum-freeze dried starch nanoparticles. *Carbohydrate Polymers, 88*(4), 1334–1341.
135. Singha, N. R., Karmakar, M., Mahapatra, M., Mondal, H., Dutta, A., Roy, C., & Chattopadhyay, P. K., (2017). Systematic synthesis of pectin-G-(sodium acrylate-Co-N-isopropylacrylamide) interpenetrating polymer network for super adsorption of dyes/M (II): Determination of physicochemical changes in loaded hydrogels. *Polymer Chemistry, 8*(20), 3211–3237.
136. Sivagangi, R. N., Madhusudana, R. K., Sudha, V. T. J., Krishna, R. K. S. V., & Lee, Y. I., (2016). Pectin/poly (acrylamide-co-acrylamidoglycolic acid) pH-sensitive semi-IPN hydrogels: Selective removal of $Cu^{2+}$ And $Ni^{2+}$, modeling, and kinetic studies. *Desalination and Water Treatment, 57*(14), 6503–6514.
137. Sivakami, M. S., Gomathi, T., Venkatesan, J., Jeong, H. S., Kim, S. K., & Sudha, P. N., (2013). Preparation and characterization of nano chitosan for treatment wastewaters. *International Journal of Biological Macromolecules, 57*, 204–212.
138. Solairaj, D., Rameshthangam, P., Muthukumaran, P., & Wilson, J., (2017). Studies on electrochemical glucose sensing, antimicrobial activity and cytotoxicity of fabricated copper nanoparticle immobilized chitin nanostructure. *International Journal of Biological Macromolecules, 101*, 668–679.
139. Spernarth, L., & Magdassi, S., (2007). Preparation of ethylcellulose nanoparticles from nano-emulsion obtained by inversion at constant temperature. *Micro & Nano Letters, 2*(4), 90–95.
140. Strom, G., Ohgren, C., & Ankerfors, M., (2013). Nanocellulose as an additive in foodstuff. *Innventia Rep., 403*, 1–25.
141. Sun, Q., Gong, M., Li, Y., & Xiong, L., (2014). Effect of retrogradation time on preparation and characterization of *Proso millet* starch nanoparticles. *Carbohydrate Polymers, 111*, 133–138.
142. Tan, Y., Xu, K., Li, L., Liu, C., Song, C., & Wang, P., (2009). Fabrication of size-controlled starch-based nanospheres by nanoprecipitation. *ACS Applied Materials & Interfaces, 1*(4), 956–959.
143. Tan, Y., Xu, K., Niu, C., Liu, C., Li, Y., Wang, P., & Binks, B. P., (2014). Triglyceride–water emulsions stabilized by starch-based nanoparticles. *Food Hydrocolloids, 36*, 70–75.
144. Tang, Z. X., Qian, J. Q., & Shi, L. E., (2007). Preparation of chitosan nanoparticles as carrier for immobilized enzyme. *Applied Biochemistry and Biotechnology, 136*(1), 77–96.

145. Tao, F., Shi, C., & Cui, Y., (2018). Preparation and physicochemistry properties of smart edible films based on gelatin–starch nanoparticles. *Journal of the Science of Food and Agriculture, 98*(14), 5470–5478.
146. Tharanathan, R. N., & Kittur, F. S., (2003). Chitin—the undisputed biomolecule of great potential. *Critical Reviews in Food Science and Nutrition, 43*(1), 63–87.
147. Thomas, D., Latha, M. S., & Thomas, K. K., (2018). Synthesis and *in vitro* evaluation of alginate-cellulose nanocrystal hybrid nanoparticles for the controlled oral delivery of rifampicin. *Journal of Drug Delivery Science and Technology, 46*, 392–399.
148. Tran, T. H., Nguyen, H. L., Hwang, D. S., Lee, J. Y., Cha, H. G., Koo, J. M., & Oh, D. X., (2019). Five different chitin nanomaterials from identical source with different advantageous functions and performances. *Carbohydrate Polymers, 205*, 392–400.
149. Triyono, K., Suhartatik, N., & Wulandari, Y. W., (2018). Nanoencapsulation of kaffir lime oil with coacervation method using Arabic gum and maltodextrin as encapsulant. *International Journal of Food and Nutrition Sciences, 3*(1), 43–48.
150. Tzoumaki, M. V., Moschakis, T., & Biliaderis, C. G., (2011). Mixed aqueous chitin nanocrystal–whey protein dispersions: Microstructure and rheological behavior. *Food Hydrocolloids, 25*(5), 935–942.
151. Tzoumaki, M. V., Moschakis, T., Kiosseoglou, V., & Biliaderis, C. G., (2011). Oil-in-water emulsions stabilized by chitin nanocrystal particles. *Food Hydrocolloids, 25*(6), 1521–1529.
152. Vadivel, M., Sankarganesh, M., Raja, J. D., Rajesh, J., Mohanasundaram, D., & Alagar, M., (2019). Bioactive constituents and bio-waste derived chitosan/xylan based biodegradable hybrid nanocomposite for sensitive detection of fish freshness. *Food Packaging and Shelf Life, 22*, 100384.
153. Veneranda, M., Hu, Q., Wang, T., Luo, Y., Castro, K., & Madariaga, J. M., (2018). Formation and characterization of zein-caseinate-pectin complex nanoparticles for encapsulation of eugenol. *LWT- Food Science and Technology, 89*, 596–603.
154. Vicente, A. A., Pinheiro, A. C., & Martins, J. T., (2019). *Book of Abstracts of the 8$^{th}$ International Symposium on Delivery of Functionality in Complex Food Systems.* Porto, Portugal. Universidade Do Minho, Departamento De Engenharia Biológica (DEB).
155. Wang, F., Yang, S., Yuan, J., Gao, Q., & Huang, C., (2016). Effective method of chitosan-coated alginate nanoparticles for target drug delivery applications. *Journal of Biomaterials Applications, 31*(1), 3–12.
156. Wang, T., Ma, X., Lei, Y., & Luo, Y., (2016). Solid lipid nanoparticles coated with cross-linked polymeric double layer for oral delivery of curcumin. *Colloids and Surfaces B: Biointerfaces, 148*, 1–11.
157. Wei, Z., Zhu, P., & Huang, Q., (2019). Investigation of ovotransferrin conformation and its complexation with sugar beet pectin. *Food Hydrocolloids, 87*, 448–458.
158. Winarti, C., Sunarti, T. C., Mangunwidjaja, D., & Richana, N., (2014). Preparation of arrowroot starch nanoparticles by butanol-complex precipitation, and its application as bioactive encapsulation matrix. *International Food Research Journal, 21*(6), 2207.
159. Wu, L. X., Qiao, Z. R., Cai, W. D., Qiu, W. Y., & Yan, J. K., (2019). Quaternized curdlan/pectin polyelectrolyte complexes as biocompatible nanovehicles for curcumin. *Food Chemistry, 291*, 180–186.
160. Wu, W., Kong, X., Zhang, C., Hua, Y., & Chen, Y., (2018). Improving the stability of wheat gliadin nanoparticles–effect of gum Arabic addition. *Food Hydrocolloids, 80*, 78–87.

161. Xu, X., Luo, L., Liu, C., & Mcclements, D. J., (2017). Utilization of anionic polysaccharides to improve the stability of rice glutelin emulsions: Impact of polysaccharide type, pH, salt, and temperature. *Food Hydrocolloids, 64*, 112–122.
162. Yan, J. K., Qiu, W. Y., Wang, Y. Y., & Wu, J. Y., (2017). Biocompatible polyelectrolyte complex nanoparticles from lactoferrin and pectin as potential vehicles for antioxidative curcumin. *Journal of Agricultural and Food Chemistry, 65*(28), 5720–5730.
163. Yan, L., Chang, P. R., Zheng, P., & Ma, X., (2012). Characterization of magnetic guar gum-grafted carbon nanotubes and the adsorption of the dyes. *Carbohydrate Polymers, 87*(3), 1919–1924.
164. Yang, P., Zhu, Z., Chen, M., Zhou, X., & Chen, W., (2019). Microwave-assisted synthesis of polyamine-functionalized carbon dots from xylan and their use for the detection of tannic acid. *Spectrochimica Acta Part A: Molecular and Biomolecular Spectroscopy, 213*, 301–308.
165. Ye, F., Miao, M., Jiang, B., Campanella, O. H., Jin, Z., & Zhang, T., (2017). Elucidation of stabilizing oil-in-water Pickering emulsion with different modified maize starch-based nanoparticles. *Food Chemistry, 229*, 152–158.
166. Zahoor, A., Sharma, S., & Khuller, G. K., (2005). Inhalable alginate nanoparticles as antitubercular drug carriers against experimental tuberculosis. *International Journal of Antimicrobial Agents, 26*(4), 298–303.
167. Zainudin, B. H., Wong, T. W., & Hamdan, H., (2018). Design of low molecular weight pectin and its nanoparticles through combination treatment of pectin by microwave and inorganic salts. *Polymer Degradation and Stability, 147*, 35–40.
168. Zeeb, B., Saberi, A. H., Weiss, J., & Mcclements, D. J., (2015). Formation and characterization of filled hydrogel beads based on calcium alginate: Factors influencing nanoemulsion retention and release. *Food Hydrocolloids, 50*, 27–36.
169. Zhai, X., Lin, D., Liu, D., & Yang, X., (2018). Emulsions stabilized by nanofibers from bacterial cellulose: New potential food-grade Pickering emulsions. *Food Research International, 103*, 12–20.
170. Zhang, X., Liu, Y., Wang, Y., Luo, X., Li, Y., Li, B., Wang, J., & Liu, S., (2019). Surface modification of cellulose nanofibrils with protein nanoparticles for enhancing the stabilization of O/W Pickering emulsions. *Food Hydrocolloids, 97*, 105180.
171. Zimet, P., & Livney, Y. D., (2009). Beta-lactoglobulin and its nanocomplexes with pectin as vehicles for Ω-3 polyunsaturated fatty acids. *Food Hydrocolloids, 23*(4), 1120–1126.

# PART II

# NANOTECHNOLOGY APPLICATIONS IN FOOD PRESERVATION AND PACKAGING

## CHAPTER 5

# NOVEL METHODS IN NANOTECHNOLOGY FOR FOOD PRESERVATION

NANGWANBIANG KURBAH and SHEENA HAORONGBAM

## ABSTRACT

Nanotechnology uses molecules, macromolecules, and atoms to formulate and technologize them into materials of unique quality and the materials thus produced are with an approximate size of 1–100 nm. These nano-materials have more than one internal structure, with or without an external dimension, from 1–100 nm scale level. This means that the technology allows the manipulation and/or observation of nano-material in nano-scale-level. The nano-materials are different in qualities which is dissimilar from the macro-scale counterparts. In the food industry, the functional bioactive compounds found are more in the range of those which deteriorate and become inactive with the passage of time, but encapsulation of these compounds with nanotechnology as a means can enhance the shelf life which reduces the deterioration process, and delivering of these compounds to reach their targeted location becomes more efficient. Nanotechnology packages are more superior to the current packaging materials with better batter strength and barrier properties due to the presence of a film called anti-microbial films which can detect the presence of pathogen.

## 5.1 INTRODUCTION

The word nano means dwarf. A nanometer is equal to a millimeter in millionth, or equal to billionth in meter [42]. Therefore, nanotechnology is

Nanotechnology Horizons in Food Process Engineering, Volume 1. Megh R. Goyal, PhD, PE, Junaid Ahmad Malik, PhD, Satish Kumar, PhD & Ritesh B. Watharkar, PhD (Eds.)

the technology that uses matter that is equal to the diameter of ½ of DNA and 1/20th of red blood cell size. This size is equivalent to the size of viruses [9]. A strand of human hair is approximately 100,000 nm, so it is easily understood how small a measurement is in nanometer [21]. The nanomaterials (NMs) in foods which have been researched the most are on in the powder form of food, and composed of a nanomaterial, which show qualities of a chemically composed powder but of bigger-sized particle [6]. The ongoing study shows the potential of nanotechnology in the food sector especially in preservation, supplementation, and packaging technology because of their novel function and application.

Nanotechnology theory was first acknowledged in 1959 by Richard Feynmana, renowned physicist, wherein it was discovered that materials can be completely changed at their molecular level [19, 28]. An academician of the University of Tokyo Norio Taniguchi, established that these atoms and molecules can be engineered in their molecular level and even change can be brought about in their application [28]. Through nanotechnology, these molecules and atoms can be controlled and manipulated in a nanoscale level between 1–100 nm [5].

Nanotechnology is the study to control the molecules at the molecular level within a nanoscale meter [7]. The application of nanotechnology also extended from medicine to mechanical, and also new nanotechnology has been discovered. Nanoparticle also acts as a carrier for the transport mechanism. They are categorized according to their structure, size, and characteristics.

An instance where food nanotechnology as a pioneered usage can be seen, when it was utilized by Louis Pasteur in Pasteurization process, which in turn, helped in the introduction of nanotechnology in food production to increase food quality. With the discovery of deoxyribonucleic acid (DNA) model by Watson and Crick, which is in the range of 2.5 nm, and increased application of nanotechnology has been seen extending from agricultural products, biotechnology, and medicines [7]. Discovery of carbon nanotubes in a nanoscale level between 1 nm, furthermore increased in discoveries and innovation in different areas of life, leading to the era of nanotechnology in the world.

This chapter focuses on the advent of nanotechnology through nanoencapsulation in probiotics, bioactive compounds, and additives in the preservation of foods.

## 5.2 CURRENT APPLICATIONS OF NANOTECHNOLOGY IN FOODS

The important factors that affect the deterioration of food are due to microbial growth; these microbes are mainly yeast, bacteria, and mold. The chemical reaction that takes place with the food such as enzymes activation and browning effects are the other factors. The deterioration can also be due to the infestation by the insects or rodents, or the imbalance of air or moisture which react with the foods [30, 44]. The most essential means to prevent the deterioration is to restrict the microbial growth. These can be controlled by preservation methods such as the use of salt, sugar, drying, and different preservatives. It can also be achieved by controlling the moisture content, air, light, etc. A generalized idea of the control measures, which can be implemented upon introduction of nanotechnology to food sector, is summarized in Figure 5.1.

The rapid discoveries and development of nanotechnology in recent years have increased its applications. The developments mainly desire to change food texture or by encapsulating bioactive compound and essential nutrients, or by control to the discharge of flavor [29].

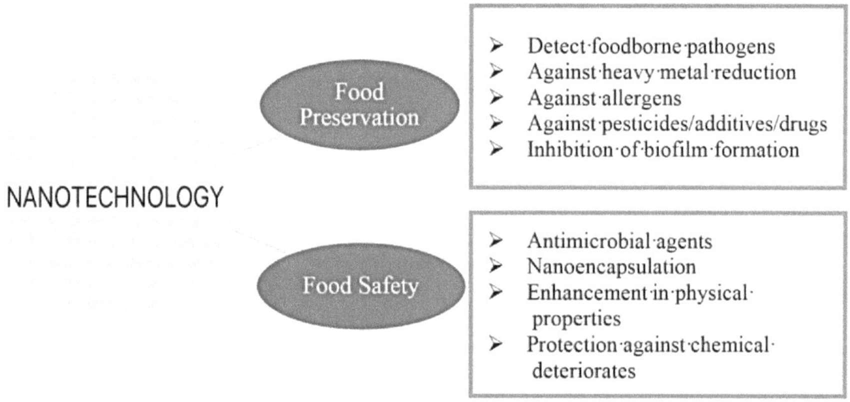


**FIGURE 5.1** Nanotechnology in food preservation, safety, and security.

*Source:* Adapted from: Bajpai et al. [4].

## 5.3 NANOENCAPSULATION

Nanoencapsulation is the process of encapsulation of desired particles or compounds on the nanometer scale within a layer of thin films [3, 35]. The encapsulation procedure consists of enclosing an essential compound such as food additives, bioactive compounds, and various flavors within the films [2]. This enclosing layer can act as a barrier to protect these essential compounds to reduce degradation and increase the absorption and bioavailability. The encapsulating layer or barrier is mostly a polymer [2]. Nanoencapsulation has a large scope in the field of cosmetics, pharmaceuticals, painting, etc. Encapsulation consists of two types: nano and microencapsulation [42].

Nanoencapsulation technique is a technology to store substances or components in a small amount and is composed of nanoemulsion, nano-composite, and nanostructure, their function which are to control the core release [37]. A few scientists have developed a unique nanotechnology that can encapsulate many essential compounds, and the encapsulated active compounds were found to be able to adsorb. Further, there has been the development of a technology that can control the release of active compounds [15, 16, 33].

## 5.4 CHARACTERIZATION OF NANOENCAPSULATION

Nanoencapsulation procedure is composed of two phases; these are the coating, membrane, or the shell which encloses the core materials and the main core material which is the desired material chosen for encapsulation as shown in Figure 5.2. The encapsulated coating materials and core materials are different in shape and size [33].

## 5.5 FORMULATION OF NANOMATERIALS (NMS)

Encapsulation material for nanotechnology is of many types; these are carbon nanotubes, liposomes, dendrimers, magnetic nanoparticles, and micelles, and they produce NMs known as nanocarriers [22]. These particles have the ability to deliver an essential compound to the target cells or organs [17, 20]. Therefore, they play an important role in delivering of essential compounds to the target sites [32].

The nanocarriers keep encapsulated particles and protect them from degradation and increase their intake by the cell [23]. The conventional nanoparticles are eliminated from the body as unwanted materials [43]. A hydrophilic nanocarrier polymer enhances its delivering process [11]. Selected coating material must have the capacity of forming a layer with a good property of flexibility, strength, impermeability, and stability. Further, the membranes must not react with the core ingredient [13]. Nanoparticles are of natural and synthetic polymers, which are gelatin, protein, chitosan (CS), PGA polylactides, PLA polylactides, PLGA polylacidecoglycolides, etc. [24].

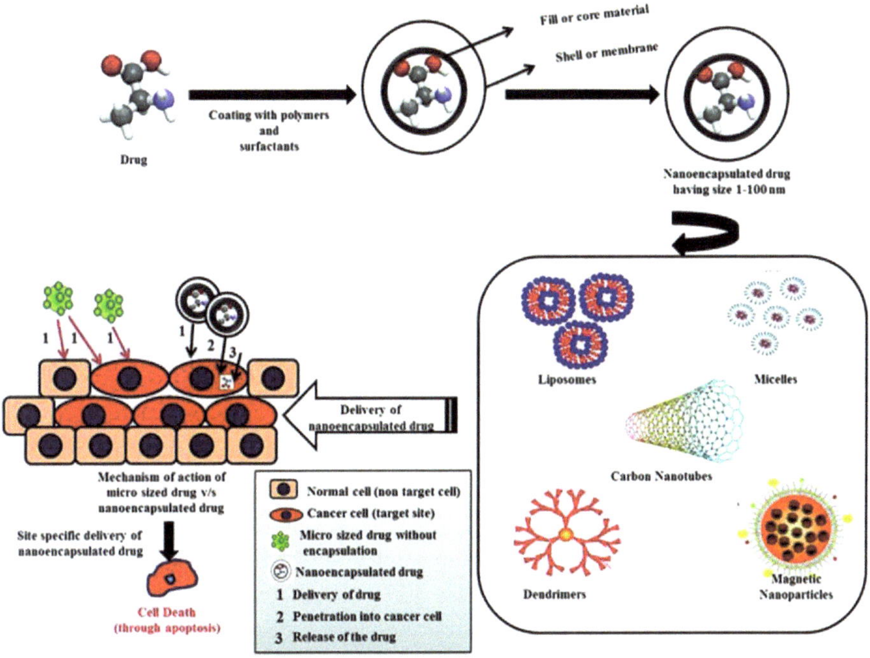


**FIGURE 5.2** Nanoencapsulation process (the drug is encapsulated within a polymer membrane and delivered to targeted cells/organs).

*Source:* Adapted from: Murthy et al. [33].

Development of nanoparticles is in the form of top-down or bottom-up approach. The top-down approach decreases the shape and size of the particles, while the bottom-up approach is produced by self-alignment which is influenced by the pH, temperature, and ion strength [1, 31, 41].

## 5.6 EVALUATION OF A NANO-PARTICLE

The evaluation of nanoparticle for their quality is measured by several physical and chemical parameters. The Fourier Transform Infra-Red (FTIR) studies the organic molecules of particles. Dynamic light scattering (DLS) is for aggregation of the nanoparticles. Scanning Colorimeter and Spectroscopy are study the interaction between particles. Zeta potential for stability EDX is study for checking the presence of another element in the particles and scanning electron microscopy (SEM) study morphology characteristics of particles [33].

## 5.7 ROLE OF NANOTECHNOLOGY IN FOOD PRESERVATION THROUGH ENCAPSULATION TECHNIQUE

Nanotechnology is the study to control physical and chemical processes at the nanoscale level. It is the leading technology in conservation and preservation of foods. The development of edible shell or membrane by scientific methodologies can provide an alternative for preservation of food [40]. For centuries, the edible-based emulsion has been used. These membranes can be from vegetable or animal oil and vegetable or animal waxes [12].

The edible nano-coatings or barrier on the encapsulated substances can give a better barrier to gas and moisture by extending the foods' storage life and controlling physicochemical properties. It can improve the retention of antioxidant, color, flavor, and stop the enzymes and browning activities even after opening the package of the foods [14, 49]. The development, production of different techniques of nanoparticles and their application have been reported along with their examples [37] which can not only provide an importance in saving the formulators, it also reduces the number of essential nutrients required and therefore are applicable in food industries for these various reasons.

Curcumin which is a natural yellow pigment of turmeric and Octenyl succinic anhydride-ε-polylysine have found a possible function as nano-emulsion to encapsulate a bioactive compound [48, 52]. Lipid-based nanoencapsulation such as liposomes, cochleates, and nano-archaeosomes encapsulates can deliver and protect the bioactive compound and also act as carrier mechanism, to control the deliveries of food nutrients [37].

A combination of α-tocopherol and a fraction of stearin which is rich in milk is encapsulated through oil in water emulsion to form sodium casein emulsion. It can protect the degradation of fats [8, 38]. Another is Zein protein, present in maize, which is bound and enclosed by lipids to prevent them from degradation. It shows to take in or consume fatty acid to form a fat layer to form membrane [34]. Encapsulation soy lecithin act as a carrier for water-insoluble nutrients and therefore increases the availability of active compound and improved stability of these compounds [36].

consumption, the application of nanotechnology in food for the provision of barrier to required food compounds and also for enhancement of functionality of these compounds can be studied and developed to provide for an enhanced food basket in the current global scenario.

## KEYWORDS


- **food preservation**
- **Fourier transform infra-red**
- **microencapsulation**
- **nanoencapsulation**
- **nanoparticles**
- **nanotechnology**


## REFERENCES


1. Augustin, M. A., & Sanguansri, P., (2009). Nanostructured materials in the food industry. *Advances in Food and Nutrition Research, 58*, 183–213.
2. Bakowska-Barczak, A. M., & Kolodziejczyk, P. P., (2011). Black currant polyphenols: Their storage stability and microencapsulation. *Industrial Crops and Products, 34*(2), 1301–1309.
3. Bajpai, A., Jadhav, K., Muthukumar, M., Kumar, S., & Srivatava, G., (2020). Use of nanotechnology in quality improvement of economically important agricultural crops. Chapter 3. In: Ghorbanpour, M., Bhargava, P., Varma, A., & Choudhary, D. K., (eds.), *Biogenic Nano-Particles and their Use in Agro-Ecosystems* (pp. 39–57). The Gateway East – Singapore: Springer International Publishers.
4. Bajpai, V. K., Kamle, M., Shukla, S., Mahato, D. K., Chandra, P., Hwang, S. K., Kumar, P., Huh, Y. S., & Han, Y. K., (2018). Prospects of using nanotechnology for food preservation, safety, and security. *Journal of Food and Drug Analysis, 26*(4), 1201–1214.
5. Bhushan, B., (2020). Frontiers in nanotribology: Magnetic storage, bio/nanotechnology, cosmetics, and bioinspiration. *Journal of Colloid and Interface Science, 577*, 127–162.
6. Calipinar, H., & Ulas, D., (2019). Development of nanotechnology in the world and nanotechnology standards in Turkey. *Procedia Computer Science, 158*, 1011–1018.
7. Chaudhry, Q., Scotter, M., Blackburn, J., Ross, B., Boxall, A., Castle, L., & Watkins, R., (2008). Applications and implications of nanotechnologies for the food sector. *Food Additives and Contaminants, 25*(3), 241–258.
8. Chellaram, C., Murugaboopathi, G., John, A. A., Sivakumar, R., Ganesan, S., Krithika, S., & Priya, G., (2014). Significance of nanotechnology in food industry. *APCBEE Procedia, 8*, 109–113.

9. Dingman, J., (2008). Guest commentary: Nanotechnology: Its impact on food safety. *Journal of Environmental Health, 70*(6), 47–50.
10. De Vos, P., Faas, M. M., Spasojevic, M., & Sikkema, J., (2010). Encapsulation for preservation of functionality and targeted delivery of bioactive food components. *International Dairy Journal, 20*(4), 292–302.
11. Ezhilarasi, P., Karthik, P., Chhanwal, N., & Anandharamakrishnan, C., (2013). Nanoencapsulation techniques for food bioactive components: A review. *Food Bioprocess Technol., 6*, 628–647.
12. Galus, S., & Kadzińska, J., (2015). Food applications of emulsion-based edible films and coatings. *Trends in Food Science & Technology, 45*(2), 273–283.
13. Ghosh, P., (2000). Hydrophilic polymeric nanoparticles as drug carriers. *Ind. J. Biochem. Biophys., 37*, 273–282.
14. Hasan, S. K., Ferrentino, G., & Scampicchio, M., (2020). Nanoemulsion as advanced edible coatings to preserve the quality of fresh-cut fruits and vegetables: A review. *International Journal of Food Science & Technology, 55*(1), 1–10.
15. Huang, Q., Given, P., & Qian, M., (2009). *Micro/Nano Encapsulation of Active Food Ingredients* (p. 328). Oxford University Press; Oxford.
16. Huang, Q., Yu, H., & Ru, Q., (2010). Bioavailability and delivery of nutraceuticals using nanotechnology. *Journal of Food Science, 75*(1), R50–R57.
17. Hughes, G. A., (2005). Nanostructure-mediated drug delivery. *Nanomed. Nanotech. Biol. Med., 1*, 22–30.
18. Hughes, J. M., Aherne, D., Bergin, S. D., O'Neill, A., Streich, P. V., Hamilton, J. P., & Coleman, J. N., (2012). Using solution thermodynamics to describe the dispersion of rod-like solutes: Application to dispersions of carbon nanotubes in organic solvents. *Nanotechnology, 23*(26), 265604.
19. Hughes, J. M., Cathcart, H., & Coleman, J. N., (2010). Dispersion and exfoliation of nanotubes with synthetic oligonucleotides: Variation of dispersion efficiency and oligo-nanotube interaction with base type. *The Journal of Physical Chemistry C, 114*(27), 11741–11747.
20. Kawashima, Y., (2001). Nanoparticulate systems for improved drug delivery. *Adv. Drug Delivery Rev., 47*, 1, 2.
21. Kingsley, J. D., Ranjan, S., Dasgupta, N., & Saha, P., (2013). Nanotechnology for tissue engineering: Need, techniques and applications. *Journal of Pharmacy Research, 7*(2), 200–204.
22. Kumari, A., Singla, R., Guliani, A., & Yadav, S. K., (2014). Nanoencapsulation for drug delivery. *EXCLI Journal, 13*, 265.
23. Kumari, A., Yadav, S. K., Pakade, Y. B., Singh, B., & Yadav, S. C., (2010). Development of biodegradable nanoparticles for delivery of quercetin. *Colloids Surf. B, 80*, 184–192.
24. Nagavarma, B., Yadav, H. K., Ayaz, A., Vasudha, L., & Shivakumar, H., (2012). Different techniques for preparation of polymeric nanoparticles: A review. *Asian J. Pharm. Clin. Res., 5*, 16–23.
25. Nedovic, V., Kalusevic, A., Manojlovic, V., Levic, S., & Bugarski, B., (2011). an overview of encapsulation technologies for food applications. *Procedia Food Science, 1*, 1806–1815.
26. Nguyen, T. T. L., Anton, N., & Vandamme, T. F., (2016). Nutraceutical compounds encapsulated by extrusion–spheronization. *New Polymers for Encapsulation of Nutraceutical Compounds,* 195–230.

27. Mahesh, S. K., Fathima, J., & Veena, V. G., (2019). Cosmetic potential of natural products: Industrial applications. Chapter 10. In: Swamy, M. K., & Akhter, M. S., (eds.), *Natural Bio-Active Compounds* (pp. 215–250). The Gateway East – Singapore: Springer International Publishers.
28. Maheshwari, P. V., & Gupta, N. V., (2012). Advances of nanotechnology in healthcare. *International Journal of PharmTech Research, 4*(3), 1221–1227.
29. Mihindukulasuriya, S. D. F., & Lim, L. T., (2014). Nanotechnology development in food packaging: A review. *Trends in Food Science & Technology, 40*(2), 149–167.
30. Mishra, U. K., (2012). Application of nanotechnology in food and dairy processing: An overview. *Pak. J. Food Sci., 22*, 23–31.
31. Mishra, B., Patel, B. B., & Tiwari, S., (2010). Colloidal nanocarriers: A review on formulation technology, types and applications toward targeted drug delivery. *Nanomed. Nanotechnol. Biol. Med., 6*, 9–24.
32. Mozafari, M. R., Flanagan, J., Matia-Merino, L., Awati, A., Omri, A., Suntres, Z. E., & Singh, H., (2006). Recent trends in the lipid-based nanoencapsulation of antioxidants and their role in foods. *J. Sci. Food Agric., 86*, 2038–2045.
33. Murthy, K. C., Monika, P., Jayaprakasha, G. K., & Patil, B. S., (2018). Nanoencapsulation: An advanced nanotechnological approach to enhance the biological efficacy of curcumin. In: Jayaprakasha, G. K., Patil, B. S., & Gattuso, G., (eds.), *Advances in Plant Phenolics: From Chemistry to Human Health* (pp. 383–405). Washington – DC: ACS Publications.
34. Padua, G. W., & Wang, Q., (2009). Controlled self-organization of zein nanostructures for encapsulation of food ingredients. In: Huang, Q., Given, P., & Qian, M., (eds.), *Micro/ Nanoencapsulation of Active Food Ingredients* (pp. 143–156). Washington – DC: ACS Publications.
35. Paredes, J. P., (2006). Written description requirement in nanotechnology: Clearing a patent thicket. *J. Pat. & Trademark Off. Soc'y, 88*, 489.
36. Peters, R., Brandhoff, P., Weigel, S., Marvin, H., Bouwmeester, H., Aschberger, K., & Mech, A., (2014). Inventory of nanotechnology applications in the agricultural, feed and food sector. *EFSA Supporting Publications, 11*(7), 621E.
37. Quintanilla-Carvajal, M. X., Camacho-Díaz, B. H., Meraz-Torres, L. S., Chanona-Pérez, J. J., Alamilla-Beltrán, L., Jimenéz-Aparicio, A., & Gutiérrez-López, G. F., (2010). Nanoencapsulation: A new trend in food engineering processing. *Food Engineering Reviews, 2*(1), 39–50.
38. Relkin, P., Yung, J. M., Kalnin, D., & Ollivon, M., (2008). Structural behavior of lipid droplets in protein-stabilized nanoemulsions and stability of α-tocopherol. *Food Biophysics, 3*(2), 163–168.
39. Reza, M. M., Johnson, C., Hatziantoniou, S., & Demetzos, C., (2008). Nanoliposomes and their applications in food nanotechnology. *Journal of Liposome Research, 18*(4), 309–327.
40. Rojas-Graü, M. A., Raybaudi-Massilia, R. M., Soliva-Fortuny, R. C., Avena-Bustillos, R. J., McHugh, T. H., & Martín-Belloso, O., (2007). Apple puree-alginate edible coating as carrier of antimicrobial agents to prolong shelf-life of fresh-cut apples. *Postharvest Biology and Technology, 2*, 254–264.
41. Sanguansri, P., & Augustin, M. A., (2006). Nanoscale materials development–a food industry perspective. *Trends Food Sci. Technol., 17*, 547–556.
42. Sekhon, B. S., (2010). Food nanotechnology: An overview. *Nanotechnology, Science and Applications, 3*, 1.

43. Storm, G., Belliot, S. O., Daemen, T., & Lasic, D. D., (1995). Surface modification of nanoparticles to oppose uptake by the mononuclear phagocyte system. *Adv. Drug Delivery Rev, 17*, 31–48.
44. Singh, P., (2018). Nanotechnology in food preservation. *Food Science, 9*(2), 435–441.
45. Vidhyalakshmi, R., Bhakyaraj, R., & Subhasree, R. S., (2009). Encapsulation "the future of probiotics": A review. *Adv. Biol. Res., 3*(3, 4), 96–103.
46. Xiaojia, H., Hua, D., & Huey-min, H., (2019). The current application of nanotechnology in food and agriculture. *Journal of Food Drug Analysis, 27*(1), 1–21.
47. Wang, X., Jiang, Y., Wang, Y. W., Huang, M. T., Ho, C. T., & Huang, Q., (2008). Enhancing anti-inflammation activity of curcumin through O/W nanoemulsions. *Food Chemistry, 108*(2), 419–424.
48. Wang, X., Li, D., Ghali, L., Xia, R., Munoz, L. P., Garelick, H., Bell, C., & Wen, X., (2016). Therapeutic potential of delivering arsenic trioxide into HPV-infected cervical cancer cells using liposomal nanotechnology. *Nanoscale Research Letters, 11*(1), 1–8.
49. Weiss, J., Takhistov, P., & McClements, D. J., (2006). Functional materials in food nanotechnology. *Journal of Food Science, 71*(9), R107–R116.
50. Wandrey, C., Bartkowiak, A., & Harding, S. E., (2010). Materials for encapsulation. In: Zuidam, N. J., & Nedovic, V., (eds.), *Encapsulation Technologies for Active Food Ingredients and Food Processing* (pp. 31–100). New York – USA: Springer International Publishers.
51. Yu, H., Huang, Y., & Huang, Q., (2010). Synthesis and characterization of novel antimicrobial emulsifiers from ε-polylysine. *Journal of Agricultural and Food Chemistry, 58*(2), 1290–1295.
52. Yu, H., & Huang, Q., (2010). Enhanced *in vitro* anti-cancer activity of curcumin encapsulated in hydrophobically modified starch. *Food Chemistry, 119*(2), 669–674.

CHAPTER 6

# POTENTIAL OF POLYMER NANO-BLENDS IN FOOD PACKAGING

SMITH V. PATIL, PRATIBHA JADHAV, and GIRISH M. JOSHI


## ABSTRACT

Remarkable milestones have been achieved in the biopolymer-based food packaging industry, which consistently focuses on the enhancement of the food packaging properties. Apart from their biodegradable nature, biopolymer-based blends are very easily available in nature and require specific processing for commercial use. Polymer nano blends are one of the rising branches of bio-based polymer food packaging where different nanoparticles are blended with the polymers to give the desired properties to the polymer films according to the target applications. Recent advances in the field of polymer nano blends and its benefits for the packaging of food products have been reviewed in this chapter. Various processes of cross-linking, their uses, and how they are useful for food packaging is disclosed in detail. Nanoparticles are the core of polymer nano blends. They are used to imbibe the desired properties to the polymer films. Its utilization in proper proportion can keep the biodegradability nature of the polymer films intact. Within the decade, the food packaging industry will be revolutionized by polymer-derived food packaging with different nanoparticles according to industrialist needs.


## 6.1 INTRODUCTION

The importance of food packaging is grown immensely in today's world. Packaging is now known to maintain the food products life. To display the

Nanotechnology Horizons in Food Process Engineering, Volume 1. Megh R. Goyal, PhD, PE, Junaid Ahmad Malik, PhD, Satish Kumar, PhD & Ritesh B. Watharkar, PhD (Eds.)

content of food and to protect them during cold storage along with transportation from various factors were important such as micro-organisms attack, synthetic foreign substance, oxygen moisture, and light. To perform these tasks, the food packaging materials should be selected properly according to the contents of a food product. It must be able to give proper physical barrier and maintain the appropriate chemical conditions for food products. For terminating oxidants from food for acquiring the palatable shell life of realistic usability and sustain the desirable food quality and freshness [30].

The permeability property should be present in food packaging system. It must have the ability to prevent penetration of water vapors, oxygen, carbon dioxide, and addition or loss of any other kind of vapors. In spite of permeability properties, it is also desirable to have good thermal properties, mechanical properties, and optical properties for food packaging covers to prevent any unwanted changes in the food product. 20% of all the plastic produced throughout the world is used as food packaging plastics including plastic bottles, cans, jars, glass, cups. After the complete utilization of plastic products, proper disposing of the plastics is very important or to reuse it if possible as they must not cause any ecological issues. However, in the recent era, despite proper planning and various sources of plastics, a growing concern on the disposal of plastic packaging materials is always present [15, 59].

Also, the removal technique is very limited for packaging units which contributes greatly to the growing concerns. These growing concerns have depleted the natural resources majorly and have raised a major question on the usability of petroleum-based packaging materials. Therefore, 'biodegradable polymer-based packaging is the best option available to face all these problems. Protein-derived nano blends is the best of the best alternative to the traditional food packaging systems. These polymer-derived Nano blends have all the desired properties and majorly they are bio-degradable in an environment in the desired amount of time [52].

The new domain is opened for the food packaging industry by the bio nanocomposites and nano blends for natural food packaging [50]. The raw material for such bio-nano blends is naturally available and polyfilm can obtain with few processing steps [65]. By the addition of the nanoparticles or by nano reinforcements it was easy to improve the property of the bio-nanocomposite according to requirements. Bionanocomposite and blends are naturally bio degradable. It was reduced the major problem of plastic pollution, which was of much interest [7].

In today's world, plastic food packaging can be replaced with the use of bio-based polymers with all the desirable properties like thermal properties, mechanical properties, and optical properties [21]. By adding perfect additives all these properties can be controlled. Also, from keeping the capacity to revolutionize the whole packaging industry the bio nanocomposites can help in the economic development of the rural areas. The raw material of bio-based polymer films can be easily available in this area and thus processed giving the livelihood for the peoples there [11].

Nanoparticle blending along with biopolymers and the formation of the polymer nano blends has created a whole new class in the packaging industry [52]. The blending of particles along with biopolymers creates a network between polymer and particles. This modified network increases the overall strength of the biopolymer and thus fine properties. Barrier properties to any sorts of gas, thermal properties to avoid any heat effect on food products, and optical properties to avoid any changes in food because of sunlight [37, 38, 42, 49]. The improved properties of biopolymers were achieved due to the high surface area of nanoparticles. It was blended with the bio-polymer systems. Rawdkuen et al. [40] observed the impact of property as a function of a nanoparticle dimensions. Various nanoparticles are discussed in this chapter.

However, there is not always a win-win situation for polymer nano blends over the traditional plastic packaging. Polymer nano blends have the huge benefit of biodegradability over plastics, but at the same time, the permeability property for the gas and barrier properties of biopolymer is not very developed with lot scope to contribute the research community. Low permeability gas properties were seen in the biopolymer due to the high hydrophilic nature of the polymer system. It has incompatibility issues with the blend, matrix phase, and nanoparticles. In this chapter authors explored, how these concerns can be terminated by ongoing research and development in this field. Not only these disadvantages were accepted as challenge to overcome but also improved mechanical and thermal properties are introduced in polymer nano blends. All the major advantages along with some of the disadvantages are discussed in detail [2, 5]. This chapter explores how polymer nano-blends are preferable for food packaging applications.

## 6.2 STRUCTURE OF PROTEINS DERIVED POLYMER NANO BLENDS

Protein is one of the important constituents of polymer nanocomposite which comes under the hetero class of polymers systems. Protein can be used as raw material because of its 3-D structure and its versatility for renewable polymer nano blends which are easily degradable. Protein consists of 20 amino acids made up of various functional groups which are responsible for its structure and orientations. Bonds present between the various functional groups confirm the structure and orientation of protein. These groups are responsible for the 3-D structure of the protein as there are many interactions between these groups which overall supports the 3-D structure. Mostly all ionic bonds of these functional groups govern for the tensile strength (TS) of the protein.

Hydrogen bonds also play an important function in protein stability at high temperatures. These hydrogen bonds are broken at high temperature. It may change the overall structure of the protein and denaturation of protein was caused. This denaturation can be avoided or terminated by increasing the bonds or amino acids in the protein structure. Therefore, the functional group present in the protein was one of the sources enhance the properties when contrasted with other packaging films [48]. The amino acids are connected by linkage known as peptide bonds. This peptide bond is responsible for the chain structure of the proteins. It was linked and acts as a one of the major sources to alter protein properties just like amino acids. Apart from the versatility of the proteins they are easily available in nature and are plentiful accessible, non-toxic in nature, and eco-friendly [34].

The protein structure can be easily altered by the choice of amino acids and functional groups present in them [17]. But some small changes like orientation are possible with the help of weak bonds such as hydrogen bonds, weak Vander walls forces. Polar and non-polar amino residue, both contains the van der walls forces [3].

Not only protein structure was decided by such forces, but also physical properties and mechanical properties are affected by them. For example, the presence of weak forces like non-covalent bonds also affects the crystalline phase of the protein structure. When the protein films compared with petroleum-based plastics demonstrated that the protein films were very unreactive toward the environment. It has very less inertia of their own. Also, the films do not affect any micro-organisms and were very defensive against them due to the presence of various hydrophilic groups. Therefore, the proteins films were one of the important raw materials for polymer nano blends and can be easily blended with various nanoparticles to obtain films of desirable properties.

## 6.3 MODIFICATION OF PROTEIN DERIVED BIO-NANOCOMPOSITES

Crosslinking of polymers has emerged as one of the best techniques to enhance the mechanical and physical properties of polymers. Hydrophobic nature of polymer film was enhanced significantly by crosslinking. Also, the thermal stability of polymer films increased. Crosslinking generally involves the treatment of the polymer film with the hydro-phobic material.

Han et al. [20] demonstrated that once the polymer undergoes cross-linking, it becomes highly permeable to any of the solvents present, thus increased the property of permeability to the full extent. Also, with the help of crosslinking, multi-layered polymer film formation was possible as it creates a bond between two of the polymer sheets [20]. Cross-connecting of proteins goes about as a guide to build up a solid intermolecular bonding, diminished polymer chain mobility, and closer molecular packaging [18].

Coltelli et al. [13] reported that the crosslinking of the polymer can overall benefit the food packaging industry. Synthetic chemicals, heating at high temperatures, enzyme action, and irradiation were the fine methods possible to introduce crosslinking in the polymers [13]. Figure 6.1 depicts the various methods of how crosslinking can be obtained in protein structure. During heating of protein, all the weak forces which include van der walls forces, hydrogen bonding are broken and new bonds are created which are the reason of crosslinks created [12].

This crosslinking of proteins enhances the overall mechanical strength of the protein structure. The 3-D structure of the protein, which was dependent on the orientation of functional groups affected by this method. However, while thermally heating the proteins almost care should be taken that the temperature does not go above the ideal preparation temperature, because in such situations, instead of introducing crosslinks, denaturation may be introduced which destroys the whole protein structure [43]. The crosslinking with the help of enzymes is also one of the emerging fields. A crosslinking enzyme such as transglutaminase is a calcium-dependent enzyme that cross-interface the structure of proteins such as wheat gelatine and milk [57]. This enzyme catalysis the peptide bonds in the protein structure which introduces the crosslinks [45]. However, this method of crosslinking is not very well established and more research work is needed in this field. Currently, the scope of this method is very limited [61].

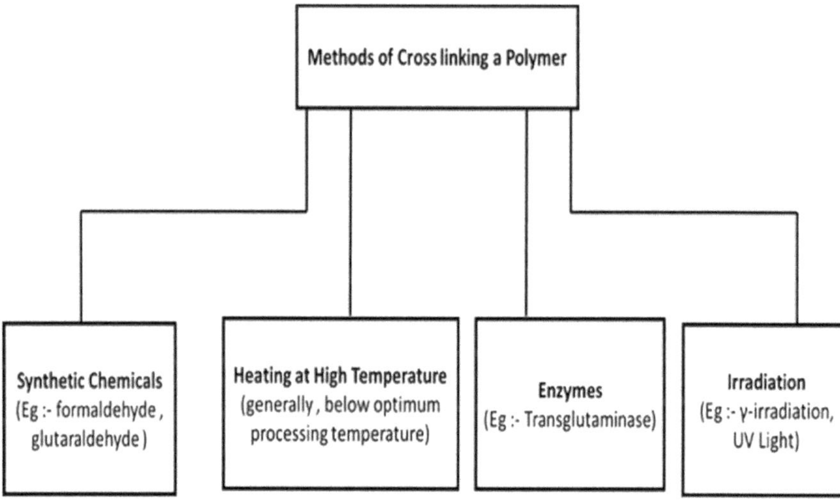


**FIGURE 6.1** Methods of crosslinking a polymer.

Synthetic compounds like NaOH, glutaraldehyde, formaldehyde, carbonyl diimidazole glycol have additionally been utilized for crosslinking of proteins. According to various studies, the use of different synthetic crosslinking agents improved the elastic nature of the polymer [62]. Also, the permeability properties, for example, oxygen barrier and $CO_2$ barrier properties are greatly enhanced. Sometimes unsaturated fatty acids are also used as crosslinking reagents as they are very hydrophobic. As permeability properties towards the gas increased the water barrier properties and extension at break properties of the polymer blend is decreased. Once the polymer is blended with the nanoparticles of the desired choice, then the use of crosslinking reagents was strengthened the whole polymer grid as new bonds would be formed throughout the grid. In this way, the appropriate use of cross-linkers can greatly enhance the polymer nano-blends grid [47].

Irradiation was one of the most preferable methods to introduce crosslinks in the polymer blends. This technique can be used in various polymers such as elastomers, thermoplastics, and TPE. This technique is used to change to a linear network of polymer chain into the 3-D network which has far more improved properties. There is a substantial increase in the branching rate and the average molecular weight is also increased by this method. Irradiation can be caused by both ionizing and non-ionizing radiation techniques. X-rays (ionizing) and UV light (non-ionizing) both are used for the crosslinking of proteins. Irradiation can affect the polymer in various forms to achieve crosslinking for example, free radicals, van der walls bonds, hydrogen bonds, and covalent bonds may be formed more polypeptide bonds between layers. This technique was the ideal choice over the synthetic compound method. Due to technique improved water barrier properties of the polymer and the elongation at break (EOB) was enhanced [12].

## 6.4 NANO-FILLERS FOR FOOD PACKAGING

The polymer blends are very efficient when compared to petroleum-based plastics and are also capable of solving all the problems in the present era. These biopolymers were capable of perfect food packaging and have many improved properties over plastics. However, there are some of the factors in the bio-based polymers which are insufficient to take a load of modern problems. Such as ineffective weight distribution over the polymer blends can sometimes exploit the polymer surface. All such problems were solved by the introduction of appropriate nanofillers in polymer blends. Selection of appropriate nano-fillers always depends on the uses of the polymer for different packaging [47]. Nanofillers are classified into two major classes, organic, and inorganic. Cellulose, chitin, zein are examples of organic nano-fillers, while carbon nanotubes, layered silicates, zinc oxide (ZnO), are inorganic nano-fillers which are incorporated in the polymer blends to enhance their properties. Figure 6.2 shows the general classification of the nano-fillers. Further, these nano-fillers are explained in detail [68].

Organic bio-based polymers sometimes contain nanoparticles which are inorganic in nature to enhance its properties. These nanoparticles are extracted from various minerals and are made crystalline line in nature so that they are properly blended with the polymers. Mostly these nanoparticles are silicate or oxide based. Despite the characterization of materials in the living body, it was difficult to emulate this self-association in the research center. A few research centers are dealing with the turn of events of various materials dependent on blends of various polymers and inorganic nanoparticles [31].

### *6.4.1 ORGANIC NANOFILLERS*

#### *6.4.1.1 CELLULOSE NANOCRYSTALS (CNCS)*

Cellulose nanocrystals (CNC's) are the most plentifully and easily accessible nanofillers. It is easy to get these nanofillers from marine animals and other plants. Cellulose nanocrystals are often known as CNC's can goes with several words worldwide such as microcrystals, nanocrystals, and micro crystallites [29, 44]. Figure 6.3 shows the process of obtaining cellulose nanocrystals (CNCs) from marine animals or plants [16]. From the marine animal's cellulose strands of cellulose fibers are extracted [8]. These strands are the raw material of cellulose nanocrystals. These strands along with the nanocrystals they also contain unwanted microfibrils. To remove this microfibril, acid hydrolysis treatment is done on the strands.

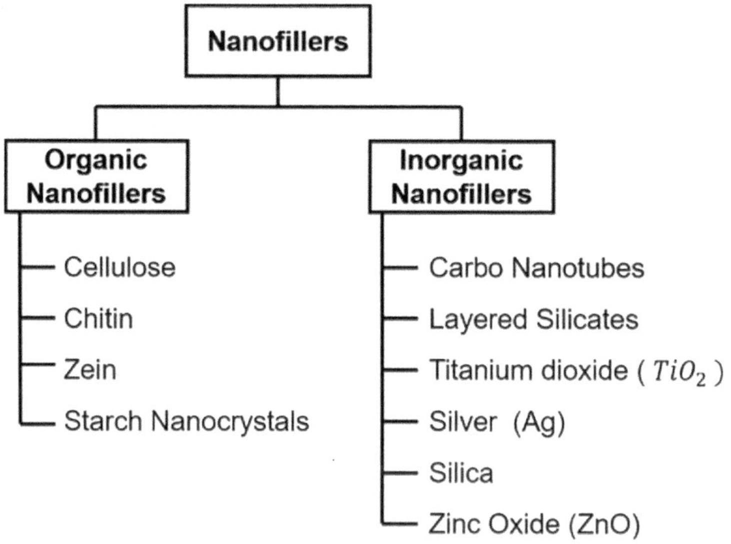


**FIGURE 6.2** Classification of nanofillers.

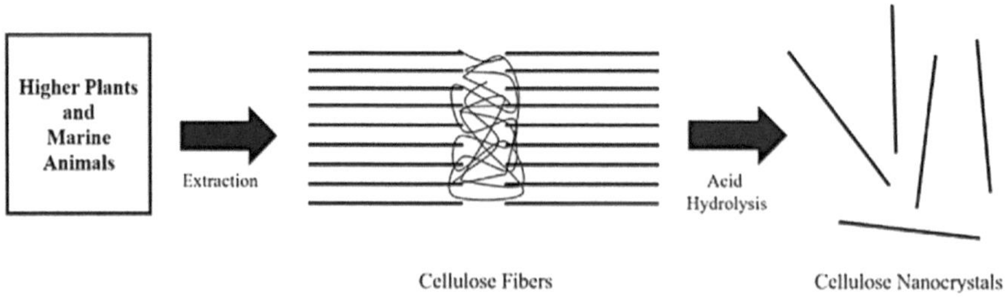


**FIGURE 6.3** Extraction of cellulose nanocrystals.

Since the microfibrils are amorphous, they dissolve in the solvent used to perform acid hydrolysis, and sharp nanocrystals are left behind [9, 46]. These crystals have a rod-like shape and breadth is in few nm. The shape of the cellulose nanocrystals depends on the purity of the raw material. Hydrochloric acid (HCL) and sulfuric acid are two of the options to perform acid hydrolysis [10, 22].

CNC's have attained as one of the most preferred choices as nanofillers for polymer blends because of its distinct nano size, high surface area, biodegradability, and rod-like compound structure. It is a known fact that the CNC's are known to greatly enhance the mechanical strength of polymer-composite. Since CNC's are lightweight, the weight of polymer nano blend is very less when compared to other films. Cellulose nanocrystals have a hydroxyl group on their surface in abundance; therefore, it is possible to make surface changes by addition of desirable functional groups according to the needs. As a function of increased concentration of sulfate ions CNC's thermal stability of the nano blend may be easily achieved. Scarcely any investigations found that all out solvency of water atoms in polymer nano blends with cellulose nanocrystal (CNC's) was decreased because of cellulose low hydrophilicity when contrasted with the greater part of another biopolymer-based network.

#### 6.4.1.2 CHITIN

Chitin raw materials are very abundant in nature. It is the second most occurring sources in nature after cellulose nanoparticles. Raw materials of the chitin are generally the biproduct of the seafood industry, which includes crab, prawns' shells. Chitin whiskers can easily obtain by boiling HCL and crab shells. While processing the chitin to incorporate it in the polymer blends was always important to treat these nanoparticles with HCL as because of it, the hydrogen-hydrogen interaction between the chitin and polymer increased [24, 36, 58]. Chitin nanoparticles are less crystalline when compared to cellulose nanoparticles, which helps to develop mechanical properties of the polymer nano blends. Also, at the low concentration of the chitin, the nanoparticle was easily dispersed into the polymer films and no conglomeration observed [23, 69].

Just like the cellulose nanoparticles, the chitin nanoparticles are also biodegradable, toxic-free, and availability in nature was very high [64]. Generally, they are used to increase the mechanical strength of various polymer blends, even at low concentrations; the mechanical strength is enhanced significantly. Along with the mechanical strength, the water barrier properties, and the permeability properties are also enhanced. Chitin has no as such improvised effect on the improvisation of the break at the elongation of bio-nano composite [35].

#### *6.4.1.3 ZEIN*

Zein nanoparticles are known for their non-toxicity and also, they are biocompatible and biodegradable. These nanoparticles are generally preferred because of their homogeneous dispersion. Also, zein nanoparticles are biocompatibilities. They can be easily blended with a vast number of polymer films. Young's modulus along with the TS has increased when zein nanoparticles is used in films. Also because of the hydrophobic nature of zein nanoparticles, they are also very helpful to increase the permeability property of the polymer blends [27, 33].

#### *6.4.1.4 STARCH DERIVED NANOCRYSTAL*

Starch derived nanocrystal raw materials are abundantly available in nature, easy to produce, and have a very low cost. Lepot et al. [26] demonstrated that the production of starch nanocrystal by simple hydrolysis of the local starch below its gelatinization temperature. The amorphous part of the starch is soluble in the solvent used and we are left with the crystalline Starch nanocrystals [26]. These starch nanocrystals are present generally in platelets form. Singular starch crystallites are difficult to acquire in light of starch platelet flocculation [1].

Starch nanoparticles were first in class to blend with the polymer system. They were used to strengthen the fillers. Even the small amount of Starch nanoparticles can enhance the elastic properties of the polymer blend according to research. Amylopectin chain molecular mobility is restricted by starch derived nanocrystals as it creates the hydrogen bonds between the starch nanocrystals and the polymer films [66]. The permeability properties of the films are also enhanced as various hydroxyl groups can frame the outside gases may attach polymer films. In this way, it can retain the water from damaging the food product inside the packaging also starch films are used to obtain the opacity properties in the polymer films which are generally obtained by the addition of the dyes [19].

### 6.4.2 INORGANIC NANOFILLERS

#### 6.4.2.1 CARBON BASED NANOTUBES

Carbon-based nanotubes which are the newly discovered form of carbon allotropes are made from graphene sheets. These sheets are cylindrically folded to obtain the hollow rod-like structure of the carbon nanotubes. These nanotubes are of two types, single-wall carbon nanotubes (SWCNTs), and multi-wall carbon nanotubes (MWCNTs). In SWCNT's, a single sheet of graphene is folded and in multiwall carbon nanotubes, a few sheets of graphene are folded. Because of this, the diameter of the SWCNTs is less compared to the MWCNTs. However, because of the exceptional high surface area and also due to adaptability of SWCNTs, while blending it with the polymer blends, it faces the problems of aggregation and because of its homogeneous dispersion of the SWCNT's is not possible. Because of the recent discoveries and progress in the nanoparticles field, huge production of carbon-based nanotubes can be easily obtained. CVD is one of the most preferred methods for the production of carbon nanotubes [60].

Carbon-based nanotubes have the extraordinary properties of the high thermal, mechanical, and electrical properties due to which it is becoming the most obvious choice for high strength food packaging systems. These nanotubes have $sp^2$ hybridized carbon atoms in their lattice, therefore when blended with the polymer blends physical and mechanical properties of the blends increase substantially. The TS of these sheets is even more than any metals. Carbon-based nanotubes are also conductive; therefore, they are also used to enhance the thermal and electrical properties of the polymer blends. If carbon-based nanotubes are well dispersed in the polymer blends, then the permeability properties were also increased with the water barrier properties [27].

However, despite such good properties of nanotubes, concerns have been raised on the toxicity. Therefore, they are not reinforced in the commercial food packaging [6].

#### 6.4.2.2 LAYERED SILICATES

The raw material of silicates is naturally available in nature and with less processability, they can be easily blended with the polymer nanocomposites. They are introduced to increase the thermal, barrier, and mechanical properties of the polymer blends. Due to high surface area of the silicates, they are used to enhance the overall TS of polymer blends. Silicates are generally of various types, but for food packaging application, only montmorillonites (MMTs) also known as MMT are used because of their high surface area. MMT has promising properties, for example, a huge perspective proportion and high surface region when contrasted with other inorganic nanoparticles [17].

Concerning the dispersion of the layered silicates in polymer blends, various types of nanocomposites can be achieved thermodynamically, which are intercalated, exfoliated, intercalated, and flocculated (Figure 6.4). All these methods contain the same amount of silicate nanoparticles inside the polymer blends, but there are differences concerning the overall properties of the polymer blends. In the first type, i.e., intercalated, most of the silicate's pates are parallel to each other and they were not very well dispersed in the polymer bends system. In the second type, intercalated, and flocculated, the layered silicates were still parallel to each other but no spacing was between them. This spacing can be filled with the ions for more bond formation. This bond formation will give more strength to the overall matrix of the polymer film and thus strengthens its TS. The third type is known as the exfoliated in which all the silicate plates are not parallel and are oriented in various directions in the polymer films. Exfoliated methods help to achieve more surface area of the silicates which helps to enhance barrier properties and the permeability properties of polymer film. In this way, all three types are equally important, and thus, by the appropriate selection, we can easily achieve the desired properties for our polymer nano blends. Besides, handling conditions, for example, high temperature and weight moreover prompted homogeneous blending of the nanofiller.

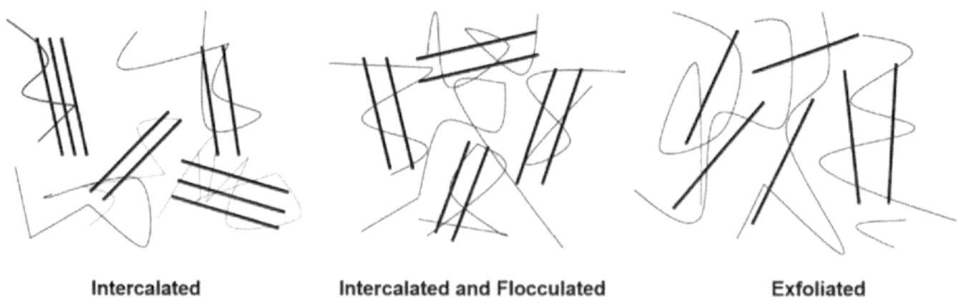


**FIGURE 6.4** Different types of silicate nanocomposites.

Li et al. reported that homogeneous dispersion of the silicates layers into the polymer film is difficult to achieve. There is generally agglomeration when silicates are blended with the polymer films which give rise to tactoid formations [28]. However, if an appropriate quantity of silicate nanoparticles was used in polymer films, then an excellent polymer nanocomposite may be obtained.

#### 6.4.2.3 TITANIUM DIOXIDE ($TIO_2$)

Titanium dioxide ($TiO_2$) is vividly used nanoparticle in the packaging industry and is in the list of the top five most-used nanoparticles in the world. They are readily available in nature in the form of $TiO_2$ as rutile and brookite minerals. $TiO_2$ nanoparticles possess the photocatalytic properties which are hardly possessed by any other nanoparticles and therefore these were in much demand. These photocatalytic properties of nanoparticles can reduce or stop the microbe growth on the food products. $TiO_2$ also possesses antimicrobial properties. It was used as an anti-microbe agent in the food packaging systems. These nanoparticles do not provide significant protection again the UV rays from the sun or any other source but they can stop food contamination by the microbes. They have the best permeability properties which can stop the oxygen and carbon dioxide or any other gases to affect the food product [67].

According to FDA, the $TiO_2$ is a toxicity-free nanoparticle and there is no such issue if mistakenly any contact is established with the food products. Because of this, they are easily used in food packaging systems to enhance barrier and mechanical properties [55].

#### 6.4.2.4 SILVER

Along with the good mechanical and physical properties of food packaging polymer Nano blends, there is also a need for good antimicrobial packaging because of the current foodborne diseases. Silver nanoparticles (AgNPs) are known to have good antimicrobial properties along with mechanical properties. Majorly these nanoparticles are used in the ready-to-eat food packaging as there more protection is needed because more water content is present in these foodstuffs [41].

Apart from the best antimicrobial properties, they also possess good mechanical and thermal properties. AgNPs are very stable at high temperatures. Due to their ease of operation, they can be easily incorporated or coated on the polymer film surface. The only problem related to the AgNPs utilization is the migration of the $Ag^+$ ions in the food kinds of stuff. Because of this, some concerns are raised on the toxicity of Ag nanoparticles [63].

#### *6.4.2.5 SILICA*

Silica dioxide is known for its water barrier properties. It exists in both crystalline and amorphous form and contains a 3-D lattice structure. These nanoparticles are spherical with dimensions in nm. They were easily dispersed in the polymer blends. Silica nanoparticles are mainly used to provide a water barrier or other liquid barrier properties to the polymer nanocomposites [54]. Because of their high surface area, they can easily restrain water molecules from interacting with the food kinds of stuff. They were simply coated over the polymer film to achieve the water barrier properties. Also, silica was recognized by the FDA and permissions have been granted to utilize added substances in food products. Therefore, they are more in demand now and their uses in food packaging field are growing day-by-day [42].

#### *6.4.2.6 ZINC OXIDE (ZNO)*

From the inorganic nanofiller group, ZnO is the most non-toxic nanofiller. It is recognized as one of the safest nanoparticles by many food organizations [66]. Along with the food packaging industry, sometimes it has also been used in food products because of its antimicrobial properties. This is also one of the five nanoparticles which are recognized as the safest nanoparticle to use in the food packaging industry. Even if any migration of the zinc ions is established may not create any issue [4].

Along with the great antimicrobial properties, it had good mechanical properties to and very stable at high temperatures. Therefore, it was recognized by various food organizations [61].

## 6.5 DIFFERENT PROPERTIES OF POLYMER NANO-BLENDS

Food Packaging demands has grown in recent years. Need to face the challenges of pollution by the use of petroleum-based packaging. Bio-based polymer nano blends are the only hope to meet all these challenges. Along with their biodegradability, also possess various properties that help to keep the food products fresh and thus increased their shell life to a great extent. However, the most desired properties for any food product would be mechanical properties and barrier properties. If the polymers fulfill the said criteria, then it was choice a packaging unit for the different types of food products. To keep the food safe during transportation and storage and to avoid the effect of environmental factors was the major contribution of the polymer blends used for food packaging. Thus, Barrier properties and mechanical properties are discussed in brief below [51]. The most desirable properties of polymer derived nano blends which the industrialist choice forward is shown in Figure 6.5.

**FIGURE 6.5** General properties of food packaging systems.

### 6.5.1 MECHANICAL PROPERTIES OF POLYMER NANO BLENDS

Mechanical properties are core of any food packaging systems. They are the most fundamental and general requirement of all the packaging units. If a polymer nano blend has good mechanical properties, then it can be used for a wide range of food packaging units. Mechanical properties were the basic framework and responsible for keeping food products safe during transportation and at storage facilities. Young's modulus of elasticity, extension at the break, mechanical strength and TS were decided the extent of the mechanical properties of the polymer nano blends [56]. Sorrentino et al. proved that high Young modulus of elasticity allows the polymer nano blends to stretch when the force is applied. If the polymer can withstand good force, it can easily protect the food product when any high force is applied during transportation. Good Mechanical strength provides a proper framework in which the food material is safe [53].

Mechanical properties for polymer nano blends are mostly dependent on the type of nanoparticles used with the polymer blends. They are greatly affected by the particle sizes, morphology, and bond formation capacity of the nanoparticles. The smaller the size of the nanoparticle, the more is the surface is and more interactions with the polymer blends. Because of this increased surface area, there will be more bond formation between the nanoparticle surface and the polymer surface [61].

Types of bonds also affect mechanical strength to a great extent. Strong bonds like covalent and ionic bonds can greatly enhance the mechanical strength of the polymer blends. Other bonds like hydrogen-hydrogen bonds, weak van der walls bonds are also equally important to maintain the orientation and 3-D shape of the polymer films [39]. If the nanoparticles are dispersed homogeneously in the polymer lattice, then the frequency of bond formation is increased, which can increase the mechanical strength of the polymer. Also, the crosslinks of polymers play an important role in increasing mechanical strength.

Good crosslinks between the polymer films and nanoparticles are known to increase the young's modulus value. Nanoparticles highly impact the core mechanical properties of polymer blends. Amorphous nature of polymer offers superior strength to them. Figure 6.6 exhibits the data about the decreasing crystallinity nature of polymer blends such as isostatic polypropylene and polypropylene [25]. The crystalline nature of polymer decreases as the percentage of spherical ZnO nanoparticles increases. In this way, by adding the appropriate percentage of nanoparticles, the mechanical properties of polymer blends can easily govern according to food packaging needs. Figure 6.7 shows the details of the melting temperature of polymer blends in the influence of the weight percentage of nanoparticles. Bar depicts that as the weight percent of nanoparticles is increased, the melting temperature of the polymer decreases [14].

Young's modulus is the measurement of the ability of a particular object to withstand changes in length under tension or compression. The higher the Young's modulus, greater is the TS of the material. Kean et al. [25] showed that as the concentration of Gelatine nanoparticles is increased, there is a significant enhancement in the Young's Modulus of polymer which increases TS of the polymer as shown in Figure 6.8. Therefore, showed the fact that all the important mechanical properties of blends are majorly enhanced by nanoparticles [25].

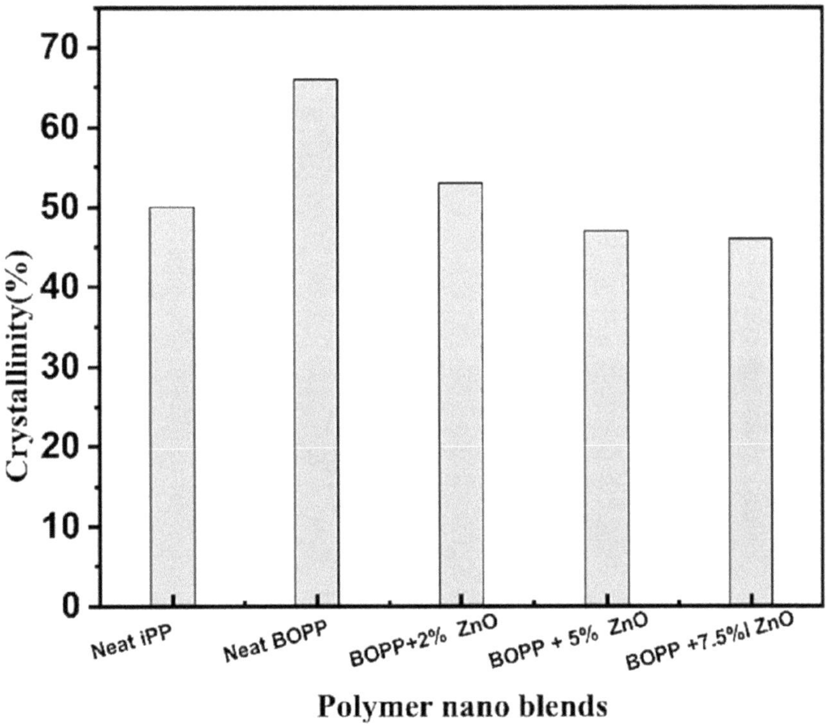


**FIGURE 6.6** Crystallinity nature of polymer nano blends.

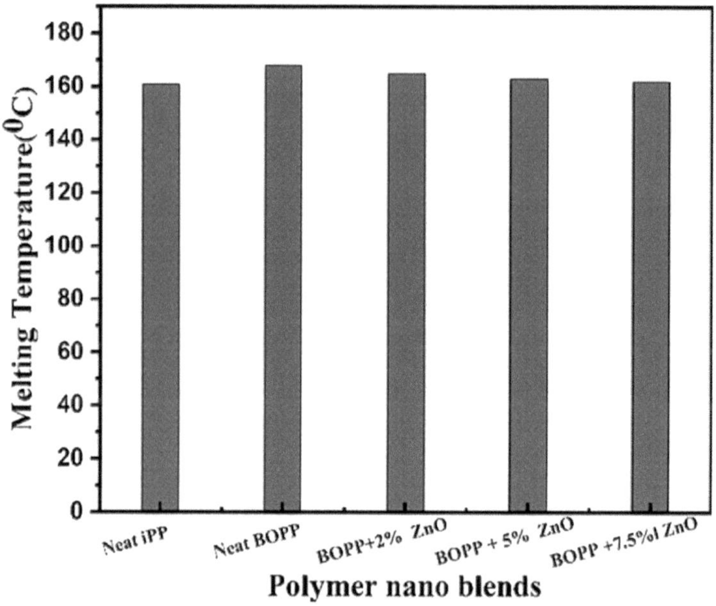


**FIGURE 6.7** Melting temperature trends of polymer nano blends.

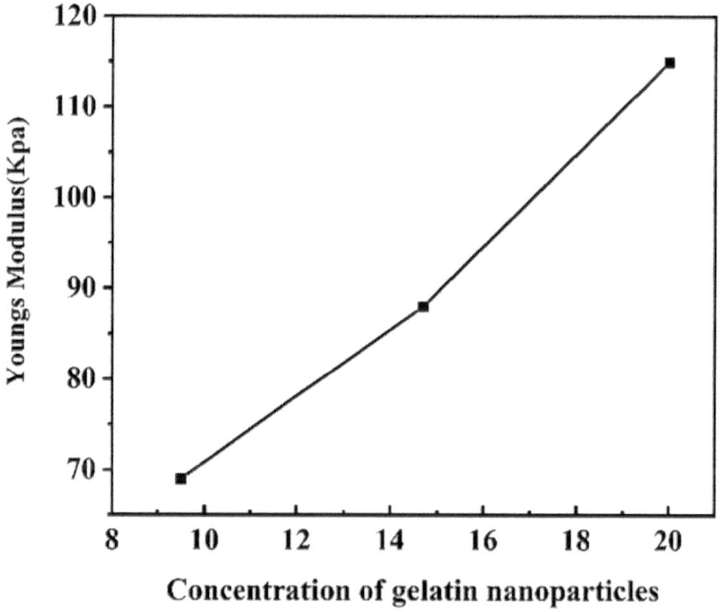


**FIGURE 6.8** Young's modulus as a function of gelatin nanoparticle loading.

### 6.5.2 BARRIER PROPERTIES OF POLYMER NANO BLENDS

The only major difference between the petroleum-based plastics and the bio-polymer nano blends is the water vapor permeability (WVP). Plastics can give the best permeability properties when compared to biopolymers. The great need for high permeability properties is yet to be accomplished in the bio-polymer food packaging industry. However, the incorporation of the nanoparticles in the polymer blends has made great progress in the permeability properties of polymer nano blends.

Nanoparticles (e.g., ZnO and $TiO_2$) have increased the water barrier properties to a great extent. If the water or any outer environmental factors are kept away from food products, then it can significantly enhance the life of the food products. These also play an important role to avoid the microorganism growth on food products. $O_2$ and $CO_2$ barrier properties of polymer nano blends are also very crucial to enhance the shelf life of food products. If the polymer has good oxygen barrier properties, then the whole oxidation process of food products can be completely avoided. This enables us to avoid the use of anti-oxidants in a food product which is also considered carcinogenic by many organizations [32].

In addition to mechanical properties, the size of nanoparticles also plays an important role in enhancing the permeability properties of polymer nano blends. The smaller the size of the nanoparticle, the greater will be interactions and bond formations with the polymer films. These bonds are the main reason to give permeability properties to the polymer blends. The higher the bonds, the higher will be the permeability power of the films. Also, the nanoparticles must be dispersed uniformly throughout the films. Due to such dispersion, it was directly shown more interaction between the nanoparticles and the polymer blends. With the blending of various nanoparticles to polymer films, all the important properties can be more enhanced in the future [36].

## 6.6 SUMMARY

The growing concerns of today's packaging industry have created a lot many problems and the only clear solution to all these problems are bio-polymer blends. Bio-based polymer blends are biodegradable, raw materials are easily available and require very little processing. The raw materials of petroleum-based plastics are limited whereas the raw materials of bio-based polymers are unlimited and easily available. Bio-polymers can act as good food packaging units as they have very enhanced gas barrier and mechanical properties.

Selections of polymers according to the need of food products can be easily done as there is a huge number of polymer blends are available, and their properties can be easily enhanced with blending of the desired nanoparticles.

The blending of nanoparticles in proper proportion enhances the mechanical properties and barrier properties of polymer blends to a great extent. Also, the biodegradability of the blends can be maintained which can solve all the modern-day food preservation issues. There are a number of nanoparticles that can be used in polymer blends to obtain desired properties. They are known to increase the TS, and optical properties of food packaging polymers. The use of nanoparticles in polymer blends will give the ability to bio-polymers to stand against petroleum-based packaging.

Also, there is very limited data available on the subject of migration of the nanoparticle ions into the food products. Some of the nanoparticles are non-toxic but many of them are toxic. Migration of ions into food products can have adverse effects on food and can also spoil the food products. Therefore, more examination on interfacing issues this particular field is required before there is the utilization of the nanoparticle in commercial food packaging units. Nevertheless, polymer nano blends are the future of the packaging industry. They have a brilliant possibility to widen the scope of food packaging materials for preserving the ideal food qualities.

## KEYWORDS

- **cellulose nanocrystals**
- **food packaging**
- **multi-wall carbon nanotubes**
- **nanofillers**
- **nanoparticles**
- **polymer nano blends**

## REFERENCES

1. Ai, F., Zheng, H., Wei, M., & Huang, J., (2007). Soy protein plastics reinforced and toughened by $SiO_2$ nanoparticles. *Journal of Applied Polymer Science, 105*(3), 1597–1604.
2. Akbari, Z., Ghomashchi, T., & Moghadam, S., (2007). Improvement in food packaging industry with biobased nanocomposites. *International Journal of Food Engineering, 3*(4), 3. doi: 10.2202/1556-3758.1120.

3. Alexandre, E. M. C., Lourenço, R. V., Bittante, A. M. Q. B., Moraes, I. C. F., & Sobral, P. J. D. A., (2016). Gelatin-based films reinforced with montmorillonite and activated with nanoemulsion of ginger essential oil for food packaging applications. *Food Packaging and Shelf Life, 10*, 87–96.
4. Angellier-Coussy, H., & Gastaldi, E., (2012). Protein/clay nano-biocomposites. In: Avérous, L., & Pollet, E., (eds.), *Environmental Silicate Nano-Biocomposites* (pp. 323–343). London - UK: Springer International Publishers.
5. Arfat, Y. A., Benjakul, S., Prodpran, T., Sumpavapol, P., & Songtipya, P., (2015). physico-mechanical characterization and antimicrobial properties of fish protein isolate/fish skin gelatin-zinc oxide (ZnO) nanocomposite films. *Food and Bioprocess Technology, 9*(1), 101–112.
6. Arora, A., & Padua, G. W., (2010). Review: Nanocomposites in food packaging. *Journal of Food Science, 75*(1), R43–49.
7. Azeredo, H. M. C. D., (2009). Nanocomposites for food packaging applications. *Food Research International, 42*(9), 1240–1253.
8. Bordes, P., Pollet, E., & Averous, L., (2009). Nano-biocomposites: Biodegradable polyester/nanoclay systems. *Progress in Polymer Science, 34*(2), 125–155.
9. Cao, X., Chen, Y., Chang, P. R., Stumborg, M., & Huneault, M. A., (2008). Green composites reinforced with hemp nanocrystals in plasticized starch. *Journal of Applied Polymer Science, 109*(6), 3804–3810.
10. Casariego, A., Souza, B. W. S., Cerqueira, M. A., Teixeira, J. A., Cruz, L., Díaz, R., & Vicente, A. A., (2009). Chitosan/clay films' properties as affected by biopolymer and clay micro/nanoparticles' concentrations. *Food Hydrocolloids, 23*(7), 1895–1902.
11. Cinelli, P., Schmid, M., Bugnicourt, E., Wildner, J., Bazzichi, A., Anguillesi, I., & Lazzeri, A., (2014). Whey protein layer applied on biodegradable packaging film to improve barrier properties while maintaining biodegradability. *Polymer Degradation and Stability, 108*, 151–157.
12. Clarinval, A. M., & Halleux, J., (2005). Classification of biodegradable polymers. Chapter 1. In: Smith, R., (ed.), *Biodegradable Polymers for Industrial Applications* (pp. 3–31). Cambridge - UK: Woodhead Publishing.
13. Coltelli, M. B., Wild, F., Bugnicourt, E., Cinelli, P., Lindner, M., Schmid, M., Weckel, V., et al., (2015). State of the art in the development and properties of protein-based films and coatings and their applicability to cellulose-based products: An extensive review. *Coatings, 6*(1), 1–59. doi: 10.3390/coatings6010001.
14. Damodaran, S., & Parkin, K. L., (2017). *Fennema's Food Chemistry* (5th edn., p. 1123). CRC Press; Boca Raton- FL.
15. Ding, P., Norton, I. T., Zhang, Z., & Pacek, A. W., (2008). Mechanical properties of gelatin-rich micro-particles. *Journal of Food Engineering, 86*(3), 307–314.
16. Dufresne, A., (2008). Polysaccharide nanocrystal reinforced nanocomposites. *Canadian Journal of Chemistry, 86*(6), 484–494.
17. Garrett, R. H., & Grisham, C. M., (1996). Part I: Molecular components of cells. In: Brooks, T., & Cole, (eds.), *Biochemistry* (3rd edn., pp. 1–375). Belmont - CA: Harcourt Brace Custom Publishers.
18. Ghanbarzadeh, B., Oleyaei, S. A., & Almasi, H., (2015). Nanostructured materials utilized in biopolymer-based plastics for food packaging applications. *Critical Reviews in Food Science and Nutrition, 55*(12), 1699–1723.
19. Hammann, F., & Schmid, M., (2014). Determination and quantification of molecular interactions in protein films: A review. *Materials, 7*(12), 7975–7996.

20. Han, Y., Li, K., Chen, H., & Li, J., (2017). Properties of soy protein isolate biopolymer film modified by graphene. *Polymers, 9*(8), 312. doi: 10.3390/polym9080312.
21. Hu, X., Cebe, P., Weiss, A. S., Omenetto, F., & Kaplan, D. L., (2012). Protein-based composite materials. *Materials Today, 15*(5), 208–215.
22. Imran, M., Revol-Junelles, A. M., Martyn, A., Tehrany, E. A., Jacquot, M., Linder, M., & Desobry, S., (2010). Active food packaging evolution: Transformation from micro- to nanotechnology. *Critical Reviews in Food Science and Nutrition, 50*(9), 799–821.
23. Jayakumar, R., Menon, D., Manzoor, K., Nair, S. V., & Tamura, H., (2010). Biomedical applications of chitin and chitosan-based nanomaterials: A short review. *Carbohydrate Polymers, 82*(2), 227–232.
24. Jin, T. Z., (2017). Current state of the art and recent innovations for antimicrobial food packaging. In: Juneja, V. K., Dwivedi, H. P., & Sofos, J. N., (eds.), *Microbial Control and Food Preservation: Theory and Practice* (pp. 349–372). New York - USA: Springer International Publishers.
25. Kean, T., & Thanou, M., (2011). Chitin and chitosan: Sources, production and medical applications. Chapter 10. In: Willams, P. A., (eds.), *Renewable Resources for Functional Polymers and Biomaterials: Polysaccharides, Proteins and Polyesters* (pp. 292–318). London - UK: The Royal Society of Chemistry Publishing.
26. Lepot, N., Van, B. M. K., Van, D. R. H., D'Haen, J., Peeters, R., Franco, D., & Mullens, J., (2011). Influence of incorporation of ZnO nanoparticles and biaxial orientation on mechanical and oxygen barrier properties of polypropylene films for food packaging applications. *Journal of Applied Polymer Science, 120*(3), 1616–1623.
27. Li, K. K., Yin, S. W., Yin, Y. C., Tang, C. H., Yang, X. Q., & Wen, S. H., (2013). Preparation of water-soluble antimicrobial zein nanoparticles by a modified antisolvent approach and their characterization. *Journal of Food Engineering, 119*(2), 343–352.
28. Li, S., & Zhao, Y., (2017). Preparation of zein nanoparticles by using solution-enhanced dispersion with supercritical $CO_2$ and elucidation with computational fluid dynamics. *International Journal of Nanomedicine, 12*, 3485–3494.
29. Li, Y., Jiang, Y., Liu, F., Ren, F., Zhao, G., & Leng, X., (2011). Fabrication and characterization of $TiO_2$/whey protein isolate nanocomposite film. *Food Hydrocolloids, 25*(5), 1098–1104.
30. Mariano, M., El Kissi, N., & Dufresne, A., (2014). Cellulose nanocrystals and related nanocomposites: Review of some properties and challenges. *Journal of Polymer Science Part B: Polymer Physics, 52*(12), 791–806.
31. Marsh, K., & Bugusu, B., (2007). Food packaging-roles, materials, and environmental issues. *Journal of Food Science, 72*(3), R39–55.
32. Meyers, M. A., Chen, P. Y., Lin, A. Y. M., & Seki, Y., (2008). Biological materials: Structure and mechanical properties. *Progress in Materials Science, 53*(1), 1–206.
33. Mittal, V., (2013). Barrier properties of renewable nanomaterials. Chapter 22. In: Dufresne, A., Thomas, S., & Pothen, L. A., (eds.), *Biopolymer Nanocomposites* (pp. 541–564). New Jersey - USA: John Wiley & Sons Inc.
34. Naveed Ul, H. A., Nadhman, A., Ullah, I., Mustafa, G., Yasinzai, M., & Khan, I., (2017). Synthesis approaches of zinc oxide nanoparticles: The dilemma of ecotoxicity. *Journal of Nanomaterials, 12*, 1–14.
35. Nur, H. Z. A., Roos, Y. H., & Kerry, J. P., (2014). Use and application of gelatin as potential biodegradable packaging materials for food products. *International Journal of Biological Macromolecules, 71*, 94–102.

36. Oymaci, P., & Altinkaya, S. A., (2016). Improvement of barrier and mechanical properties of whey protein isolate based food packaging films by incorporation of zein nanoparticles as a novel bionanocomposite. *Food Hydrocolloids, 54,* 1–9.
37. Ozcalik, O., & Tihminlioglu, F., (2013). Barrier properties of corn zein nanocomposite coated polypropylene films for food packaging applications. *Journal of Food Engineering, 114*(4), 505–513.
38. Pandey, J. K., Kumar, A. P., Misra, M., Mohanty, A. K., Drzal, L. T., & Singh, R. P., (2005). Recent advances in biodegradable nanocomposites. *Journal of Nanoscience and Nanotechnology, 5*(4), 497–526.
39. Pavlidou, S., & Papaspyrides, C. D., (2008). A review on polymer–layered silicate nanocomposites. *Progress in Polymer Science, 33*(12), 1119–1198.
40. Rawdkuen, S., Sai-Ut, S., & Benjakul, S., (2010). Properties of gelatin films from giant catfish skin and bovine bone: A comparative study. *European Food Research and Technology, 231*(6), 907–916.
41. Rhim, J. W., & Ng, P. K., (2007). Natural biopolymer-based nanocomposite films for packaging applications. *Critical Reviews in Food Science and Nutrition, 47*(4), 411–433.
42. Rhim, J. W., Wang, L. F., & Hong, S. I., (2013). Preparation and characterization of agar/silver nanoparticles composite films with antimicrobial activity. *Food Hydrocolloids, 33*(2), 327–335.
43. Rhim, J. W., Wang, L. F., Lee, Y., & Hong, S. I., (2014). Preparation and characterization of bio-nanocomposite films of agar and silver nanoparticles: Laser ablation method. *Carbohydrate Polymers, 103*, 456–465.
44. Rostamzad, H., Paighambari, S. Y., Shabanpour, B., Ojagh, S. M., & Mousavi, S. M., (2016). Improvement of fish protein film with nanoclay and transglutaminase for food packaging. *Food Packaging and Shelf Life, 7*, 1–7.
45. Ruiz, E., Romero, I., Moya, M., Cara, C., Vidal, J. D., & Castro, E., (2013). Dilute sulfuric acid pretreatment of sunflower stalks for sugar production. *Bioresource Technology, 140*, 292–298.
46. Schmid, M., Sängerlaub, S., Wege, L., & Stäbler, A., (2014). Properties of transglutaminase crosslinked whey protein isolate coatings and cast films. *Packaging Technology and Science, 27*(10), 799–817.
47. Shakeri, A., & Radmanesh, S., (2013). Preparation of cellulose nanofibrils by high-pressure homogenizer and zein composite films. *Advanced Materials Research, 829*, 534–538.
48. Shankar, S., Teng, X., Li, G., & Rhim, J. W., (2015). Preparation, characterization, and antimicrobial activity of gelatin/ZnO nanocomposite films. *Food Hydrocolloids, 45*, 264–271.
49. Silva, K. S., Garcia, C. C., Amado, L. R., & Mauro, M. A., (2015). Effects of edible coatings on convective drying and characteristics of the dried pineapple. *Food and Bioprocess Technology, 8*(7), 1465–1475.
50. Sinharay, S., & Bousmina, M., (2005). Biodegradable polymers and their layered silicate nanocomposites: In greening the 21st century materials world. *Progress in Materials Science, 50*(8), 962–1079.
51. Siracusa, V., Rocculi, P., Romani, S., & Rosa, M. D., (2008). Biodegradable polymers for food packaging: A review. *Trends in Food Science & Technology, 19*(12), 634–643.
52. Song, F., Tang, D. L., Wang, X. L., & Wang, Y. Z., (2011). Biodegradable soy protein isolate-based materials: A review. *Biomacromolecules, 12*(10), 3369–3380.

53. Sorrentino, A., Gorrasi, G., & Vittoria, V., (2007). Potential perspectives of bio-nanocomposites for food packaging applications. *Trends in Food Science & Technology, 18*, 84–95.
54. Tharanathan, R. N., (2003). Biodegradable films and composite coatings: Past, present and future. *Trends in Food Science & Technology, 14*(3), 71–78.
55. Tian, H., (2011). Processing and properties of soy protein/silica nanocomposites fabricated in situ synthesis. *Journal of Composite Materials, 46*(4), 427–435.
56. Tran, Q. H., Nguyen, V. Q., & Le, A. T., (2018). Corrigendum: Silver nanoparticles: Synthesis, properties, toxicology, applications and perspectives. *Advances in Natural Sciences: Nanoscience and Nanotechnology, 9*(4), 049501. doi: 10.1088/2043-6254/aad12b.
57. Vieira, M. G. A., Da Silva, M. A., Dos, S. L. O., & Beppu, M. M., (2011). Natural-based plasticizers and biopolymer films: A review. *European Polymer Journal, 47*(3), 254–263.
58. Wang, J. S., Zhao, M. M., Yang, X. Q., Jiang, Y. M., & Chun, C., (2007). Gelation behavior of wheat gluten by heat treatment followed by transglutaminase crosslinking reaction. *Food Hydrocolloids, 21*(2), 174–179.
59. Wang, W., Du, Y., Qiu, Y., Wang, X., Hu, Y., Yang, J., Cai, J., & Kennedy, J. F., (2008). A New green technology for direct production of low molecular weight chitosan. *Carbohydrate Polymers, 74*(1), 127–132.
60. Weber, C. J., Haugaard, V., Festersen, R., & Bertelsen, G., (2002). Production and applications of biobased packaging materials for the food industry. *Food Additives and Contaminates, 19*, 172–177.
61. Weiss, J., Takhistov, P., & McClements, D. J., (2006). Functional materials in food nanotechnology. *Journal of Food Science, 71*(9), R107–R116.
62. Wihodo, M., & Moraru, C. I., (2013). Physical and chemical methods used to enhance the structure and mechanical properties of protein films: A review. *Journal of Food Engineering, 114*(3), 292–302.
63. Winkler, H., Vorwerg, W., & Schmid, M., (2015). Synthesis of hydrophobic whey protein isolate by acylation with fatty acids. *European Polymer Journal, 62*, 10–18.
64. Xiao, D., & Zhong, Q., (2011). *In vitro* release kinetics of nisin as affected by tween 20 and glycerol co-encapsulated in spray-dried zein capsules. *Journal of Food Engineering, 106*(1), 65–73.
65. Ye, M., Neetoo, H., & Chen, H., (2008). Effectiveness of chitosan-coated plastic films incorporating antimicrobials in inhibition of listeria monocytogenes on cold-smoked salmon. *International Journal of Food Microbiology, 127*(3), 235–240.
66. Zhao, R., Torley, P., & Halley, P. J., (2008). Emerging biodegradable materials: Starch- and protein-based bio-nanocomposites. *Journal of Materials Science, 43*(9), 3058–3071.
67. Zheng, H., Ai, F., Wei, M., Huang, J., & Chang, P. R., (2007). Thermoplastic soy protein nanocomposites reinforced by carbon nanotubes. *Macromolecular Materials and Engineering, 292*(6), 780–788.
68. Zhou, J. J., Wang, S. Y., & Gunasekaran, S., (2009). Preparation and characterization of whey protein film incorporated with $TiO_2$ nanoparticles. *Journal of Food Science, 74*(7), N50–56.
69. Zubair, M., & Ullah, A., (2020). Recent Advances in protein derived bionanocomposites for food packaging applications. *Critical Reviews in Food Science and Nutrition, 60*(3), 406–434.

CHAPTER 7

# ROLE OF INORGANIC FILLERS IN NANOCOMPOSITES FOR FOOD PACKAGING

GARGI D. PATIL, PRATIBHA JADHAV, and GIRISH M. JOSHI

## ABSTRACT

Currently, demand for food with longer shelf life and bare processing has increased. Therefore, antimicrobial, antifungal, ultra-violet resistant, moisture resistant, and antioxidant properties of the packaging material are desired. Hence, nanocomposites are used massively in food packaging applications. These properties depend not only on the type of nanoparticles, but also on various factors such as filler orientations, synthesis method and fabrication of nanocomposites used, degree of mixing, and adhesion between the fillers and the matrix. Inorganic nanofillers such as silver, zinc oxide (ZnO), silicon dioxide ($SiO_2$), titanium dioxide ($TiO_2$), kaolinite, montmorillonite (MMT), and copper owing to their antimicrobial property against gram-positive and gram-negative bacteria and superior mechanical and barrier properties gained extensive demand in food packaging. It was incorporated into synthetic as well as natural polymers. This chapter overviews how actually these fillers affect the physical properties of the formed composites and how actually it is better than conventional polymers, health, and environmental effects.

## 7.1 INTRODUCTION

The last decade witnessed enormous attention with respect to research in the field of nanocomposites with polymer as a host matrix system. Survey

Nanotechnology Horizons in Food Process Engineering, Volume 1. Megh R. Goyal, PhD, PE, Junaid Ahmad Malik, PhD, Satish Kumar, PhD & Ritesh B. Watharkar, PhD (Eds.)

reported that the market of nano-enabled packaging has been grown at a rate of 12.7% with a worth of $15 from the year 2014 to 2020. Packaging serves various purposes such as shielding against temperature rise, shock, and vibrations, oxygen, carbon dioxide, dirt, moisture [26, 34, 71]. It acts successfully for storage of foodstuffs. Apart from that, it was used for labeling various nutritional aspects of food. It acts as a marketing aspect. Efficient and good designing of packaging attracts more consumers. Besides, the packaging material should be easily sealable, light in weight, cost effective, and protect food from extreme climatic conditions of heat, moisture, and cold [75].

Currently people were more interested to prefer the ready-to-use, ready-to-cook food. Hence it was in more demand for packaging material with longer shelf life with low concentration of preservatives is enlarging day by day [119]. Nanocomposite deals with nano-sized particles of the order of 1 nm to 100 nm reinforced into a polymer matrix. This reinforcement found to carve out superior mechanical, chemical, optical, electrical, and barrier performance. Food deterioration can happen during handling, processing, storage, and distribution [22, 70, 81].

Food packaging is packaging for food items. A package provides protection, tamper resistance to food stored in it. This domain of nanotechnology explores various intelligent/smart and active packaging which is of storage interest for handling of foodstuffs [72]. Intelligent packaging takes into consideration of maintaining the food quality by monitoring the environment with the use of bio-sensors. It thus alerts the consumer about the safety of food. Some of the food quality indicators include $O_2$, $CO_2$, humidity, Ph detectors [82, 111]. The reinforcement of nanostructures has markedly lifted the ability of packaging materials, both by improving the barrier properties for water vapor and gas and by enhancing tensile strength (TS), elongation at break (EOB), young's modulus, and thermal stability [16, 19, 129]. Active packaging uses nanoparticles, for a better interaction of environment and food. With the alteration in environment, it according alters its characteristics ultimately increasing its shelf life [32, 33, 124]. However, its impact on human health due to migration is limited. Present legislations do not really cover-up this entirely. This is so far the most growing field of nanotechnology.

In nanocomposites, nanofiller was induced into the polymer matrix, which may be non-biodegradable such as polypropylene, nylon. As well as biodegradable naturally occurring biopolymers such as starch, cellulose, pectin, chitosan (CS), chitin, casein, whey, collagen, gelatin, and some chemically synthesized polymer such as poly(ε-caprolactone) (PCL), poly-lactic acid (PLA), poly (vinyl alcohol), (PVOH), poly (butylene succinate)

(PBS), poly (glycolic acid) (PGA). All these can be used for food packaging applications into which organic or inorganic nanofillers can be reinforced forming nanocomposites [117].

This chapter focuses on inorganic nanofillers. Inorganic nanofillers can be divided into the three categories referring to Figure 7.1, metallic nanoparticles include silver, gold, copper, and iron and non-metallics include selenium nanoparticles. The use of metal oxides such as magnesium oxide (MgO), zinc oxide (ZnO), iron oxide, and titanium dioxide ($TiO_2$) are augmented in the food packaging industry. Also, Silicon dioxide ($SiO_2$) nanoparticles are also exploited significantly in food packaging. Apart from one-dimensional filler, several two-dimensional fillers such as clays of montmorillonite (MMT), hectorite, kaolinite, these fillers are also in demand currently due to their platelet shape and increased barrier properties.

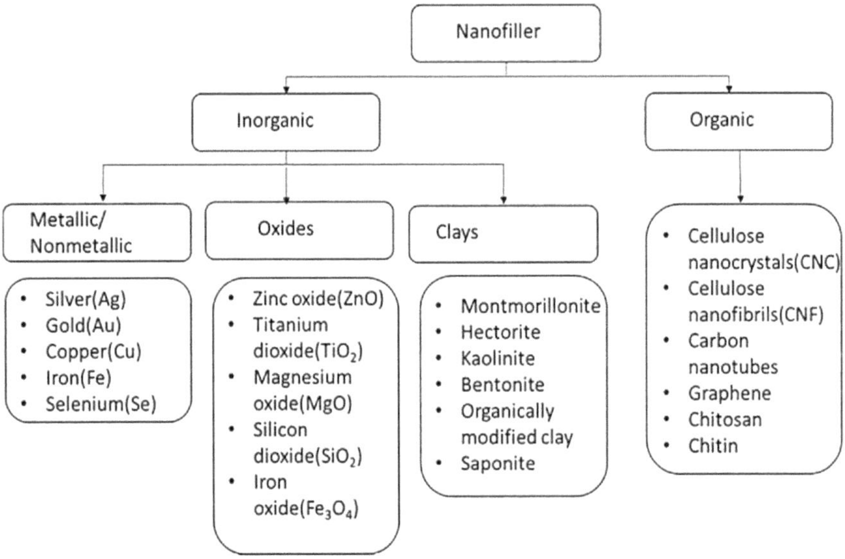


**FIGURE 7.1** Classification of nanofillers.

This chapter focuses on how actually nanoparticle as an individual contributes to increasing its physical properties. It overviews how actually the nanofillers influence on mechanical, barrier, and antimicrobial properties of composites. Various effects due to nano entity loading and the alteration in their properties are addressed. At last, the toxicological evaluation of nanoparticles on health and environment is discussed.

## 7.2 IMPACT OF VARIOUS INORGANIC FILLERS ON DIFFERENT NANOCOMPOSITES: LITERATURE SURVEY (TABLE 7.1)

**TABLE 7.1** Influence of Inorganic Nanofillers on Antimicrobial, Barrier, and Mechanical Properties

| System | Fillers | Effect on Properties | Applications | References |
|---|---|---|---|---|
| Carboxymethyl cellulose (CMC) | Montmorillonite (MMT) + ε-poly-(lysine) (ε-PL) | Improved barrier against UV and water vapor, increased tensile strength, good antibacterial activities against *Staphylococcus aureus*, *Escherichia coli* and antifungal against *Rhizopus oligosporus* and *Botrytis cinerea.* | Antimicrobial food packaging | [48] |
| Chitosan + polyacrylic acid | Copper hydrogel | Improved antibacterial activity, antifungal activity | Food and plant packaging | [38] |
| Gelatin | Grapefruit seed extract + $TiO_2$ nanoparticles | Improved water barrier, UV resistant properties, antibacterial properties against *Escherichia coli* and *Listeria monocytogenes* | Antioxidant, antimicrobial active food packaging material | [47] |
| Poly (butylene adipate co-terephthalate) (PBAT) | $TiO_2$ nanoparticles | Improved tensile strength, decreased oxygen permeability, excellent antimicrobial activity against *Escherichia coli* and *Staphylococcus aureus* | Antimicrobial food packaging | [45] |

**TABLE 7.1** *(Continued)*

| System | Fillers | Effect on Properties | Applications | References |
|---|---|---|---|---|
| Poly (ethylene oxide) | Montmorillonite nanoparticles | Increased moisture barrier and oxygen barrier properties | Active food packaging | [43] |
| Poly (L-lactic acid) (PLLA) | ZnO and $SiO_2$/ZnO nanoparticles | Antimicrobial activity against *Escherichia coli.* | Antibacterial food packaging | [42] |
| Poly (vinyl alcohol) (PVA) + chitosan (CS) | ZnO-$SiO_2$ nanoparticles | Antimicrobial activity *Staphylococcus aureus*, *Escherichia coli.* | Ecofriendly-antibacterial food packaging | [37] |
| Polylactic acid (PLA) | Nanocellulose fibrils (carrot cellulose nano-fibrils (CCNF)) + Silver nanoparticles | Increased mechanical properties and hydrophilicity, antibacterial activity against *Escherichia coli* and *Bacillus cereus* | Antibacterial food packaging | [46] |
| Polymethyl methacrylate (PMMA) | Titanium dioxide nanotubes (TNTs) | Improved antimicrobial properties | Antimicrobial food packaging | [39] |
| Polycaprolactone (PCL) + Poly (L-lactic acid) | Phycocyanin/ polyvinyl alcohol (PVA) nanoparticles | Improved oxygen barrier properties. | Antioxidant food packaging | [40] |
| Polyethylene films (PE) | Zinc oxide | Antimicrobial activity against *Staphylococcus aureus*, *Escherichia coli*, increased barrier against UV and increased tensile strength | Antimicrobial food packaging | [44] |
| Polyhydroxy butyrate (PHB) + polycaprolactone (PCL) | Silica having N-halamine | Antibacterial activity against *Staphylococcus aureus*, *Escherichia coli.* | Antibacterial food packaging | [41] |

## 7.3 MECHANICAL PROPERTIES

The packaging material for some special purposes, e.g., juice bottles and cans require certain toughness, TS, young's modulus, and plasticity [3, 95, 112]. As a consequence, mechanical properties are given utmost importance in packaging technology. Several possible techniques and methods to improve the mechanical strength of food packaging are being studied by researchers all over the globe. In earlier century, it was noticed that, even with the tiny addition of nanoparticle (0–5% wt.), the flexural along with impact strength increased dramatically. This suggested that the morphological interaction between the nanofillers and the polymer matrix within the nanocomposite greatly influences the mechanical properties of the reinforced material [62]. As by incorporating them density of composite per unit volume increases the propagation of polymerization and the dynamics of cross-linking is swayed by the presence of nanofiller.

Considering the polymeric matrix being semi-crystalline in nature, several researchers proved change in morphology as soon as nanofillers are added. While if matrix was amorphous in nature, its interaction with nanoparticle enthalpy and entropic effect observed. Wang et al. [125, 127] demonstrated the increased modulus of PBAT/OMMT from 85 MPa to 156 MPa with the simultaneous rise in clay loading and the further increase in clay load did not significantly affect the young's modulus due to increase in crystallinity [35, 64]. Figure 7.2 shows the young's modulus of different polymers over the loading of ZnO nanoparticles. Salehi Vaziri et al. [100] showed that stretch of the composite is related with the fracture mechanism. Moreover, justified that because of the silica nanoparticles young's modulus of pure polystyrene (PS) increased [100]. An enhancement in the TS and Young modulus of PHBV-CS biopolymer blend with the increase in ZnO-Ag nanoparticle content to 5% wt./wt. due to even dispersion of nanofillers [21, 50, 115, 130].

Nanofillers tend to increase the entropy as a result of which they are repelled by polymer chains. Due to this, the polymeric chains around the fillers were weakened, leading to the formation of low-density regions [79]. Low density regions in turn lower the strength of the composite. The effect is worsened if the filler content increases as the inter-filler distance decreases and the low-density chains try to spread out lowering the elastic modulus of the composite. Filler-matrix enthalpic interactions such as covalent bonds and van der walls radiate stress transfer across the surface of composite. If these interactions are stronger, a group of fillers act as singe large filler and impart characteristic property to the composite viz., elastic strength as it is proportional to number of filler-filler interaction [5, 52, 67]. For as much as weak interactions are considered, the formation of aggregate lowers the overall strength of composite [14, 48, 61]. Figure 7.3 shows the influence of nanoparticles contributing to the overall enhancement of nanofiller.

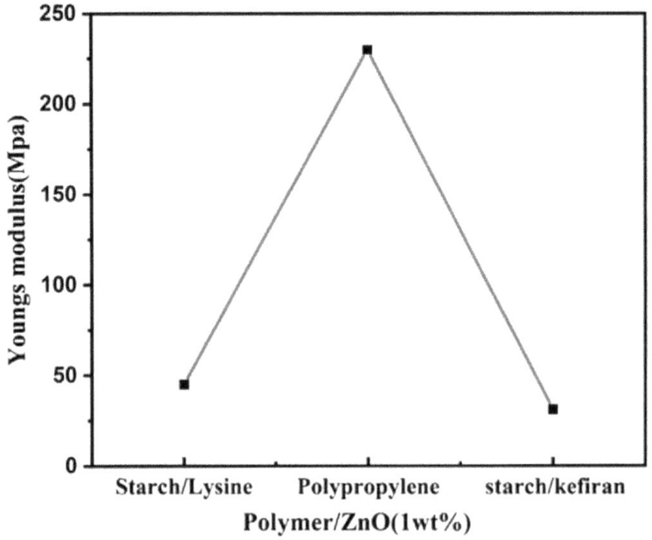


**FIGURE 7.2** Effect of ZnO nanoparticles on Youngs modulus.

Another important aspect for mechanical improvement was the surface-to-volume ratio. It was an indication of the interfacial area of the filler polymer matrix. This factor is influenced by the shape of the nanofiller. If the filler was spherical in shape, the surface-to-volume ratio is in direct proportion to volume fraction and is inversely related to the radius of the filler. This concludes that, with the decrease in the size of the nanoparticles, the interfacial area increases with the increase in volume fraction of filler due to the layer called 'interfacial layer' around the nanoparticles. And as the interfacial area increases, the chain interactions between the polymer and the nano-filler subsequently rises leading to an increase in stiffness [24, 110]. But after a certain value, it starts to decrease because of the transfer of residual stress [59]. It is also being seen that the viscosity increases as the size of the filler decreases. There is no such direct relationship but it is the interfacial area which influences the viscosity [2].

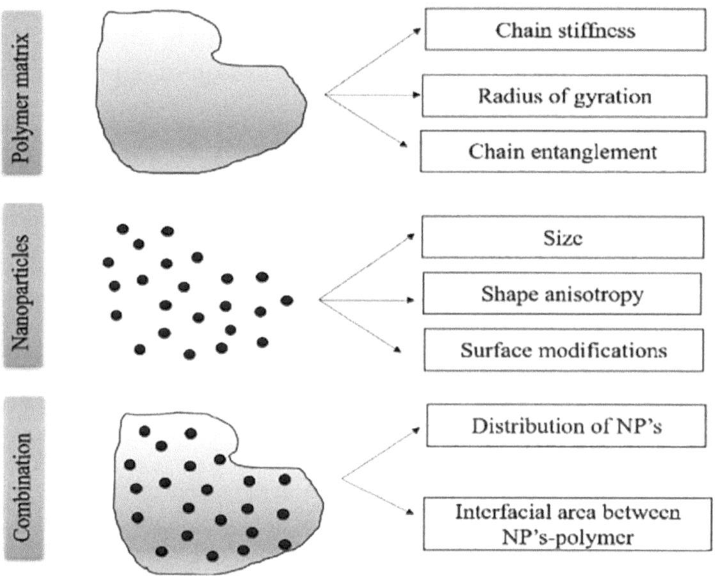


**FIGURE 7.3** Factors affecting mechanical properties of nanocomposites.

Shape of the filler is also equally important. Cylindrical fillers have greater surface-to-volume ratio than spherical filler. These fillers being anisotropic in nature, they can achieve better mechanical properties even with low loadings [52, 54, 98]. But, if cylinders are long, with the increase in aspect ratio, the filler dispersion into the matrix decreased. This limitation can be cured by modifying the filler using different strategies [27]. The effect of aspect ratio on nanocomposites overall properties can be better understood by considering nano-clays. With the high aspect ratio, the stiffness of composite material is also high [10, 97].

With the increase in volume fraction, and decrease in nanofiller size, fracture toughness is also increased [20, 52]. Numerous toughening mechanism on the basis of the energy dissipate they have been predicted to understand the matrix-filler interactions [41, 55, 90]. One of which is debonding in which separation of nanoparticles occur during the deformation process which may occur due to stress initiation and formation of voids [68, 101]. Another mechanism is crack deflection. Due to the presence of nanoparticles, acting as a hurdle in the path of crack propagation, the crack gets deflected following a torturous path. Therefore, the surface reflected rough and the toughness was increased. Other toughening mechanisms that are caused as a result of absorption of energy were void growth and shear banding.

In case of layered composites, particularly intercalated, gallery failure was an important mechanism that should be taken into consideration. This gives an explanation as to why exfoliation does not help in improving mechanical properties to that extent [10, 104]. EOB is another important mechanical property of polymer which affect due to the addition of nanofillers. Figure 7.4 shows the change in EOB of various polymer matrices due to the addition of (1 wt.%) ZnO nanoparticles.

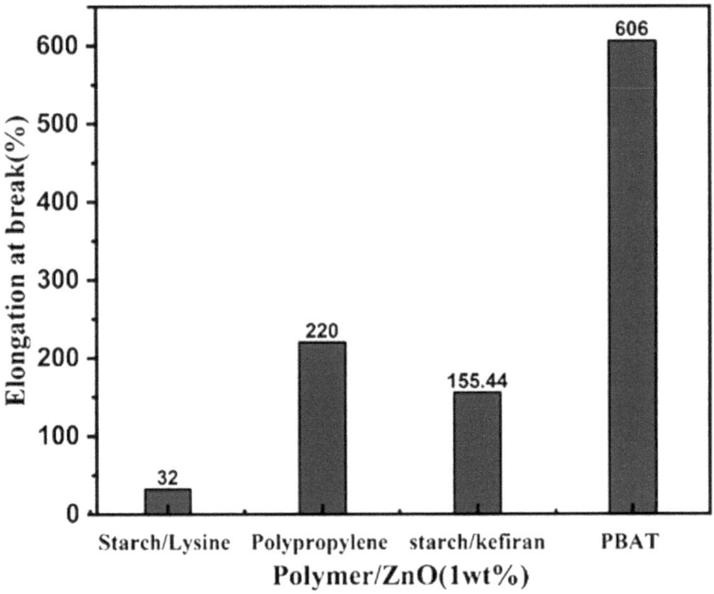


**FIGURE 7.4** Elongation at break vs. ZnO composites.

## 7.4 BARRIER PROPERTIES

The main concern of the food industry is an increment of shelf life of foodstuffs. This can easily be achieved if the barrier properties are strong of the packaging film. This would even lessen the production cost as the need for adding preservatives would vanish [21, 85, 88]. The food packaging technologist has always been seeking for the methods to enlarge the barrier fulfillment. Oxygen from the environment permeates through the packaging material and causes rancidity and unavoidable oxidation of the foodstuff leading to the loss of flavor as well as aroma. Water containing food such as juices, sauces, jams are more prone to oxidative deterioration. It was also reported to cause a loss of various nutrients such as vitamin C and can lead to the growth of microorganisms. Hence it is important to understand how actually oxygen permeability is affected with the use of nanofiller (Figure 7.5) [77]. Dry foods such as spices, powders, and snacks are easily damaged when in contact with water. Other aqueous foods such as juices and beverages should be protected so as not to evaporate. Hence along with water barrier, it is also important to achieve UV barrier as it could degrade essential nutrients, aroma, and flavors.

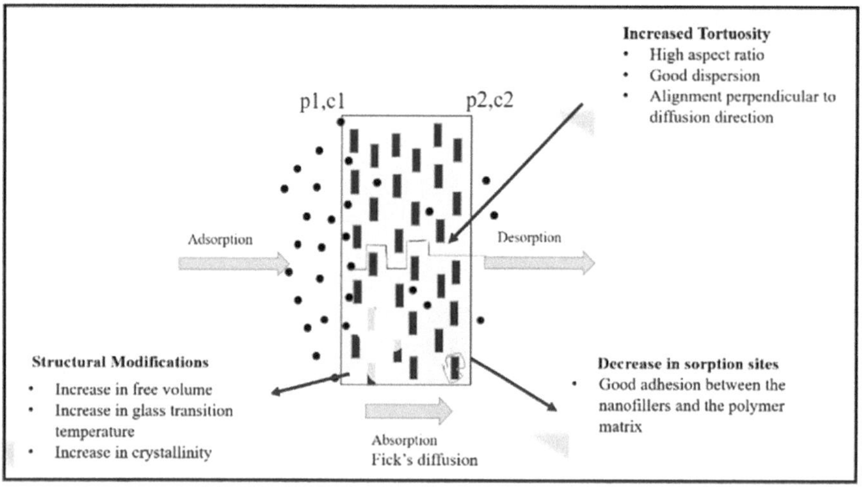


**FIGURE 7.5** Effect of nanofillers on barrier property: causes of decreased permeability.

Barrier properties are governed by diffusion in which the penetrant is adsorbed on the surface of film, and then absorbed into the film due to concentration difference and then finally desorbed out of the film [13, 25, 74]. Diffusion is based on Fick's law which states that the permeant travels on the account of concentration gradient across a film of a certain thickness. The flux is also triggered by the presence of pressure gradients and is directly related to permeability coefficient as follows:

$$J = p\frac{(p1 - p2)}{l} \tag{1}$$

The increased barrier properties are result of decrease in permeability. This was caused due to the decrease in diffusivity as well as solubility given as follows:

$$P = D \times S \tag{2}$$

where; P, D, S are the permeability, diffusivity, and solubility coefficients, respectively.

The $CO_2$ and $O_2$ are influenced majorly by diffusion coefficient following 'Fickian diffusion' whereas water vapor by solubility coefficient following 'non-Fickian diffusion.' This is because water exhibits hydrogen bonding and hydrophilic polymer matrices tend to increase the solubility coefficient. Hence, there is a need to systematically design system considering all these parameters.

Nanoparticles, due to their impermeable characteristics do not allow the gas to move. It leads to increase the tortuosity which has the direct relationship with the gas permeability. If they are exfoliated and oriented perpendicular to the solute gases, its performance is even multiplied. It was the main key challenge in this field of nanotechnology because they tend to align parallelly. The presence of filler results in the immobility of the side polymer chains close to nanofillers. It was often termed as the formation of interphase whose properties are distinct from that of the bulk. The interfacial region has a low diffusion coefficient due to low free volume known as, 'interfacial void.' The experiment results showed that due to the formation of soft regimes around the platelets, caused due to immobilization of polymer segments in the vicinity of filler, the free volume holes decreased as the volume fraction in the constrained region increases [66, 92].

A relationship in his model that clearly says that, as the volume fraction (φ) of the filler increases, the effective solubility ($S_e$) ultimately decreases. The equation can be stated as:

$$\frac{S_e}{S_p} = (1 - \varphi) \tag{3}$$

where; $S_p$ is the solubility of the penetrant [84].

Several experiments proved that the permeability of the composites decreases to a certain extent. After that, agglomeration effect dominates and the permeability increases. In most of the cases, 5% wt./wt. volume was considered to be ideal. The effect of shape on permeability of $O_2$, $CO_2$ and $H_2O$ are reviewed with the help of 100 different literature sources. It was observed that the relative permeability for nanoplatelets drops more readily than the spheres and cylinder/elongated fillers in spite of low volume fraction. This type of behavior may be attributed to torturous path and aspect ratio. The analysis further tailored that platelet less need the permeability to 10 folds.

The addition of iso-dimensional fillers does not affect the permeability. It was also proved that platelet/elongated filler when oriented perpendicular to the flux achieved least permeability. But at the same time, it is very challenging to attain a regular arrangement. Methods of synthesis may be useful for retaining them, e.g., solution intercalation and in-situ polymerization. Other factors contributing are reduction in free volume caused due to crystallinity due to increase in the glass transition temperature. Similarly, studied why is there nonmonotonic variation after certain volume fraction, particularly to 5% wt./wt. in most of the cases. This could be again attributed to the formation of agglomerates [99].

Two-dimensional platelet fillers owing to their high aspect ratio found to decrease the VWP and oxygen permeability to a great extent [96, 116]. It was very important to know and understand different morphologies of nanocomposites aggregated, intercalated, and exfoliated. In aggregated process, the tactoids of clay are dispersed well throughout the matrix. Polymer chains do not interfere inside the gallery of tactoids. This type of morphology is no longer nanocomposites [107]. While, in intercalation, the polymers chains interact with single layers, but the tactoids remain intact. Similarly, in exfoliation single layers are completely dispersed, and no trace of tactoids can be found. It was the most desired morphology as it increased the diffusion path due to high aspect ratio. After a certain volume fraction, agglomeration starts and hence the permeability reduces [58].

Recently, LbL (layer-by-layer) is being used to improve the barrier performance. The thin film with nanostructures is deposited precisely where exfoliation is not possible [46]. Various clays such as MMT, mica, vermiculite, owing to the presence of charges on its surface, are used extensively in LbL [160]. Due to the biodegradability, PLA have gained wide potential in food packaging as compared to petroleum-based transparent polymers. It has been used along with clay which showed almost 60% reduction in OTR, which monitored the effect of LbL process for such composites. Electrostatic force of attraction was developed with the help CS intercalated between MMT. The resulting composite exhibited 2-fold times reduced oxygen permeability and 20% reduced VWP than the pure polymer itself [108].

Many eminent scientists have come up with models which could more precisely manifest the governing equations of barrier properties of nanocomposites and bring off outcome closer to what is obtained experimentally. Nevertheless, all the models validated the experimental results that came to a same conclusion of: increased filler's aspect ratio, aligning the fillers perpendicular to the direction of diffusion, increasing the adhesion between the matrix and the filler as it is directly related with the crystallinity (free volume), modifying the surface of filler and improved dispersion of filler [28].

Pioneer of those was Neilson who first proposed the concept of tortuosity. It was assumed that the nanofillers are impermeable. Thus, the penetrating gases will travel through the slits between the fillers which followed a torturous path, thus increasing the diffusion path. As a result, permeability decreases. Model was often known as 'torturous model.' It was demonstrated that tortuosity is mostly influenced by the orientation patterns of the filler as well as its aspect ratio [126]. The effect of form and orientation on tortuosity was taken into account. Also, predicted equations for the same which lets us better understand various aspects and relationships between properties. According to equations, platelets which are placed perpendicular to the diffusion's direction, the tortuosity should increase with the increase in the filler volume fraction while the thickness of the fillers is decreased. The relationship can be given as;

$$f = \left(\frac{l_p}{2t_p}\right)\varphi \tag{4}$$

Similar expressions have been given for spheres as well as cylinder, which tells us how crucial aspect ratio and shape play in minimizing the barrier properties. This equation also indicates that if the length of the filler increases, the permeability substantially decreases. In some of the models, Neilson's model was modified with the introduction of order parameter which depicts information about the orientation of filler.

$$f = 1 + \frac{l_p}{3t_p}\varphi\left(S + \frac{1}{2}\right) \tag{5}$$

The direct effect of aspect ratio in model and took into account of shape along with orientation, type of alignment, effect of wiggles. Under the assumption of filler content being low ($\varphi \ll 1$) but they overlap on each oriented perpendicularly ($\varphi\alpha \gg 1$) [44, 84]. Tortuosity was given as:

$$f = 1 + \frac{\alpha^2\varphi^2}{4(1-\varphi)} \tag{6}$$

The accounts for the wiggling effect of filler which led to increased effective diffusion path, whereas tell us about increased filler volumes reducing effective cross-sectional area for diffusion. This was further revalidated by Aris who came up with a new form of Cussler's equation which even accounted for "necking" and resistance due to the silts in between the platelets [15, 30, 123].

All these models were predicted under the assumption that the filler molecules are oriented perpendicular to the direction of diffusion. They did not consider any other arrangement. A model with zein/bentonite nanocomposite observed through TEM image that due to the high aspect ratio and fillers volume fraction, casting led the filler to be parallel to the suspension and hence finally perpendicular to the direction of the diffusion. This, in turn found to decrease to permeability due to the decrease in solubility [23]. This result was totally contradictory when compared to other literatures, which covey that random orientation is prominent in increasing the barrier performance.

Later on, it is considered plasticization as well as solubility factors and manifested relative permeability in terms of aspect ratio, volume fraction, solubility factors, immobility factors and plasticizing factors [17]. The barrier properties of PBAT/OMMT nanocomposites that were well fitted against water vapor were tested [64] and similarly, relations were developed in terms of relative diffusivities [105, 128]. The process did not show dependence on diffusion coefficients. Hence, prepared an analytical model along with simulations that could also account for diffusion in oriented and arbitrarily oriented composites [42, 43, 63].

It is also worth noticing that permeability was unrestrained of the polymer type. It depends solely on the nature and type of filler used. But, not true in case of solubility for water permeation. If the polymer is hydrophilic, plasticization effects were acted upon and water diffusivity is increased. There exponential relationship between diffusion and water concentration. Various authors examined plasticization effects and found that these effects do not work to 4% wt./wt. of clay content and start to increased henceforth in case of intercalated morphology [8, 9]. If the polymer was hydrophobic nature, water clustering haven reported around the filler which has a positive effect on water permeability [36, 113].

$TiO_2$, $Fe_2O_3$, $SiO_2$ and $Al_2O_3$ spherical particles along with PLLA is a matrix to study the light and barrier properties. DSC analysis had shown that crystallization increases with the addition of nanoparticles. It was proved that $SiO_2$ and $Al_2O_3$ owing to their isotropic dispersion as compared to $Fe_2O_3$ and $TiO_2$ would perform better as a barrier shield. Although, it was seen that the hydrophobicity of PLLA was increased with the use of $TiO_2$, which indicated that $TiO_2$/PLLA performs best against water. This could be attributed to neutral zeta potential value. Others had either positive or negative zeta potential values. Water being amphoteric in nature interacts with these formed charges, thus decreasing hydrophobicity. But sometimes they may form water cluster and thus decrease permeability of water [69].

PVA/vermiculite composite exhibited exceptionally good oxygen barrier property, but increased moisture permeability. PVA exhibits hydrogen bonding with water forming clusters. It was obviously, decreased the water permeability. But this effect was reversible as water again can break its bond allowing $O_2$ to diffuse into the packaging film [7, 57]. Another cause affecting the water permeability was agglomeration of nanoparticles. It is mainly caused due to insufficiency in attaining dispersion of nanofillers and poor adhesive forces between the filler and the matrix. It was even influenced by the type of synthesis method, e.g., due to lack of shear forces in melt processing. Decrease in WVTR and WVP of PVA/starch/glycerol bionanocomposites with increase in halloysite nanotubes is usually observed. AFM (atomic force microscopy) analysis foreseen the formation of aggregates in the polymer system [4].

It was also the type of gas that determines the permeability characteristics of the nanocomposites. Owing to the kinetic diameters of gases, the one with greater kinetic diameter is less probable to permeate through the film. Amongst $CO_2$, $N_2$ and $O_2$, the kinetic diameter follows the order of $N_2 < O_2 < CO_2$. It means that $CO_2$ can easily pass through the composite due to availability of free volume followed by $O_2$ and $N_2$ respectively.

Amongst all the nanofillers, layered fillers are considered to be most effective against all sorts of barriers. Layered phyllosilicates from smectite group containing MMT, kaolinite, saponite, and hectorite were mostly used in food packaging, particularly MMT. MMT is a 2:1 type phyllosilicate clay with aluminum oxide ($Al_2O_3$) being placed in between silicon oxide ($SiO_2$) forming a platelet like structure. This gives MMT its characteristically high aspect ratio. Due to its high gas barrier performance, it was the most used clay nanofiller. The aluminum ions form negative charges on the surface of the filler, and these are balanced by cations such as $Na^+$ present in the gallery, which was seen to have a negative impact of gas permeation due to the formation of agglomerates. This limitation is cured by the filler surface modification as these cations have the capability to interchange with organic ions. It was often known as organic modification and hence the name, "organically modified MMT" most commonly used nowadays. It was the most crucial step in achieving homogeneous dispersion. In some cases, compatible agents were added to counterbalance the hydrophilicity of the clay with the hydrophobicity of the matrix.

The method of synthesis also affects the barrier properties. Three ways of synthesizing clay nanocomposites first was melted intercalated. It was simply involved heating the mixture of filler and the matrix to soften nature. It was the most commonly used method due to its feasibility and economically. In-situ polymerization suggested polymerization reaction to produce exfoliated platelets with the use of catalyst, heat or irradiation, monomer. This method was limited due to unavailability of suitable monomers. Solution intercalation deals with solvent; the clays are differentiated as single platelets and exfoliated. Its suspension was then adsorbed on polymer surface after mixing. After that, the solvent was evaporated. During this phase, changes in orientation are expected to happen.

## 7.5 ANTIMICROBIAL PROPERTIES

### *7.5.1 MECHANISM OF ACTION*

Food products such as fish and meat, poultry, milk products such as cheese, breads, and vegetables, fruits are more probable of microbial ingress. By incorporating antimicrobial agents, it inhibits foodborne pathogens which affect robustly on the foodstuff under packaging, thereby preventing deterioration and hence increasing the shelf life of it [86]. The concept of response to stimuli was broadened. The active substances such as antimicrobial agents are released when some sort of biological activity of microorganism was seen [78].

Due to the increasing demand of active antimicrobial food packaging, various scientists have investigated its diversity of materials, methods of synthesis, and mechanism of action. Mostly, metals such as silver, gold, cooper, iron, and some metal oxides such as ZnO, $TiO_2$, $SiO_2$, calcium oxide (CaO) are used for inducing antimicrobial action into the matrix due to the increased surface-to-volume ratio. They work on the basis of some of the other sort of mechanisms. Once these nanoparticles enter the cell, they inhibit the protein synthesis processes. This causes lack of nutrition to cells leading to apoptosis.

#### 7.5.1.1 SILVER (AG) NANOPARTICLES

Silver has been utterly used in food preservation. Even, the sweets were coated with silver layer to prevent its contamination against foodborne pathogens. It was also used in the medical appliances and equipment. Dentist use its coating over teeth to avoid tooth decay caused by microorganisms. This property of silver is also explored in food packaging. According to Ag's antibacterial performance it acts below 10 nanometers attributing to AgNPs. While, above 10 nm, antimicrobial action was due to silver ions [78]. AgNPs may easily attack and permeate through the cell membrane as compared to its ions. The actual possible antimicrobial mechanism routes are illustrated in the following diagram (Figure 7.6).

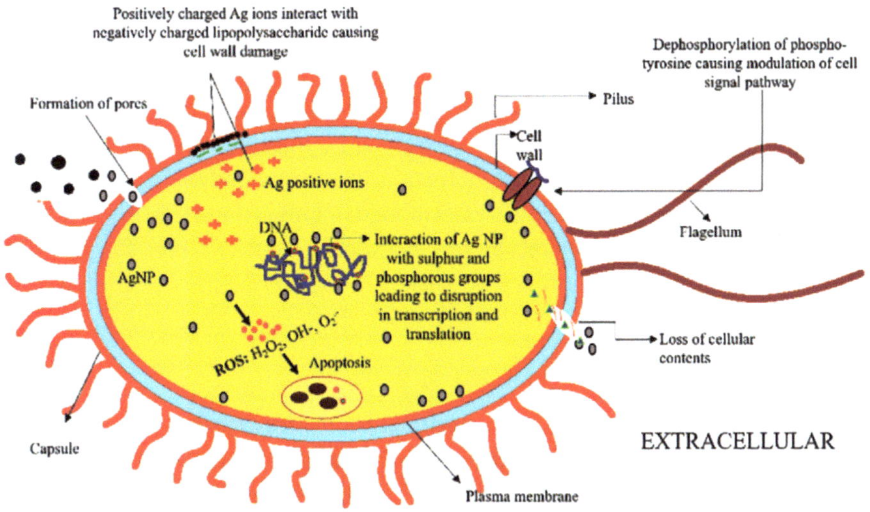


**FIGURE 7.6** Antimicrobial mechanism of action of silver nanoparticles inside bacterial cell.

The wholeness of the bacterial cell membrane is disturbed due to AgNPs and they are able to permeate through it. This causes the formation of pores within the cell wall which bring about leakage of cellular constituents and finally cell death. Due to the interference of AgNPs in the chain reactions (respiratory), the AgNPs combine with the sulfhydryl groups. This causes damage to proteins due to peroxidation of lipids as well as DNA due to oxidation. The phosphotyrosine are dephosphorylation stimulated due to the presence of AgNPs. Tyrosine phosphorylation was responsible for the formation of several enzymes such as RNA polymerase and UDP glucose dehydrogenase.

Dephosphorylation inhibits the formation of these enzymes responsible for cell's signaling pathway. Lack of nutrients to cell then ultimately leads to cell death. The nanoparticles tie up to the DNA containing phosphorous and sulfur causing disruption in the DNA's transcription and translation process. Thus, it causes damage to cell leading lack of cell division and to death. Under aerobic environment AgNPs release positively charged reactive $Ag^+$ ions which react with negatively charged cell membrane causing potential damage to it. The ions also interact negatively with other cell components. The presence of dissolved oxygen triggers the AgNPs to act as a catalyst and generate reactive oxygen species (ROS) such as superoxide radical ($O_2^-$), hydroxyl radical ($OH^-$) and hydrogen peroxide ($H_2O_2$). These radicals may react with and may cause damage to cell membrane and cellular organelles, DNA, proteins, thus impair the normal functioning of cells, leading to mutations and apoptosis.

Silver nanoparticles (AgNPs) acts more prominently on gram-negative bacterial membrane with because of thinner peptidoglycan layer than gram-positive bacteria as it has a thick cell wall. But some studies found this result contrary to their experiments. Formulated a bio-composite consisting of polyvinyl alcohol (PVA) electro-spunned nanofibers of silver for preservation of fruits such as lemon and strawberry. The composites exhibited better antimicrobial properties for gram-positive bacteria. This combination was best suited for *S. aureus* [31].

Anthocyanin rich purple corn extract, along with acting as a coloring agent in food was seen to have antimicrobial properties. Chitosan/Ag NP composite with purple corn extract. This composite exhibited an excellent barrier against UV light, $O_2$, and water. The TS was increased by 15.30% when compared with pure CS film and 11% is increased compared to Chitosan/AgNP film. It was proved excellent antibacterial property against *E. coli, L. monocytogenes, Salmonella, S. aureus* [89]. Several other researches explored the antimicrobial activity of silver [12, 93].

#### *7.5.1.2 TITANIUM DIOXIDE ($TIO_2$) NANOPARTICLES*

$TiO_2$ is a photocatalytic substance found mostly in toothpastes. Its anatase form with 3.2 eV is most commonly used, which produces highly reactive species. The electron-hole pairs are generated when $TiO_2$ is exposed to UV forming $H_2O_2$ and $OH^-$ radicles [76]. These radicles destroy the metabolic activities within the bacterial cell [87]. Some researches proved that the use of metals nanoparticles such as gold, silver, platinum along with $TiO_2$ can tremendously increase the antimicrobial performance even without the use of UV irradiation [76] and developed PBAT/$TiO_2$/Ag bio-composites film using solvent casting. At 5%wt./wt., the TS was elevated to 2 folds. EOB was also increased.

WVP and OTR also decreased to a large extent. PBAT and PBAT/$TiO_2$ did not show any antimicrobial action on *E. coli* and *S. aureus*. With the involvement of Ag, the colonies reduced significantly and this synergetic effect of Ag along the $TiO_2$ reinforced into PLA. The composites showed good inhibition against *L. monocytogenes* and *E. coli* [65]. Also adding $H_2O_2$ directly with $TiO_2$ ultimately triggers the photocatalytic reaction increasing the bacteriostatic effect [73, 114]. $TiO_2$, being a photocatalytic substance release hydroxyl radicles and reactive oxygen species (ROS) when are exposed to UV radiation. It causes the oxidation to the essentials of the cell such as phospholipids, hence damaging the cell membrane. Exploring the antimicrobial activity of modified $TiO_2$ with anatase and LDPE (low density polyethylene) through extrusion process, these composites were illuminated with UV light. The survival rate of the microorganisms such as *Pseudomonas* spp. and *Rhodotorula mucilaginosa* ceased significantly [18].

#### 7.5.1.3 ZINC OXIDE (ZNO) NANOPARTICLES

ZnO has a wide bandgap of 3.37 eV. This property allowed ZnO nanoparticles to be used in various industries such as pharmaceutical, energy, electronics as well as food packaging. Due to the presence of a positive charge, it acts as good on gram-negative bacteria. It was due the gram-negative bacteria's cell wall was thin when compared with gram-positive ones due to the presence of peptidoglycan. PLA/ZnO nanocomposites were investigated with increasing content of nanoparticles to 10 wt.%. The composites were seen to reduce *S. aureaus* and *E. coli* colonies. This action of ZnO was attributed to the oxidation action of $H_2O_2$ formed from ZnO. The $H_2O_2$ and ROS reacted with the cytoplasmic contents leading to cell death and after 3 wt.%, degradation in barrier and mechanical properties [56, 80].

ZnO doped Neodymium nanoparticles against *beta-lactamases*. It proved inhibition against *E. coli* and *Klebsiella pneumoniae* [47]. The films hold strong antibacterial activity against *Staphylococcus aureus* and *Escherichia coli* because of the increase in the nanoparticle content [1, 102]. It should also be noted that as the size of nanoparticle decreases, its biocidal activity elevates.

### 7.5.1.4 OTHER NANOPARTICLES

Copper is an essential metabolic constituent for all the living organisms. But, when its concentration increases beyond a certain limit, it may cause potential hazard. It has been utilized by people since ages for sterilizing water and for treating wounds, infections, and boils [38]. Its use has also fount crucial food packaging. Arfat et al. [11] studied bio-nanocomposite made up of gelatin derived from fish skin impregnated with Ag-Cu. UV barrier and TS increased with the decrease in EOB. Moreover, it showed excellent antibacterial property against *S. typhimurium* and *L. monocytogenes* [11].

Selenium nanoparticles exhibited an antimicrobial property due to the formation of ROS as in case of AgNPs thus causing damage to cell constituents. Selenium nanoparticles were found to be excellent free radical scavenger in recent studies. Reinforced selenium nanoparticles into an adhesive were further used in layers with LDPE and PET. The composites exhibited good antioxidant property until 1 month. No migration of selenium nanoparticles inside the food has been recorded. Such types of packaging can be useful for dry fruits, meats, ready-to-use veggie mixtures [120, 121].

The bactericidal effects of gold nanoparticles (AuNPs) on a variety of bacteria such as *Klebsiella pneumoniae, Pseudomonas aeruginosa, Bacillus subtilis,* and *Escherichia coli* were evaluated by researchers. Here, *Pseudomonas aeruginosa* was found to exhibit the highest inhibition [109]. The mechanism concerned with AuNPs deals with rendering the metabolism of cell due to reduction in ATP (adenine triphosphate) synthase and prevention of assembling of ribosomal subunit for tRNA (transport ribonucleic acid) [51]. Studies have found that AuNP's inhibits antibacterial action against *Klebsiella pneumoniae, Staphylococcus aureus, Escherichia coli,* and antifungal action against *Candida albicans, Aspergillus niger* and *Aspergillus flavus* [83, 86].

While synthesizing a hybrid composite of PSMA/$SiO_2$/$Al_2O_3$ with the use of the sol-gel method, the hybrid showed excellent effect against *Bacillus cereus* and *Escherichia coli* [91]. The synergetic effect of $SiO_2$ was also explored which hampered *P. aeruginosa, E. coli, S. aureus,* and *S. epidermidis* [58]. The antimicrobial properties of other nanoparticles are demonstrated in Table 7.2.

**TABLE 7.2** Antimicrobial Properties of Nanoparticles

| Nanoparticles | Antibacterial | Antifungal | Antiviral |
|---|---|---|---|
| Silver | *E. coli, Staphylococcus aureus, Staphylococcus epidermidis, Pseudomonas aeruginosa* | *Aspergillus niger, Candida albicans, Alternaria alternata, Penicillium brevicompactum* | HIV-1, hepatitis B virus |
| Copper | *E. coli, Staphylococcus aureaus, Salmonella typhimurium, Listeria monocytogenes* | *Candida species, Saprolegnia* sp., *Fusarium* sp. | Influenza A virus |
| Gold | *E. coli, Pseudomonas aeruginosa* | *Candida* species, *P. graminis* | Bluetongue virus |
| $TiO_2$ | *Pseudomonas aeruginosa, Bacillus subtilis, Enterobacter aerogenes,* | *Candida albicans, P. graminis* | Newcastle virus (NDV), entero-bacter aerogenes |
| ZnO | *Streptococcus mutans, Streptococcus pyogenes, Vibrio cholerae, Shigella flexneri,* and *Salmonella typhi, A. baumannii* | *B. cinerea, P. expansum., Colletotrichum gloeosporioides* | H1NI influenza virus, Chikanguniya virus |

Some essential oils such as cinnamaldehyde, clove oil, have also been exploited to its antimicrobial mechanism of action. The oils basically, in their protonated form come across high pH form and gets dissociated releasing charges in the form of protons and electrons. They cannot pass across the plasma membrane, thus damaging the bacterial cell [78]. Impregnated C30B clay (5% wt./wt.) with PLA along with thymol and cinnamaldehyde did not show much effect on mechanical, thermal, and barrier properties due to plasticizing effects. But it proved to be efficient against *E. coli* and *S. aureus* [122]. C30B was also found to show antibacterial effect against *L. monocytogenes* [94]. Carvacrol due to its volatility found limitations. MMT is an active carrier for carvacol with LDPE as a matrix. These composites were found to have excellent antibacterial action against *E. coli* along with increased thermal stability [78].

With the use of solution casting, prepared PBAT/$SiO_2$ composites and investigated its characteristics. It was found that the composite significantly improved its mechanical properties. TS increased from 33.75 MPa to 58.01 MPa. A similar trend was observed with EOB. OTR was reduced appreciably due to torturous path. It showed prominent inhibition against *E. coli* and

*S. aureus* [118]. The use of biopolymers had also been explored [103]. While impregnating copper nanoparticles with cellulose acetate, experimental data showed that the composite exhibited good antibacterial activity against a variety of bacteria such as *Staphylococcus aureus, Candida albicans, Pseudomonas aeruginosa, Aspergillus niger* at 2 wt.%. At higher loadings, its antimicrobial activity retards, but mechanical property improves [6]. $TiO_2$ due to its UV blocking capacity, prevents food from oxidation caused due to irradiation. This helps in preserving food with increase in its shelf life [29].

## 7.6 EFFECTS AND FUTURE PROSPECTS

### 7.6.1 EFFECTS OF NANOCOMPOSITES ON HUMAN HEALTH

Release of nanomaterials (NMs) from composites materials was very complex. Apart from protecting the packaging material, it can possess possible hazard to environment as well as to the health of human being. In defined controlled releasing packaging as a packaging which controls the release of certain substances over a period of times with the concern of safety, threats, and preservation of quality of food inside the packaging material. Achieving the controlled release of active packaging including aspects of barrier, mechanical, and antimicrobial is one of the most crucial challenges faced in this field of nanotechnology.

Migration assessment and toxicity analysis of any nanoparticles should be carried out before it comes to market. It depends on factors such as its size, its diffusivity and solubility into the polymer matrix, its volume fraction. Migration depends not only on nanoparticles, but also on food pH value, temperature, mechanical properties of the packaging film. Increased surface-to-volume was most probable for migration. The present legislation on nanoparticles does not fully cover the safety aspects of the nanoparticles. The EU, which is the central authority for nanoparticles have given risk assessment, quantity to be used in nanocomposites in their articles. A recent study showed that $TiO_2$ can even activate migration of ethylene glycol into the food causing threat to the consumer [39]. Therefore, proper designing of the material should be done, considering its risk factor on health and environment as well as achieving a balance between the microbial, $O_2$, $CO_2$, and $H_2O$ ingress [53].

### 7.6.2 EFFECT OF NANOCOMPOSITES ON ENVIRONMENT

Biodegradability is the biggest concern today due large-scale use of plastics. The materials that do not degrade remain as trash and settle into the lakes and oceans, contaminating the aquatic biodiversity. If the water is potable, excessive accumulation of plastic can also affect the public health and safety. Hydrophilic NMs could easily interact which water with a probability of migration affecting aquatic life. This is not fully understood in any of the literatures [60]. Hence more emphasis should be given on developing biodegradable polymers such as Starch, Cellulose, PHA, and PLA.

Apart from contributing to mechanical, thermal, antimicrobial, and barrier properties, the nano-reinforcements could affect negatively on biodegradation as well [49]. SEPS9 (sepiolites9) clay was incorporated into the PCL and PLA matrix. The reinforcements delayed the degradation process. It may be attributed to the presence of antimicrobial property of nanoparticles, which hampers the attack of microorganisms in the soil to slow the process of biodegradation process. The changes in crystallinity of the composites could also be a probable reason. But the exact eco-toxicological evaluation is not clear so far. It needs to be addressed properly. Also, the photochemical and other harmful reactions such as combustion and redox reactions when disposed in the soil should be evaluated and modeled are an urgent need of the food packaging industry [106]. Besides that, bio-based packaging must also ensure that the physical and barrier properties are restored.

## 7.7 SUMMARY

Redesigning conventional polymers into an improved nanocomposite is trending. Understanding its exact mechanism of action is very important as it renders us information about the variation of properties with a small change in the system. Moreover, much perfect analysis and models need to be predicted. In a nutshell, the size of nanoparticles greatly influences their morphological and physical properties. Decrement in size improved mechanical, barrier as well as antimicrobial properties. Besides, the shape of nanofillers helps in increased aspect ratio, which was most crucial for barrier properties. Layered and cylindrical nanofiller tend to decrease the diffusion process. An increase in the content of nanofillers was found to affect negatively on all the particles due to aggregation, crystallization, increase in sorption sites allowing ingress of gases and water. It also leads to the formation of cracks degrading the mechanical properties. The nature of nanoparticles is also important. Hydrophilic nature attracts water permeation. The nature of the polymer matrix is equally important. The migration of optimized nanofillers is important for exhibiting its antibacterial action, but if exceeded above a certain limit, it may cause hazards to the environment and human health. Thorough eco-toxicological should be done to know its exact action. Unfortunately, very little literature addressed this issue due to the lack of availability of knowledge in this domain. Further challenges of nanosafety for health must be investigated for worthwhile applications of nanocomposites.

## KEYWORDS

- **antimicrobial agents**
- **calcium oxide**
- **food packaging**
- **inorganic filler**
- **nanocomposites**
- **nanoparticles**

## REFERENCES

1. Aahil, P. M., Thomas, P. S., & Thomas, S., (2012). Effect of organoclay on the gas barrier properties of natural rubber nanocomposites. *Polymer Composites, 33*(4), 524–531.
2. Abbas, M., Buntinx, M., Deferme, W., & Peeters, R., (2019). Biopolymer/ ZnO nanocomposites for packaging applications: A review of gas barrier and mechanical properties. *Nanomaterials, 9*(10), 1494.
3. Abd El-Aziz, M. E., Morsi, S. M. M., Salama, D. M., Abdel-Aziz, M. S., Abd, E. M. S., Shaaban, E. A., & Youssef, A. M., (2019). Preparation and characterization of chitosan/polyacrylic acid/copper nanocomposites and their impact on onion production. *International Journal of Biological Macromolecules, 123*, 856–865.
4. Abdullah, Z. W., Dong, Y., Han, N., & Liu, S., (2019). Water and gas barrier properties of polyvinyl alcohol (PVA)/starch (St)/ glycerol (Gl)/halloysite nanotube (HNt) bionanocomposite films: Experimental characterization and modelling approach. *Composites Part B: Engineering, 174*(5), 107033. doi: 10.1016/j.compositesb.2019.107033.
5. Abdulrazzaq, N. S., Jafarzadeh, K. T. S., Pourhajibagher, M., Behroozibakhsh, M., Masaeli, R., & Bahador, A., (2018). Evaluation of antimicrobial properties of conventional poly(methyl methacrylate) denture base resin materials containing

hydrothermally synthesized anatase $TiO_2$ nanotubes against cariogenic bacteria and candida albicans. *Iranian Journal of Pharmaceutical Research, 17*, 161–172.
6. Abou-Yousef, H., Saber, E., Abdel-Aziz, M. S., & Kamel, S., (2018). Efficient alternative of antimicrobial nanocomposites based on cellulose acetate/Cu-NPs. *Soft Materials, 16*(3), 141–150.
7. Al-Tayyar, N. A., Youssef, A. M., & Al-Hindi, R. R., (2020). Antimicrobial packaging efficiency of ZnO-$SiO_2$ nanocomposites infused into PVA/CS film for enhancing the shelf life of food products. *Food Packaging and Shelf Life, 25*, 100523. doi: 10.1016/j.fpsl.2020.100523.
8. Alexandre, B., Langevin, D., Médéric, P., Aubry, T., Couderc, H., Nguyen, Q. T., Saiter, A., & Marais, S., (2009). Water barrier properties of polyamide 12/montmorillonite nanocomposite membranes: Structure and volume fraction effects. *Journal of Membrane Science, 328*(1, 2), 186–204.
9. Alix, S., Follain, N., Tenn, N., Alexandre, B., Bourbigot, S., Soulestin, J., & Marais, S., (2012). Effect of highly exfoliated and oriented organoclays on the barrier properties of polyamide 6 based nanocomposites. *The Journal of Physical Chemistry C, 116*(8), 4937–4947.
10. Almasi, A., Silani, M., Talebi, H., & Rabczuk, T., (2015). Stochastic analysis of the interphase effects on the mechanical properties of clay/epoxy nanocomposites. *Composite Structures, 133,* 1302–1312.
11. Arfat, Y. A., Ahmed, J., Hiremath, N., Auras, R., & Joseph, A., (2017). Thermo-mechanical, rheological, structural and antimicrobial properties of bionanocomposite films based on fish skin gelatin and silver-copper nanoparticles. *Food Hydrocolloids, 62,* 191–202.
12. Arfat, Y. A., Ahmed, J., & Jacob, H., (2017). Preparation and characterization of agar-based nanocomposite films reinforced with bimetallic (Ag-Cu) alloy nanoparticles. *Carbohydrate Polymers, 155*, 382–390.
13. Aris, R., (1986). On a problem in hindered diffusion. *Archive for Rational Mechanics and Analysis, 95*(2), 83–91.
14. Bansal, A., Yang, H., Li, C., Cho, K., Benicewicz, B. C., Kumar, S. K., & Schadler, L. S., (2005). Quantitative equivalence between polymer nanocomposites and thin polymer films. *Nature Materials, 4*(9), 693–698.
15. Baschetti, M. G., Doghieri, F., & Sarti, G. C., (2001). Solubility in glassy polymers: correlations through the nonequilibrium lattice fluid model. *Industrial & Engineering Chemistry Research, 40*(14), 3027–3037.
16. Ben, D. F., Dil, E. J., Tabatabaei, S. H., Mighri, F., & Ajji, A., (2019). Effect of nanoclay orientation on oxygen barrier properties of LBL nanocomposite coated films. *RSC Advances, 9*(3), 1632–1641.
17. Bharadwaj, R. K., (2001). Modeling the barrier properties of polymer-layered silicate nanocomposites. *Macromolecules, 34*(26), 9189–9192.
18. Bodaghi, H., Mostofi, Y., Oromiehie, A., Zamani, Z., Ghanbarzadeh, B., Costa, C., Conte, A., & Del, N. M. A., (2013). Evaluation of the photocatalytic antimicrobial effects of a $TiO_2$ nanocomposite food packaging film by *in vitro* and *in vivo* tests. *LWT - Food Science and Technology, 50*(2), 702–706.
19. Brune, D., & Bicerano, J., (2002). Micromechanics of nanocomposites: Comparison of tensile and compressive elastic moduli, and prediction of effects of incomplete exfoliation and imperfect alignment on modulus. *Polymer, 43,* 369–387.

20. Burgos-Mármol, J. J., & Patti, A., (2017). Unveiling the impact of nanoparticle size dispersity on the behavior of polymer nanocomposites. *Polymer, 113*, 92–104.
21. Cabedo, L., & Gamez-Pérez, J., (2018). Inorganic-based nanostructures and their use in food packaging. Chapter 2. In: Cerqueira, M. A. P. R., Lagaron, J. M., Castro, L. M. P., & Soares, A. A. M. O., (eds.), *Nanomaterials for Food Packaging* (pp. 13–45). Cambridge – USA: Elsevier Publishing.
22. Cameron, S. J., Hosseinian, F., & Willmore, W. G., (2018). A current overview of the biological and cellular effects of nanosilver. *International Journal of Molecular Sciences, 19*(7), 2030. doi: 10.3390/ijms19072030.
23. Cerisuelo, J. P., Gavara, R., & Hernández-Muñoz, P., (2015). Diffusion modeling in polymer-clay nanocomposites for food packaging applications through finite element analysis of TEM images. *Journal of Membrane Science, 482*, 92–102.
24. Cheng, S., Xie, S. J., Carrillo, J. Y., Carroll, B., Martin, H., Cao, P. F., Dadmun, M. D., et al., (2017). Big effect of small nanoparticles: A shift in paradigm for polymer nanocomposites. *ACS Nano, 11*(1), 752–759.
25. Choudalakis, G., & Gotsis, A. D., (2009). Permeability of polymer/clay nanocomposites: A review. *European Polymer Journal, 45*(4), 967–984.
26. Cooksey, K., (2006). *Food Packaging, Principles and Practices* (2nd edn., p. 550). Gordon Robertson; CRC, Taylor and Francis; Boca Raton, FL.
27. Crosby, A. J., & Lee, J. Y., (2007). Polymer nanocomposites: The "nano" effect on mechanical properties. *Polymer Reviews, 47*(2), 217–229.
28. Cui, Y., Kumar, S., Rao, K. B., & Van, H. D., (2015). Gas barrier properties of polymer/clay nanocomposites. *RSC Advances, 5*(78), 63669–63690.
29. Duncan, T. V., (2011). Applications of nanotechnology in food packaging and food safety: barrier materials, antimicrobials and sensors. *Journal of Colloid and Interface Science, 363*(1), 1–24.
30. Duran, N., Duran, M., De Jesus, M. B., Seabra, A. B., Favaro, W. J., & Nakazato, G., (2016). Silver nanoparticles: A new view on mechanistic aspects on antimicrobial activity. *Nanomedicine, 12*(3), 789–799.
31. Kowsalya, E., Mosa Christas, K., Balashanmugam, P., & Rani, J. C. (2019). Biocompatible silver nanoparticles/poly(vinyl alcohol) electrospun nanofibers for potential antimicrobial food packaging applications. *Food Packaging and Shelf Life, 21*, 100379. doi: 10.1016/j.fpsl.2019.100379.
32. Enescu, D., Cerqueira, M. A., Fucinos, P., & Pastrana, L. M., (2019). Recent advances and challenges on applications of nanotechnology in food packaging. a literature review. *Food and Chemical Toxicology, 134*, 110814. doi: 10.1016/j.fct.2019.110814.
33. Falla, W. R., Mulski, M., & Cussler, E. L., (1996). Estimating diffusion through flake-filled membranes. *Journal of Membrane Science, 119*(1), 129–138.
34. Faraz, M. I., Besseling, N. A. M., Korobko, A. V., & Picken, S. J., (2014). Structure-property relationships and modeling of the mechanical properties of a high-temperature resistant thermoset nanocomposite. *Composites Part B: Engineering, 56*, 9–14.
35. Fertig, R. S., & Garnich, M. R., (2004). Influence of constituent properties and microstructural parameters on the tensile modulus of a polymer/clay nanocomposite. *Composites Science and Technology, 64*(16), 2577–2588.
36. Follain, N., Belbekhouche, S., Bras, J., Siqueira, G., Marais, S., & Dufresne, A., (2013). Water transport properties of bio-nanocomposites reinforced by luffa cylindrica cellulose nanocrystals. *Journal of Membrane Science, 427*, 218–229.

37. Fornes, T. D., & Paul, D. R., (2003). Modeling Properties of nylon 6/clay nanocomposites using composite theories. *Polymer, 44*(17), 4993–5013.
38. Fortunati, E., Puglia, D., Armentano, I., Valdés, A., Ramos, M., Juárez, N., Garrigós, M. C., & Kenny, J. M., (2017). Multifunctional antimicrobial nanocomposites for food packaging applications. Chapter 8. In: Grumezescu, A. M., (ed.), *Food Preservation Nanotechnology in the Agri-Food Industry* (pp. 265–303). Cambridge – USA: Academic Press, Elsevier Publishing.
39. Garcia, C. V., Shin, G. H., & Kim, J. T., (2018). Metal oxide-based nanocomposites in food packaging: Applications, migration, and regulations. *Trends in Food Science & Technology, 82*, 21–31.
40. Gill, Y. Q., Song, M., & Abid, U., (2019). Permeation characterization and modelling of polyethylene/clay nanocomposites for packaging. *Polymer Bulletin, 77*(7), 3749–3765.
41. Gómez-Del, R. T., Rodríguez, J., & Pearson, R. A., (2014). Compressive properties of nanoparticle modified epoxy resin at different strain rates. *Composites Part B: Engineering, 57*, 173–179.
42. Greco, A., (2014). Simulation and modeling of diffusion in oriented lamellar nanocomposites. *Computational Materials Science, 83*, 164–170.
43. Greco, A., & Maffezzoli, A., (2013). Two-dimensional and three-dimensional simulation of diffusion in nanocomposite with arbitrarily oriented lamellae. *Journal of Membrane Science, 442*, 238–244.
44. Gusev, A. A., & Lusti, H. R. J. A. M., (2001). Rational design of nanocomposites for barrier applications. *Advanced Materials, 13*(21), 1641–1643.
45. Guzman De, V. R., & Miravete, A., (2007). Mechanical model to evaluate the effect of the dispersion in nanocomposites. *Acta Materialia, 55*(9), 3025–3031.
46. Hagen, D. A., Foster, B., Stevens, B., & Grunlan, J. C., (2014). Shift-time polyelectrolyte multilayer assembly: Fast film growth and high gas barrier with fewer layers by adjusting deposition time. *ACS Macro Letters, 3*(7), 663–666.
47. Hameed, A. S., Karthikeyan, C., Ahamed, A. P., Thajuddin, N., Alharbi, N. S., Alharbi, S. A., & Ravi, G., (2016). *In vitro* antibacterial activity of ZnO and Nd-doped ZnO nanoparticles against ESBL producing *Escherichia Coli* and *Klebsiella Pneumoniae. Scientific Reports, 6*, 24312. doi: 10.1038/srep24312.
48. Harton, S. E., & Kumar, S. K., (2008). Mean-field theoretical analysis of brush-coated nanoparticle dispersion in polymer matrices. *Journal of Polymer Science Part B: Polymer Physics, 46*(4), 351–358.
49. He, Y., Fei, X., & Li, H., (2019). Carboxymethyl cellulose-based nanocomposites reinforced with montmorillonite and E-poly- L -lysine for antimicrobial active food packaging. *Journal of Applied Polymer Science, 137*(18), 48782. doi: 10.1002/app.48782.
50. Hetrick, E. M., Shin, J. H., Paul, H. S., & Schoenfisch, M. H., (2009). Anti-biofilm efficacy of nitric oxide-releasing silica nanoparticles. *Biomaterials, 30*(14), 2782–2789.
51. Hoseinnejad, M., Jafari, S. M., & Katouzian, I., (2018). Inorganic and metal nanoparticles and their antimicrobial activity in food packaging applications. *Critical Reviews in Microbiology, 44*(2), 161–181.
52. Hu, H., Onyebueke, L., & Abatan, A., (2010). Characterizing and modeling mechanical properties of nanocomposites-review and evaluation. *Journal of Minerals and Materials Characterization and Engineering, 9*(4), 275. doi: 10.4236/jmmce.2010.94022.

53. Huang, J. Y., Li, X., & Zhou, W., (2015). Safety assessment of nanocomposite for food packaging application. *Trends in Food Science & Technology, 45*(2), 187–199.
54. Hull, M. S., (2003). The file: Agency, authority, and autography in an Islamabad bureaucracy. *Language & Communication, 23*(3, 4), 287–314.
55. Ji, X. L., Jing, J. K., Jiang, W., & Jiang, B. Z., (2002). Tensile modulus of polymer nanocomposites. *Polymer Engineering and Science, 42*(5), 983–993.
56. Kim, I., Viswanathan, K., Kasi, G., Sadeghi, K., Thanakkasaranee, S., & Seo, J., (2019). Poly(lactic acid)/ZnO bionanocomposite films with positively charged ZnO as potential antimicrobial food packaging materials. *Polymers, 11*(9), 1427. doi: 10.3390/polym11091427.
57. Kim, J. M., Lee, M. H., Ko, J. A., Kang, D. H., Bae, H., & Park, H. J., (2018). Influence of food with high moisture content on oxygen barrier property of polyvinyl alcohol (PVA)/vermiculite nanocomposite coated multilayer packaging film. *Journal of Food Science, 83*(2), 349–357.
58. Koh, H. C., Park, J. S., Jeong, M. A., Hwang, H. Y., Hong, Y. T., Ha, S. Y., & Nam, S. Y., (2008). Preparation and gas permeation properties of biodegradable polymer/layered silicate nanocomposite membranes. *Desalination, 233*(1–3), 201–209.
59. Kontou, E., & Anthoulis, G., (2007). The effect of silica nanoparticles on the thermomechanical properties of polystyrene. *Journal of Applied Polymer Science, 105*(4), 1723–1731.
60. Koo, J. H., (2016). *Fundamentals, Properties, and Applications of Polymer Nanocomposites*. (1st edn., p. 716). Cambridge University Press.
61. Krishnamoorti, R., & Vaia, R. A., (2007). Polymer nanocomposites. *Journal of Polymer Science Part B: Polymer Physics, 45*(24), 3252–3256.
62. Kumar, S. K., & Krishnamoorti, R., (2010). Nanocomposites: Structure, phase behavior, and properties. *Annual Review of Chemical and Biomolecular Engineering, 1*, 37–58.
63. Li, C., Jiang, T., Wang, J., Peng, S., Wu, H., Shen, J., Guo, S., Zhang, X., & Harkin-Jones, E., (2018). Enhancing the oxygen-barrier properties of polylactide by tailoring the arrangement of crystalline lamellae. *ACS Sustainable Chemistry & Engineering, 6*(5), 6247–6255.
64. Li, J., Lai, L., Wu, L., Severtson, S. J., & Wang, W. J., (2018). Enhancement of water vapor barrier properties of biodegradable poly(butylene adipate-co-terephthalate) films with highly oriented organomontmorillonite. *ACS Sustainable Chemistry & Engineering, 6*(5), 6654–6662.
65. Li, W., Zhang, C., Chi, H., Li, L., Lan, T., Han, P., Chen, H., & Qin, Y., (2017). Development of antimicrobial packaging film made from poly(lactic acid) incorporating titanium dioxide and silver nanoparticles. *Molecules, 22*(7), 1170. doi: 10.3390/molecules22071170.
66. Liang, C. Y., Uchytil, P., Petrychkovych, R., Lai, Y. C., Friess, K., Sipek, M., Mohan, R. M., & Suen, S. Y., (2012). A comparison on gas separation between pes (polyethersulfone)/MMT (Na-Montmorillonite) and PES/$TiO_2$ mixed matrix membranes. *Separation and Purification Technology, 92*, 57–63.
67. Lin, X., Yin, M., Liu, Y., Li, L., Ren, X., Sun, Y., & Huang, T. S., (2018). Biodegradable polyhydroxybutyrate/poly-E-caprolactone fibrous membranes modified by silica composite hydrol for superhydrophobic and outstanding antibacterial application. *Journal of Industrial and Engineering Chemistry, 63*, 303–311.

68. Liu, H., & Brinson, L., (2008). Reinforcing efficiency of nanoparticles: A simple comparison for polymer nanocomposites. *Composites Science and Technology, 68*, 1502–1512.
69. Lizundia, E., Vilas, J. L., Sangroniz, A., & Etxeberria, A., (2017). Light and gas barrier properties of plla/metallic nanoparticles composite films. *European Polymer Journal, 91*, 10–20.
70. Malhotra, B., Keshwani, A., & Kharkwal, H., (2015). Antimicrobial food packaging: Potential and pitfalls. *Frontiers in Microbiology, 6*, 611. doi: 10.3389/fmicb.2015.00611.
71. Mania, S., Cieslik, M., Konzorski, M., Swiecikowski, P., Nelson, A., Banach, A., & Tylingo, R., (2020). The synergistic microbiological effects of industrial produced packaging polyethylene films incorporated with zinc nanoparticles. *Polymers (Basel), 12*(5), 1198. doi: 10.3390/polym12051198.
72. Marchiore, N. G., Manso, I. J., Kaufmann, K. C., Lemes, G. F., Pizolli, A. P. D. O., Droval, A. A., Bracht, L., et al., (2017). Migration evaluation of silver nanoparticles from antimicrobial edible coating to sausages. *LWT - Food Science and Technology, 76*, 203–208.
73. Meichtry, J. M., Dillert, R., Bahnemann, D. W., & Litter, M. I., (2015). Application of the stopped-flow technique to the $TiO_2$-heterogeneous photocatalysis of hexavalent chromium in aqueous suspensions: Comparison with $O_2$ and $H_2O_2$ as electron acceptors. *Langmuir, 31*(22), 6229–6236.
74. Mesbah, A., Zaïri, F., Boutaleb, S., Gloaguen, J. M., Naït-Abdelaziz, M., Xie, S., Boukharouba, T., & Lefebvre, J. M., (2009). Experimental characterization and modeling stiffness of polymer/clay nanocomposites within a hierarchical multiscale framework. *Journal of Applied Polymer Science, 114*(5), 3274–3291.
75. Minelli, M., Baschetti, M. G., Doghieri, F. J. J. O. M. S., (2009). Analysis of modeling results for barrier properties in ordered nanocomposite systems. *Journal of Membrane Science, 327*(1–2), 208–215.
76. Mohr, L. C., Capelezzo, A. P., Baretta, C. R. D. M., Martins, M. A. P. M., Fiori, M. A., & Mello, J. M. M., (2019). Titanium dioxide nanoparticles applied as ultraviolet radiation blocker in the polylactic acid biodegradable polymer. *Polymer Testing, 77*, 105867. doi: 10.1016/j.polymertesting.2019.04.014.
77. Molinaro, S., Cruz-Romero, M., Sensidoni, A., Morris, M., Lagazio, C., & Kerry, J. P., (2015). Combination of high-pressure treatment, mild heating and holding time effects as a means of improving the barrier properties of gelatin-based packaging films using response surface modeling. *Innovative Food Science & Emerging Technologies, 30*, 15–23.
78. Mousavi, K. A., Hashemi, S. M. B., & Limbo, S., (2018). Antimicrobial agents and packaging systems in antimicrobial active food packaging: An overview of approaches and interactions. *Food and Bioproducts Processing, 111*, 1–19.
79. Nair, S. B., Alummoottil, N. J., & Moothandasserry, S. S., (2017). Chitosan-konjac glucomannan-cassava starch-nanosilver composite films with moisture-resistant and antimicrobial properties for food-packaging applications. *Starch - Stärke, 69,* 1, 2.
80. Nunes, M. R., De Souza, M. C. M., De Lima, V. A. P., Da Rosa, C. G., Noronha, C. M., Maciel, M., & Barreto, P. M., (2018). Antioxidant and antimicrobial methylcellulose films containing *Lippia alba* extract and silver nanoparticles. *Carbohydrate Polymers, 192*, 37–43.

81. Odegard, G. M., Clancy, T. C., & Gates, T. S., (2005). Modeling of the mechanical properties of nanoparticle/polymer composites. *Polymer, 46*(2), 553–562.
82. Ortiz-Duarte, G., Pérez-Cabrera, L. E., Artés-Hernández, F., & Martínez-Hernández, G. B., (2019). Ag-chitosan nanocomposites in edible coatings affect the quality of fresh-cut melon. *Postharvest Biology and Technology, 147*, 174–184.
83. Pagno, C. H., Costa, T. M., De Menezes, E. W., Benvenutti, E. V., Hertz, P. F., Matte, C. R., Tosati, J. V., et al., (2015). Development of active biofilms of quinoa (*Chenopodium Quinoa* W.) starch-containing gold nanoparticles and evaluation of antimicrobial activity. *Food Chemistry, 173*, 755–762.
84. Picard, E., Vermogen, A., Gerard, J., & Espuche, E., (2007). Barrier properties of nylon 6-montmorillonite nanocomposite membranes prepared by melt blending: Influence of the clay content and dispersion stateconsequences on modeling. *Journal of Membrane Science, 292*(1, 2), 133–144.
85. Picken, S. J., Vlasveld, D. P. N., Bersee, H. E. N., Özdilek, C., & Mendes, E., (2008). Structure and mechanical properties of nanocomposites with rod- and plate-shaped nanoparticles. Chapter 4. In: Knauth, P., & Schoonman, J., (eds.), *Nanocomposites: Ionic Conducting Materials and Structural Spectroscopies* (Vol. 10, pp. 143–173). Berlin – Germany: Springer International Publishing.
86. Piruthiviraj, P., Margret, A., & Krishnamurthy, P. P., (2015). Gold nanoparticles synthesized by *Brassica oleracea* (broccoli) act as antimicrobial agents against human pathogenic bacteria and fungi. *Applied Nanoscience, 6*(4), 467–473.
87. Pleskova, S. N., Golubeva, I. S., & Verevkin, Y. K., (2016). Bactericidal activity of titanium dioxide ultraviolet-induced films. *Materials Science and Engineering C: Materials for Biological Applications, 59*, 807–817.
88. Qiao, R., & Catherine, B. L., (2009). Simulation of interphase percolation and gradients in polymer nanocomposites. *Composites Science and Technology, 69*(3, 4), 491–499.
89. Qin, Y., Liu, Y., Yuan, L., Yong, H., & Liu, J., (2019). Preparation and characterization of antioxidant, antimicrobial and pH-sensitive films based on chitosan, silver nanoparticles and purple corn extract. *Food Hydrocolloids, 96*, 102–111.
90. Quaresimin, M., Schulte, K., Zappalorto, M., & Chandrasekaran, S., (2016). Toughening mechanisms in polymer nanocomposites: From experiments to modeling. *Composites Science and Technology, 123*, 187–204.
91. Ramesh, S., Sivasamy, A., Rhee, K. Y., Park, S. J., & Hui, D., (2015). Preparation and characterization of maleimide-polystyrene/$SiO_2$-$Al_2O_3$ hybrid nanocomposites by an in situ sol-gel process and its antimicrobial activity. *Composites Part B: Engineering, 75*, 167–175.
92. Rath, S. K., Sudarshan, K., Bhavsar, R. S., Kharul, U. K., Pujari, P. K., Patri, M., & Khakhar, D. V., (2016). Characterizing the nanoclay induced constrained amorphous region in model segmented polyurethane-urea/clay nanocomposites and its implications on gas barrier properties. *Physical Chemistry Chemical Physics, 18*(3), 1487–1499.
93. Regiel-Futyra, A., Kus-Liskiewicz, M., Sebastian, V., Irusta, S., Arruebo, M., Kyziol, A., & Stochel, G., (2017). Development of noncytotoxic silver-chitosan nanocomposites for efficient control of biofilm-forming microbes. *RSC Advances, 7*(83), 52398–52413.
94. Rhim, J. W., Lee, S. B., & Hong, S. I., (2011). Preparation and characterization of agar/clay nanocomposite films: The effect of clay type. *Journal of Food Science, 76*(3), N40–48.

95. Riahi, Z., Priyadarshi, R., Rhim, J. W., & Bagheri, R., (2021). Gelatin-based functional films integrated with grapefruit seed extract and $TiO_2$ for active food packaging applications. *Food Hydrocolloids, 112*, 106314.
96. Rokbani, H., Daigle, F., & Ajji, A., (2018). Combined effect of ultrasound stimulations and autoclaving on the enhancement of antibacterial activity of ZnO and $SiO_2$/ZnO nanoparticles. *Nanomaterials (Basel), 8*(3), 59–73.
97. Roy, S., Shankar, S., & Rhim, J. W., (2019). Melanin-mediated synthesis of silver nanoparticle and its use for the preparation of carrageenan-based antibacterial films. *Food Hydrocolloids, 88*, 237–246.
98. Saber-Samandari, S., & Afaghi-Khatibi, A., (2007). Evaluation of elastic modulus of polymer matrix nanocomposites. *Polymer Composites, 28*(3), 405–411.
99. Sadeghi, M., Mehdi, T. M., Ghalei, B., & Shafiei, M., (2013). Preparation, characterization, and gas permeation properties of a polycaprolactone based polyurethane-silica nanocomposite membrane. *Journal of Membrane Science, 427*, 21–29.
100. Salehi, V. H., Abadyan, M., Nouri, M., Omaraei, I. A., Sadredini, Z., & Ebrahimnia, M., (2011). Investigation of the fracture mechanism and mechanical properties of polystyrene/silica nanocomposite in various silica contents. *Journal of Materials Science, 46*(17), 5628–5638.
101. Schmatz, D. A., Costa, J. A. V., & Morais, M. G. D., (2019). A novel nanocomposite for food packaging developed by electrospinning and electrospraying. *Food Packaging and Shelf Life, 20*, 100314. doi: 10.1016/j.fpsl.2019.100314.
102. Shahmohammadi, J. F., & Almasi, H., (2016). Morphological, physical, antimicrobial and release properties of ZnO nanoparticles-loaded bacterial cellulose films. *Carbohydrate Polymers, 149*, 8–19.
103. Sharma, R., Jafari, S. M., & Sharma, S., (2020). Antimicrobial bio-nanocomposites and their potential applications in food packaging. *Food Control, 112*, 107086. doi: 10.1016/j.foodcont.2020.107086.
104. Shemesh, R., Goldman, D., Krepker, M., Danin-Poleg, Y., Kashi, Y., Vaxman, A., & Segal, E., (2015). Ldpe/clay/carvacrol nanocomposites with prolonged antimicrobial activity. *Journal of Applied Polymer Science, 132*, 41261. doi: 10.1002/app.41261.
105. Sorrentino, A., Tortora, M., & Vittoria, V., (2006). Diffusion behavior in polymer-clay nanocomposites. *Journal of Polymer Science Part B: Polymer Physics, 44*(2), 265–274.
106. Souza, V. G. L., & Fernando, A. L., (2016). Nanoparticles in food packaging: Biodegradability and potential migration to food: A Review. *Food Packaging and Shelf Life, 8*, 63–70.
107. Spencer, P. E., & Sweeney, J., (2009). Modelling of polymer clay nanocomposites for a multiscale approach. Chapter 15. In: Karger-Kocsis, J., & Fakirov, S., (eds.), *Nano- and Micromechanics of Polymer Blends and Composites* (pp. 545–578). Munich – Germany: Carl Hanser Verlag.
108. Svagan, A. J., Akesson, A., Cardenas, M., Bulut, S., Knudsen, J. C., Risbo, J., & Plackett, D., (2012). Transparent films based on PLA and montmorillonite with tunable oxygen barrier properties. *Biomacromolecules, 13*(2), 397–405.
109. Syed, B., Prasad, N. M. N., & Satish, S., (2016). Endogenic mediated synthesis of gold nanoparticles bearing bactericidal activity. *Journal of Microscopy and Ultrastructure, 4*(3), 162–166.
110. Szymanska-Chargot, M., Chylinska, M., Pieczywek, P. M., Walkiewicz, A., Pertile, G., Frac, M., Cieslak, K. J., & Zdunek, A., (2020). Evaluation of nanocomposite made of

polylactic acid and nanocellulose from carrot pomace modified with silver nanoparticles. *Polymers, 12*(4), 812; doi: 10.3390/polym12040812.

111. Tan, B., & Thomas, N. L., (2016). A review of the water barrier properties of polymer/clay and polymer/graphene nanocomposites. *Journal of Membrane Science, 514*, 595–612.
112. Teck, K. Y., Min, B., & Won, K. K., (2014). General characteristics of packaging materials for food system. Chapter 2. In: Han, J. H., (eds.), *Innovations in Food Packaging* (pp. 13–35). San Diego – California: Academic Press, Elsevier Publishing.
113. Tenn, N., Follain, N., Soulestin, J., Crétois, R., Bourbigot, S., & Marais, S., (2013). Effect of nanoclay hydration on barrier properties of PLA/montmorillonite based nanocomposites. *The Journal of Physical Chemistry C, 117*(23), 12117–12135.
114. Tian, F., Zhu, R., Song, K., Ouyang, F., & Cao, G., (2016). Synergistic photocatalytic degradation of phenol using precious metal supported titanium dioxide with hydrogen peroxide. *Environmental Engineering Science, 33*(3), 185–192.
115. Tsai, J., & Sun, C. T., (2016). Effect of platelet dispersion on the load transfer efficiency in nanoclay composites. *Journal of Composite Materials, 38*(7), 567–579.
116. Tsiantis, A., & Papathanasiou, T. D., (2018). An evaluation of models and computational approaches for the barrier properties of coatings containing flakes of high aspect ratio. *Journal of Coatings Technology and Research, 16*(2), 521–530.
117. Vanderroost, M., Ragaert, P., Devlieghere, F., & De Meulenaer, B., (2014). Intelligent food packaging: The next generation. *Trends in Food Science & Technology, 39*(1), 47–62.
118. Venkatesan, R., & Rajeswari, N., (2016). Preparation, mechanical and antimicrobial properties of $SiO_2$/poly(butylene adipate-co-terephthalate) films for active food packaging. *Silicon, 11*(5), 2233–2239.
119. Venkatesan, R., & Rajeswari, N., (2017). $TiO_2$ nanoparticles/poly(butylene adipate-co-terephthalate) bionanocomposite films for packaging applications. *Polymers for Advanced Technologies, 28*(12), 1699–1706.
120. Vera, P., Canellas, E., & Nerin, C., (2018). New antioxidant multilayer packaging with nanoselenium to enhance the shelf-life of market food products. *Nanomaterials, 8*(10), 837. doi: 10.3390/nano8100837.
121. Vera, P., Echegoyen, Y., Canellas, E., Nerín, C., Palomo, M., Madrid, Y., & Cámara, C., (2016). Nano selenium as antioxidant agent in a multilayer food packaging material. *Analytical and Bioanalytical Chemistry, 408*(24), 6659–6670.
122. Villegas, C., Arrieta, M. P., Rojas, A., Torres, A., Faba, S., Toledo, M. J., Gutierrez, M. A., et al., (2019). PLA/organoclay bionanocomposites impregnated with thymol and cinnamaldehyde by supercritical impregnation for active and sustainable food packaging. *Composites Part B: Engineering, 176,* 107336. doi: 10.1016/j.compositesb.2019.107336.
123. Wakeham, W. A., & Mason, E. A., (1979). Diffusion through multiperforate laminae. *Industrial & Engineering Chemistry Fundamentals, 18*(4), 301–305.
124. Wang, G. T., Liu, H. Y., Yu, Z. Z., & Mai, Y. W., (2008). Evaluation of methods for stiffness predictions of polymer/clay nanocomposites. *Journal of Reinforced Plastics and Composites, 28*(13), 1625–1649.
125. Wang, H. W., Zhou, H. W., Peng, R. D., & Mishnaevsky, L., (2011). Nanoreinforced polymer composites: 3D FEM modeling with effective interface concept. *Composites Science and Technology, 71*(7), 980–988.

126. Wang, J., & Pyrz, R., (2004). Prediction of the overall moduli of layered silicate-reinforced nanocomposites-part II: Analyses. *Composites Science and Technology, 64*(7, 8), 935–944.
127. Wang, Y., (2003). 2-D nano-scale finite element analysis of a polymer field. *Composites Science and Technology, 63*(11), 1581–1590.
128. Yadav, S., Mehrotra, G. K., & Dutta, P. K., (2021). Chitosan-based ZnO nanoparticles loaded gallic-acid films for active food packaging. *Food Chemistry, 334*, 127605. doi: 10.1016/j.foodchem.2020.127605.
129. Yildirim, S., Rocker, B., Pettersen, M. K., Nilsen-Nygaard, J., Ayhan, Z., Rutkaite, R., Radusin, T., et al., (2018). Active packaging applications for food. *Comprehensive Reviews in Food Science and Food Safety, 17*(1), 165–199.
130. Zare, M., Namratha, K., Ilyas, S., Hezam, A., Mathur, S., & Byrappa, K., (2019). Smart fortified PHBV-CS biopolymer with ZnO-Ag nanocomposites for enhanced shelf life of food packaging. *ACS Applied Materials and Interfaces, 11*(51), 48309–48320.

CHAPTER 8

# NANOFORMULATION: A PROMISING APPROACH TOWARDS FOOD FORTIFICATION

JYOTIKA DHANKHAR, PREETI KUNDU, and RITA MEHLA

## ABSTRACT

Nanoformulations are one of the latest delivery systems for food fortificants in form of micronutrients or non-nutrient bioactive components. The interest for food fortification has increased lately as internationally recognized organizations have claimed that nearly 2 billion people in the world are suffering from different micronutrient deficiency diseases. Food fortification is considered as effective and reasonable health strategy to alleviate the deficiency diseases as it involves the addition or supplementation of beneficial nutrients to the food matrix. Conventional fortification programs have mainly focused on the elimination of diseases with the use of a higher concentration of active ingredients for eradication of deficiency disease; however, nowadays, focus has shifted towards lowering dietary intakes and achieving higher bioavailability of nutrients. Hence, there is a paramount need for suitable delivery vehicle for active ingredients as their fortification pose many problems like physical or chemical instability, low bioavailability, incompatibility with food matrix and unpleasant taste. Nanoformulations are mostly synthesized using bottom-up or top-down process. Nanoformulations often address the challenge by encapsulating the fortificant in appropriate loading vehicle for fortification of desired food matrix. Utilization of nanoformulation-based delivery system with appropriate design may overcome the limitations associated with the fortification of food matrices. Effective design and fabrication of nanoformulation for delivery of the active ingredient in the food

Nanotechnology Horizons in Food Process Engineering, Volume 1. Megh R. Goyal, PhD, PE, Junaid Ahmad Malik, PhD, Satish Kumar, PhD & Ritesh B. Watharkar, PhD (Eds.)

system provides benefits like protection against biochemical and microbial degradation, retention of sensory attributes like texture, flavor, mouthfeel, and overall appearance; improved bioavailability and storage stability of fortified product.

## 8.1 INTRODUCTION

The food fortification approach involves the intentional addition of nutrient to a food system to alleviate the limitations associated with nutrient intake and related disorders, to stabilize the levels of nutrients within a diet, to reestablish the nutrients lost in processing operations of foods or to attract health-conscious consumers who are interested in healthy diet. The General Principles Codex, laid down with regard to addition of Essential Nutrients to Foods refers to the word fortification as incorporation of one or many important nutrients in a diet, for the improvement of established insufficiency of one or more nutrients, for the specific groups or the whole population in larger context [20]. The General Principles Codex recommends that the undertaking of such a fortification program for a particular group of people or entire population should be based on the display of particular symptoms or clinical evidence generated due to nutrient inadequacy in a regular diet.

Although food fortification program in past was aimed at mitigating the risk of nutrient deficiency diseases such as anemia, goiter, pellagra, beriberi, and rickets, the current focus is on achieving lower dietary intakes with higher bioavailability instead of treating epidemiological condition. The present approach for fortification programs requires the incorporation of fortification agent to achieve the desirable levels in staple foods, effective methods of monitoring, and latest technology for efficacious delivery.

Food fortification with nutrients and bioactive components is a complicated task because there is a requirement of suitable encapsulation system for fortifying agents that would retain the intact form of component, and uphold the organoleptic and physicochemical properties under different processing and storage conditions of food matrix [3, 18, 100]. Additionally, the micronutrient of interest selected for food fortification must remain bioavailable and unreactive within the food system. This condition often leads to major fortification problems as many nutrients are highly absorbable but have negative effects on sensory attributes while others are poorly absorbable with no adverse effect on the food system. Iron fortification is an example of such problem as among the different salts that are used for iron fortification, ferrous sulfate ($FeSO_4$) is well absorbable but results in off-flavor and unpleasant discoloration while other salts exhibit better stability with poor absorption rate [49].

Similarly, fat-soluble vitamins like vitamin A and E pose problems for food fortification as they lack stability under normal conditions of storage and processing. The vitamins are highly susceptible to destruction by sunlight, reactive oxygen species (ROS), free radicals and transition metals [26, 29, 91], and lack solubility in water making them unsuitable for incorporation in aqueous medium like beverages and drinks [66].

Over the past few years, novel technologies like nanoformulations have been explored for the efficient delivery of nutrients in different food matrices [102]. Smart nanostructured materials can transfer desired compounds to the desired organs in lowered doses and in a controlled release method to diminish the negative effects associated with conventional methods with higher dosage. In general, nanoformulations address the important issues faced with traditional fortification approaches for instance higher dosage, lower bioavailability, and toxicity. With the latest advance of nanoformulations systems, there is a wide range of nanomaterials (NMs) available for food fortifications like biomolecules-based micelles, liposomes, nanostructured micronutrients, nanoparticles, and nanoemulsions. Nanoformulation-based platforms are systems meant for delivery of nutrient compounds are composed of nanoformulations with submicron size ($d<500$ nm) and are mainly characterized by a high surface area to volume ratio.

This chapter deals with several methods of fortification of different micronutrients and vitamins. Some of the fortificants and methods of their incorporation in commonly used nanoformulations have also been discussed.

## 8.2 MICRONUTRIENTS AND VITAMINS FOR FORTIFICATION

Fortification is an approach that has long been implemented productively to correct nutrient deficiencies and their associated diseases. In recent years, different governmental bodies, regulatory agencies, and other related authorities have focused their interest more towards improvisation of health rather than dealing with deficiency diseases as in the past century [121]. It has been projected that approximately 2 billion people globally suffer from micronutrient deficiency which represents almost about 7.7% of total ailments [65].

### 8.2.1 IRON

The major part of the iron is present as hemoglobin in erythrocytes in human body for oxygen transportation to the tissues. Iron deficiency anemia is the most common dietary ailment that affects the general wellbeing of population across the world [51]. Myoglobin, an important protein found in the muscle tissue of all mammals, contains iron as its important constituent for oxygen supply to the muscles [28]. Furthermore, iron is an essential synthesis of hormones and the connective tissues involved in normal cellular activities [24]. The different forms in which iron exists in diet are heme and nonheme [23]. The vegetarian diets contain only nonheme iron, whereas non-vegetarian diets contain both heme and nonheme iron [6, 51].

Iron is usually lost in small amounts in humans through the gastrointestinal tract, urine, feces, and skin. A circulating peptide hormone, hepcidin, has been implicated in maintaining the iron levels in the body [117].

The recommended dietary allowance (RDA) varies from 8 mg/d to 11 mg/d for males while in case of females it varies from 8 mg/d to 18 mg/d depending upon the age group. The higher RDA is recommended for pregnant females to the levels of 27 mg/d, while in the case of vegetarians, the recommended intake is 1.8 times greater than for non-vegetarians. Compared to other milks, bioavailability of iron is higher in human milk quantities; however, amount is inadequate to meet the requirements of an infant [117]. Many developed countries like the United States and Canada, permit the iron fortification of different types of flour obtained from cereals [2]. In a study carried out for two years in South Africa, the occurrence of anemia in females decreased drastically upon consumption of the curry powder fortified with NaFeEDTA (sodium ferric ethylenediaminetetraacetate) [8]. While in another study carried out in China, the consumption of soy sauce supplemented with iron (as NaFeEDTA) was found to be effective in the alleviation of anemia due to iron in subjects [68].

### 8.2.2 CALCIUM

Calcium is an important mineral required for carrying out essential functions in the human body [27]. Calcium is naturally found in green leafy vegetables, milk, cheese, and other dairy products. Calcium improves bone health that has become a serious public concern nowadays and women, especially suffer from poor bone health post-menopause. Moreover, osteoporosis has also affected human population and symptoms include restricted absorption of calcium and vitamin D, a changing lifestyle, unhealthy food habits, and some genetic factors. Vitamin D is responsible for maintaining the levels of calcium in the body. Calcium deficiency may result into reduction in blood clot formation, weak bones and teeth, and reduced body activities [73]. The recommended intake concentrations for calcium range between 871 and 1,266 mg per day for men whereas 748 and 968 mg per day for females [21].

### 8.2.3 SELENIUM

Selenium (Se) is a trace element necessary for cellular function and is an important component of several enzymes. Selenium is a constituent of selenoprotein glutathione (GSH) peroxidase that helps in maintaining the reduced status of cell and oxidative damage due to free radicals is prevented. Moreover, it regulates thyroid function, immune system, and sperm motility [35, 102, 119]. Major dietary sources for selenium are foods from plants grown in selenium-rich soil [31]. Selenium deficiency is characterized by hypothyroidism, goiter, cretinism, and mental slowing. The recommended daily allowance for selenium is 60 μg per day for adults [33].

### 8.2.4 ZINC

Zinc is an essential trace mineral required for the normal functioning of immune system. It is involved in various cellular metabolic pathways and plays a role in the catalytic activity of many enzymes. It regulates the protein synthesis, cell multiplication, DNA synthesis and wound healing [83]. The natural food sources containing zinc are whole grains, beans, dairy products, nuts, poultry, and seafood. Zinc deficiency can cause delayed healing of wounds, mental lethargy, developmental delays, impaired immune functioning, and taste abnormalities [41, 69]. The average recommended dietary intake for zinc is 11 mcg/day and 8 mcg/day for men and woman, respectively.

### 8.2.5 VITAMIN A

Vitamin A is an important fat-soluble vitamin that helps in the modulation of cell growth, reproduction, and epithelial cell integrity. It helps in proper working of the immune system and maintenance of vision in human beings. Vitamin A requirements are commonly met by foods of animal origin like milk products, fish, meat, and foods of plant origin containing carotenoids that act as precursors of vitamin A [50, 89, 88]. The common aspects that cause vitamin A deficiency include substandard nutritional stature, a diet with insufficient vitamin A sources and infection of diseases like diarrhea and measles. Vitamin A deficiency affects the visual function and can cause night blindness and xerophthalmia [120]. The recommended daily allowance for vitamin A is 900 and 700 μg retinol activity equivalent/day for men and women respectively and tolerable upper intake levels should be less than 3,000 μg in adults to prevent toxicity [108].

### 8.2.6 VITAMIN D

Vitamin D is a fat-soluble vitamin that is produced endogenously due to influence of ultraviolet (UV) radiations of sunlight. It has a unique metabolism due to its synthesis in the skin. Vitamin D stimulates the absorption of calcium and minimizes the risk of osteoporosis. The deficiency of vitamin D is associated with the occurrence of skeletal diseases such as rickets in children and osteomalacia in adults. It is naturally present in mostly in seafoods like fatty fish and fish liver oils the best-known sources [21, 111]. Vitamin D is also present in egg yolks, beef liver, cheese, and some mushroom varieties in considerably low quantity [15]. RDA for vitamin D is normally fixed at a level of 10–20 μg or 400–800 IU.

### 8.2.7 FOLATE (VITAMIN $B_9$)

Folate is an essential water-soluble vitamin required for the metabolism of DNA and RNA required for growth and division. Folate plays a similar role like cobalamin in the production and metabolism of protein. Folate occurs naturally in a broad range of foods such as leafy green vegetables, fruits, grains, nuts, beans, eggs, peas, poultry, meat, seafood, and dairy products [64]. The main sources of folic acid are brussels sprouts, spinach, liver, yeast, and asparagus [16, 110]. The recommended daily allowance for folate is 400 mcg per day for adults.

### 8.2.8 COBALAMIN (VITAMIN $B_{12}$)

Cobalamin belongs to water-soluble category of vitamins and cobalt is an integral component of the vitamin. Cobalamin is produced by microbes and occurs in several chemical forms in animal foods including milk, fish, cheese, poultry, eggs, and meats [75]. The deficiency of cobalamin can lead to serious growth retardation in infants and young children [64]. Other deficiency disorders include improper functioning of the immune system, brain injury, megaloblastic anemia, and high level of homocysteine in blood which is a risk factor for heart diseases. Cobalamin dose recommended for daily intake is 2.4 mcg/day for people over 14 years.

## 8.3 DIFFERENT APPROACHES FOR SYNTHESIS OF NANOFORMULATIONS FOR FORTIFICATION

The nanoformulations generally display high bioavailability and good encapsulation efficiency. They present a useful mechanism for maintenance of structure and chemical composition of a wide variety of different nutrients while their colloidal stability as well as large size surface area enables better incorporation and delivery properties.

Owing to different methods of synthesis, nanoformulations can be subdivided into two main groups, namely, nanostructures that undergo the self-assembly processes (such as amphiphilic systems) and another category that are synthesized typically by energy-intensive methods like microfluidization, sonication, etc.

### 8.3.1 METHOD OF SYNTHESIS FOR NANOFORMULATIONS FOR MINERALS

Minerals in nanoformulated forms are manufactured by means of either top-down or bottom-up methods. In top-down processes, a solid substance is disintegrated into minor units, usually by use of high temperatures or by automated methods [57]. In general, the processes include (high-energy) high-pressure homogenization, microfluidization, ball milling, or grinding. For example, seashells were ground to generate nano-$CaCO_3$ powder [60]. The subdivision has also been carried out by dry cryo-nanonization [17], or by means of pulsed air-flow pulverizer [48]. Nanopowders manufactured by top-down techniques generally display broader particle size range and a non-uniform surface. Moreover, powders are more prone to contamination because of abrasion from milling balls or grinding wall. The synthesis of nanosized formulations using milling techniques is based on simple and conventional approaches which are energy-intensive and require longer milling durations, due to which it seems less suitable [104].

In the bottom-up approach, nanoformulations are synthesized from the aggregation of specific atoms in the liquid or gas phase. Studies have shown that the bottom-up approach is a better strategy for designing nanoformulations compared to top-down processes as they result in a more uniform size range or structure of nanoformulations [96]. For synthesis of nanoformulations, the solution or gas is brought to supersaturation, by variation of temperature or pH, or by addition of essential reactants [1]. The size range and structure of nanoformulations in formulations can be cautiously monitored by choosing suitable process conditions. Processes involving liquid phase like wet chemical precipitation have been utilized to synthesize nanoformulations of iron pyrophosphate [90, 113], Fe hydroxide [78, 82], Ca carbonate [9, 13], and hydroxyapatite [92]. However, for manufacture at extensive levels, liquid phase procedures are not suitable since huge quantity of liquid waste is created and many process steps are involved.

The techniques using gas phase (aerosol-based) approach like flame spray pyrolysis (FSP) or flame-assisted spray pyrolysis (FASP) [106] have also been used to synthesize nanoformulations containing iron, zinc, and/or calcium [44, 45, 55, 56, 81, 86, 124]. In the FSP/FASP process, acetylene-based support flame or methane gas causes atomization of organic metal compound into fine spray, and nanoformulations are formed by solution vaporization and crystallization followed by amalgamation, compressed liquefication, coalescence, and agglomeration of units [105]. These techniques are beneficial compared to fluid or solid form-based processes as they involve reduced effluents, simplified process factors, convenient to use, and good quality nanoformulations of increased purity are synthesized. Furthermore, FSP can be scaled up with high output [116]. Although, in certain cases, non-uniform powders can be formed [14, 101]. Spray drying has also been successfully used for the development of Ca phosphate nanoformulations with different Ca: P ratios [103].

### 8.3.2 SYNTHESIS METHOD FOR NANOFORMULATIONS FOR VITAMINS

In general, encapsulation of a fat-soluble vitamin in nanoformulation is typically done by emulsification in combination with spray drying. Nanoformulations usually with particle sizes <100 nm display increased bioavailability because of increased surface area, and higher shelf stability since coalescence is prevented by Brownian movement of the particles in motion [72]. Vitamin E nanoformulations have been synthesized by numerous passes through a microfluidizer which displayed more clarity and could be used in formulation of clear beverages [18]. Horn and Rieger [47] described the synthesis of nanohydrosols with higher β-carotene bioavailability in contrast to crystalline nanoformulations in milled forms.

Liposomes, or nanosized bilayer phospholipid vesicles, have been explored as potential candidates for delivery of vitamins as they have the potential to incorporate both hydrophobic and hydrophilic materials, and they are able to ensure protection, enhance bioavailability, ensure sustained-release, and increase storage stability. The exclusive properties of liposomes in nanoformulations make them a promising framework for efficient delivery of fortificants [54]. Vitamins such as cobalamin, α-tocopherol and ergocalciferol have been incorporated in liposomes using ultrasound-assisted technique and thin-film hydration method with encapsulation of 56%, 76%, and 57%, respectively [4].

Nanoformulations containing vitamin D with size range (< 200 nm) were synthesized by emulsification using medium-chain triglycerides (MCT) and Tween 80 with were assessed to be relatively stable at ambient temperatures and the use of emulsifier (sodium dodecyl sulfate) increase their thermal stability [37]. Nanoformulations containing vitamin E with a size range of about 30 nm could be synthesized with high-pressure homogenization using natural emulsifiers like Quillaja saponin, and lecithin [40]. In the studies carried out for fabricating nanoformulations for incorporation of vitamin D, different coating material like gum Arabic (468 nm), maltodextrin (266 nm), whey protein concentrate (190 nm) and soy protein isolate (132 nm) were used, the NF prepared with soy protein isolate revealed the smallest droplet size and highest encapsulation efficiency (85%) at 4% concentration by weight [97].

## 8.4 CHARACTERIZATION OF NANOFORMULATIONS

Various physical and chemical properties of nanoformulations have been extensively studied by different characterization techniques. The techniques including Fourier transform infrared spectroscopy (FTIR), scanning electron microscopy (SEM), transmission electron microscopy (TEM), X-ray diffraction (XRD), nuclear magnetic resonance (NMR) spectroscopy, UV-Vis spectroscopy (UV-Vis), Brunauer–Emmett–Teller (BET) and particle size analysis are of primary importance to reveal the morphological as well as structural characteristics of nanoformulations.

SEM is a commonly used techniques for morphological studies of nanoformulations. Morphology has a considerable influence over most of the properties of nanoformulations, so the characterization of morphological features always gains immense attention. SEM is based on the electron scanning principle and uses a focused high energy beam of electrons over a surface to generate an image. It provides information about the external morphology and chemical composition of nanoformulations [53]. However, sample preparation is cumbersome and could dehydrate sample leading to sample shrinkage and dried samples tend to aggregate, which presents additional artifacts.

TEM is a microscopy technique which operates on electron transmittance principle. In this method, a beam of electrons is passed through an ultrathin specimen or a suspension, and an image is produced due to interaction between electrons and sample. TEM reveals the internal fine and crystalline structure due to transmittance through samples and also gives information about two or more-layer materials [36, 53].

XRD is a widely used characterization technique which finds application in the analysis of structural characteristics of nanoformulations as it gives detailed information about the chemical composition, lattice parameters, crystallinity, and nature of phase of nanoformulations [76]. A rough idea regarding particle size can also be obtained by XRD via Debye Scherer formula for XRD measurement of broadening of the most intense peak of a particular sample [76]. However, it may be difficult to measure the structural parameters correctly in the case of amorphous materials and nanoformulations with a size below 3 nm as they can influence the XRD diffractogram [53].

NMR is a rapid analytical spectroscopic technique which uses the electromagnetic radiation in the radio wave range and is primarily utilized for the computational and physical analysis of the structure of formulated NMs. NMR is typically useful as a characterization technique for diamagnetic or antiferromagnetic nanoformulations and provides information on surface properties by analyzing the interaction between ligands and surface of these nanoformulations but is not convenient to characterize ferrimagnetic or ferromagnetic materials [67].

FTIR is among the tremendously used analytical techniques that help in determining the structural features of nanoformulations. It measures the absorption of electromagnetic radiation in the infrared region and the infrared absorption bands gives information about molecular structures and interactions [10]. In a study of NMs, FTIR spectra revealed the existence of the specific groups at the exterior of the nickel nanoformulations having a hexagonal crystal structure and also investigated the characteristic modes of these groups involved in bonding [109].

Being a simplistic and economical method, UV-Vis is tremendously utilized for studying the optical attributes of nanoscale materials. This characterization method relies on the measurement of the intensity of scattering and absorption of a beam of light after it passes through a sample. In this method, the intensity of light reflected from a reference material is compared with the intensity of reflected light of the sample. The optical characteristics of nanoformulations are easily affected by some factors like concentration, size, shape, and agglomeration state. Thus, UV-Vis is a remarkable tool to investigate the nanoscale materials [52].

BET is a simple, accurate, and rapid analysis technique for surface area characterization. It monitors the absorption of gas molecules on the surface of nanoscale materials and is broadly used to evaluate specific surface area of nanoformulations [67]. Surface area and porosity are critical factors that regulate the *in vivo* application of nanoparticles. Gas adsorption tests based on the Brunauer Emmett-Teller theory and Barrett-Joyner-Halenda analysis are used for evaluation of surface area and pore size, respectively. Nanoformulations with particles having large surface area and porosity are likely to adsorb nonspecific protein or could cover polymer on the exterior of these biomolecules, hence leaving them inadequate during *in vivo* experiments. Besides the shape and size of the particles in nanoformulations, the charge is also a significant parameter for formulating greatly stable nanosystems for effective encapsulation of fortifying agent.

Dynamic light scattering (DLS) is a standardized technique that can be used for particle size analysis of nanoformulations dispersed in colloidal solutions in nanometer range. DLS is based on the measurement of light scattering in a suspension and determines the mean particle size and size range distribution of nanoformulations. DLS is regarded as a facile and useful technique in which no multiple scattering effect occur because of relatively low nanoparticle concentration used [52]. DLS can be combined with zeta potential measurement which determines the stability of colloidal dispersions, thus providing a more complete characterization [7]. However, misinterpretation of results may take place due to nonuniformity in size and thus could lead to variations in size-based properties like charge density, surface area, etc. [46].

Polydispersity index (PDI) is a dimensionless parameter that is used to designate particle size range of nanoformulation based on molecular mass distribution of contributing particles. The range of PDI values varies between 0 and 1. In case nanodispersion displays low values close to 0, it represents particles of uniform particle size of monodisperse nature while a value of 1 depicts broad size distribution consisting of heterogeneous population of particles.

PDI signifies an important parameter in attaining perfect correlation between particle size and molecular weight distribution, which leads to determination of precise size range. In general, the procedure of synthesis of nanoformulation involves the complex procedures which cannot be precisely monitored and hence homogeneous population cannot be always expected. This leads to fabrication of nanoformulation with particles having a range of shapes and sizes, causing variation in different batches. Nevertheless, good optimization and close control over process parameters may result in monodisperse formulation. Moreover, preliminary formulation of sample components is crucial in carrying out DLS analysis as minor impurity with extraneous material like dust could lead to incorrect measurements [12].

## 8.5 STABILITY OF FOOD FORTIFICANTS IN NANOFORMULATIONS

The stability of nanoformulations is governed by the potential of engineered formulation to maintain their physicochemical attributes for a given storage period. The stability of the fortificant is crucial as they are incorporated into a nanoformulation through different engineering procedures. The incorporation and stability could be due to their arrangement within nanoformulation because the fortificants may not chemically change, but could physically vary their orientation during storage or under altered functional conditions encompassing light, polarity, pH, etc. In case, the nanoformulations have been engineered to act as delivery vehicle of fortificant, the overall understanding needs to be developed for the stability and interaction of the encapsulating material and fortifying agent incorporated within. For the purpose of synthesis of nanoformulation, an important issue to be addressed is fine-tuning of the fortificant/nanoformulation physicochemical properties to realize adequate stability as well as potential to shield the fortificant and permit it to perform, i.e., discharge the fortificant at target sites.

Besides being highly bioavailable, food fortificants should also not diminish the nutritional and organoleptic qualities of the target food matrix. Fortification compounds should not substantially reduce the shelf life of fortified food and should exhibit inertness towards assimilation of other nutrients along with the absence of toxicity. Some quick screening tests are required to examine the reactivity of forificants before adding them to food. The food matrices having large amount of polyphenols for example fruit yogurts and chocolate milk are problematic to fortify as they become susceptible to color changes in the proximity of iron mineral [79].

Santillan-Urquiza et al. [93] investigated the influence of fortification of inulin-coated nanoformulations on the physiological and rheological characteristics of yogurt throughout the storage period of 28 days. When fortifying liquid foods, evaluation of physical stability is also significant in addition to sensory stability. Van Leeuwen et al. [113] studied the effect of salt addition and dialysis on the stabilization of colloidal ferric pyrophosphate suspension and also explained the aggregation of these nanoformulations with regard to ionic strength and interaction potentials. Other than iron, fortification of calcium in food may also pose a challenge due to its requirement in high concentration which can develop chalky mouthfeel and texture. Moreover, the release of calcium ions alters the pH of fortified food [107].

Similarly, fortification with vitamins is also challenging in fulfilling the organoleptic attributes of a food matrix. The functionality of nanoformulation is significantly dependent on particle size. In addition to improvisation of sensory characteristics, particle size can also substantiate better mixing in dry powder applications. In the case of liquid food applications, for the production of opaque beverages, particles having more than 100 nm size are used while smaller particles that do not scatter light can be used in transparent products with low turbidity [18, 71, 87]. The color-changing attribute of carotenoids from yellow-orange-red is also a function of particle size that facilitates their use as natural food colorants [47].

The high surface area of nanoformulations results in high total surface energy, making the system thermodynamically unstable and hence agglomeration occurs *perse* in the system for minimization of surface energy [122]. Agglomeration may lead to destabilization of nanoformulations causing them to settle quickly or cream. In general, the issue is addressed by the incorporation of stabilizers to the formulation. For the purpose of application of stabilizers in nanosystem, the choice is dependent on the potential of stabilizers to allow wetting of the particle surface and present a hurdle for inhibition of nanoformulations to participate in agglomeration [34, 112].

The important phenomenon through which colloidal suspensions achieve stability in both hydrophilic and lipophilic systems: electrostatic repulsion and steric stabilization [84, 77]. The aforementioned phenomenon may be accomplished by incorporation of ionic and nonionic stabilizers into the nanosystem. The frequently added ionic surfactants in hydrophilic system are lecithin, sodium dodecyl sulfate, docusate sodium, and sodium lauryl sulfate [39]. The nonionic stabilizers used in a lipophilic system include Tween 80, polyethylene (PE) glycol and modified cellulose polymers.

## 8.6 TOXICITY OF NANOFORMULATIONS

With regard to the issue related to toxicity of nanoformulations, there has been as such no proof of adverse effects of administration of nanoformulations in the studies carried out till date, however, the long-standing effects on humans still need to be studied extensively [58]. The mechanisms that could lead to deleterious effects have been attributed to formation of toxic components like free radicals, ROS; ROS induced apoptosis and inflammatory substances with their genesis being related to changes in reagents and polymer substance used in synthesis of nanoformulations [19]. Since the data on toxicity of nanoformulations is inconsistent; more attention should be given to toxicological studies. Owing to different issues related to the safety of nanoformulations, reliable experiments should be carried out to understand the interaction between delivery vehicle and biological media where final release of fortificant takes place [114].

For the purpose of evaluating toxicity, different physicochemical attributes that are commonly investigated include particle size, size distribution, morphology, agglomeration, dispersibility, solubility, surface reactivity, surface area to volume ratio and physicochemical properties [5, 11]. The assessment of nanoformulations for probable toxicity is usually carried out using in vitro assays such as cell viability assay, inflammatory assay, and oxidative stress assay. The routine cell viability tests used to estimate the toxicity of nanoformulations are done using MTT [3-(4,5-dimethylthiazol-2-yl)-2,5-diphenyltetrazolium bromide] assays; LDH (lactate dehydrogenase) assay for assessing membrane integrity. Other parameters are tested by electron microscopy for analyzing intracellular fate of nanoparticles, immunotoxicity biomarkers for studying apoptosis. The inflammatory biomarkers have also been used to investigate the negative change that take place during the fabrication of nanoformulations include enzyme-linked immunosorbent assay (ELISA), chemokines IL-8, TNF-α, and IL-6 [61].

Comet assay is commonly used to assess the impact of oxidative stress on DNA that could have ensued by the reactive nanoformulations. The toxicity of synthesized nanoparticles has been demonstrated to result in DNA a strand break that is studied suitably by comet assay. By using the alkaline type of the comet assay, breaks in DNA strand and the sites available for alkali attacks are exposed and oxidatively impaired DNA is studied using the enzyme formamidopyrimidine glycosylase. Oxidative stress is also studied by monitoring the extent of lipid peroxidation and GSH levels. ROS have been studied to result in oxidative damage of cell membranes which can be assessed malondialdehyde or other thiobarbituric acid reactive substances. Such analytical methods have been comprehensively used to reveal the potential of NMs to initiate lipid peroxidation in cell membranes, in particular carcinoma cells [94].

The investigation of damage due to oxidative stress that could have ensued during engineering of nanoformulations is reliably done by monitoring levels of lipid peroxidation, DNA damage, protein carbonylation or substantial decrease in antioxidant activity assessed by levels of superoxide dismutase and GSH levels [25]. In the studies for nanotoxicology, it has been observed that depletion of levels of superoxide dismutase takes place, or GSH is oxidized by oxidative stress to result in the formation of oxidized glutathione (GSSG). The estimation of ratio of GSH and its disulfide oxidative product GSSG is either carried out by high performance liquid chromatography or by colorimetric tests. The colorimetric method involves evaluation of the concentration of GSH subsequent to reaction with 5,5'-dithio-bis(2-nitrobenzoic acid) (DTNB) [98].

Despite the number of studies presently being focused on understanding the mechanisms governing nanoformulation toxicity, yet the data generated to date has been inconsistent. There are many aspects that need to be monitored while assessing the toxicity of nanostructures like the steps involved from their synthesis to storage and from delivery point to final application point. The approaches currently being used to envision the probable toxic status have been under constant scrutiny and criticism by a segment of researchers. Since the toxicological studies are based on evidence generated through in vitro and in vivo experiments, within which cell assays and test animals are subjected to very high concentrations to evoke toxicity that mostly, lead to improbable results that cannot be used for meaningful interpretation for human usage as therapeutic doses are administered at very low levels [74]. Therefore, such kind of studies mostly leads to skeptical and uninterested consumer. Needless to say, there is a pressing need to rely on a uniform approach for evaluation of toxicity of nanoformulation based on long term studies as well as practical dosage with well-defined route of administration for reaching their full potential.

## 8.7 CHALLENGES AND ISSUES RELATED TO NANOFORMULATIONS OF FOOD FORTIFICANTS

Recently, much attention has been paid towards the application of nanosized mineral compounds as nutritional supplements for fortification of food. However, there is scanty information available about the fortune of nanoparticle after oral administration [70]. On incorporation into food matrices, nanoformulations may aggregate or dissolve depending upon their external environment and properties resulting in alteration of some morphological and structural features over time, which is often ignored in research [22, 71]. This can lead to a discrete exposure of nanomaterial before and after getting incorporated to food and subjected to the gut [96]. Therefore, characterization of nanomaterial is required to evaluate its properties by validated techniques. The complex nature of the food matrix makes it challenging to detect the nanoformulations [72, 106]. Hence, appropriate analytical techniques for example IR spectroscopy, UV-Vis, NMR, and XRD techniques are needed [59, 62, 123].

Some inert nanoformulations such as $TiO_2$ and $SiO_2$ can be harmful because their nondigestibility can result into accumulation of these NMs in the body [38, 42, 118]. But nanofortificants are designed to get fully solubilize in an acidic medium, thus minimizing the risk of accumulation. Even though, there are concerns about the partial solubility of some nanoformulations that could lead to oxidative damage [43]. In spite of the proven nontoxicity, pulmonary exposure to nanoformulations is not considered safe [115]. To reduce this, some processing steps like spray drying can be performed which ensure safe processing of nanoformulations without affecting their bioavailability [30].

Cost is also a major obstacle in the translation of academic research into industrial production. Food processing industry favors the use of minerals of low cost [49, 85]. Therefore, the production cost of nanosized fortificants should not dominate their advantages. Similar to minerals, synthesis, and fortification of vitamins should be economical and cost-effective. The demand for utilization of natural components and clean-label products is increasing. In addition, products should be labeled considering the cultural and health aspects, for example, "animal-free" or "allergen-free." However, industry is cautious in commercialization of nanoformulations as it is likely to entail labeling issues from regulatory bodies. Thus, researchers should sustain their efforts towards unfolding the advantages and safety of nanosized formulations. The legislative authorities must promote the nano-innovations and take essential steps in the direction of alliance between food corporations and academic world.

## 8.8 SUMMARY

The significance of balanced nutrition encompassing essential nutrients, for example, vitamins or minerals, in appropriate amount required for the health of humans and animals is undeniable. From the recent research, it may be established that vitamins and mineral in nanoformulations form have immense potential for food fortification. Furthermore, recently scientists have been emphasizing on growing importance of the intake of fortified food products to prevent deficiency diseases. For the purpose of regular consumption, fortification of food products (e.g., bread, biscuit, cake, butter, and yogurt) or beverages (e.g., milk, juice) with particular fortificant, the most suitable nanoformulation could be chosen and used. The fortification of food products with such nanoformulations ensures the stability of the active component in the formulation [80]. However, there are many issues concerning with respect to applications of fortificants in nanoformulations [32].

The safety issues essentially need to be addressed, since at present there are numerous lacunae in the concept of nanoparticle safety and toxicity. Nanoformulations have high bioavailability, as they carry large surface area, however, it could very well lead to toxicity in nanoform than observed in intact form. Furthermore, more information is necessary for understanding the interaction of nanofortified compounds with constituents of the food system and changes after consumption. Food corporations may be reluctant to use fortificants in nanoformulations since their cost, safety, and "nano-labeling" are challenging issues for their promotion in the market. Therefore, increased attention should be paid to more research and in-depth study with regard to their safety, risk factors associated with their uses, and probable adverse effects to humans with the application of sophisticated scientific studies.

## KEYWORDS


- **bioavailability**
- **deficiency diseases**
- **fortification**
- **medium chain triglycerides**
- **micronutrients**
- **nanoformulation**


## REFERENCES


1. Altavilla, C., & Ciliberto, E., (2011). Inorganic nanoparticles: Synthesis, applications, and perspectives: An overview. In: Altavilla, C., & Ciliberto, E., (eds.), *Inorganic Nanoparticles: Synthesis, Applications, and Perspectives*. Boca Raton – Florida: CRC Press.
2. Andrew, T. G., Huchzermeyer, K. D. A., Mbeha, B. C., Nengu, S. M., & Arthur, J. R., (2015). Scientific opinion on dietary reference values for iron. *EFSA Journal, 13*(10), 4254.
3. Augustin, M. A., & Sanguansri, L., (2015). Challenges and solutions to incorporation of nutraceuticals in foods. *Annual Review in Food Sciences and Technology, 6*, 1.1–1.15.
4. Babazadeh, A., Ghanbarzadeh, B., & Hamishehkar, H., (2017). Phosphatidylcholine-rutin complex as a potential nanocarrier for food applications. *Journal of Functional Foods, 33*, 134–141.
5. Bahadar, H., Maqbool, F., Niaz, K., & Abdollahi, M., (2016). Toxicity of nanoparticles and an overview of current experimental models. *Iranian Biomedical Journal., 20*(1), 1–11.
6. Baker, R. D., & Greer, F. R., (2010). Diagnosis and prevention of iron deficiency and iron-deficiency anemia in infants and young children (0–3 years of age). *Pediatrics, 126*, 1040–1050.
7. Baldassarre, F., Cacciola, M., & Ciccarella, G., (2015). A predictive model of iron oxide nanoparticles flocculation tuning Z-potential in aqueous environment for biological application. *Journal of Nanoparticle Research, 17*(9), 1–21.
8. Ballot, D. E., MacPhail, A. P., Bothwell, T. H., Gillooly, M., & Mayet, F. G., (1989). Fortification of curry powder with NaFe(111)EDTA in an iron-deficient population:

Report of a controlled iron-fortification trial. *The American Journal of Clinical Nutrition, 49*(1), 162–169.
9. Bentov, S., Weil, S., Glazer, L., Sagi, A., & Berman, A., (2010). Stabilization of amorphous calcium carbonate by phosphate-rich organic matrix proteins and by single phosphoamino acids. *Journal of Structural Biology, 171*, 207–215.
10. Blanco-Andujar, C., (2014). *Sodium Carbonate Mediated Synthesis of Iron Oxide NPs to Improve Magnetic Hyperthermia Efficiency and Induce Apoptosis*. Ph.D. Thesis; University College, London.
11. Borm, P. J., Robbins, D., Haubold, S., Kuhlbusch, T., Fissan, H., Donaldson, K., Schins, R., et al., (2006). The potential risks of nanomaterials: A review carried out for ECE-TOC. *Particle and Fibre Toxicology, 3*, 11.
12. Boyd, R. D., Pichaimuthu, S. K., & Cuenat, A., (2011). New approach to inter-technique comparisons for nanoparticle size measurements; using atomic force microscopy, nanoparticle tracking analysis and dynamic light scattering. *Colloids and Surfaces A: Physicochemical and Engineering Aspects, 387*(1–3), 35–42.
13. Boyjoo, Y., Pareek, V. K., & Liu, J., (2014). Synthesis of micro and nano-sized calcium carbonate particles and their applications. *Journal of Materials Chemistry, 2*, 14270–14288.
14. Buesser, B., & Pratsinis, S. E., (2012). Design of nanomaterial synthesis by aerosol processes. *Annual Review of Chemical and Biomolecular Engineering, 3*(3), 103–127.
15. Calvo, M. S., Whiting, S. J., & Barton, C. N., (2004). Vitamin D fortification in the United States and Canada: Current status and data needs. *The American Journal of Clinical Nutrition, 80*, 1710S–1716S.
16. Carmel, R., Green, R., Rosenblatt, D. S., & Watkins, D., (2003). Update on cobalamin, folate, and homocysteine. *Hematology, 2003*(1), 62–81.
17. Chen, B., McClements, D. J., & Decker, E. A., (2013). Design of foods with bioactive lipids for improved health. *Annual Review in Food Sciences and Technology, 4*, 35–56.
18. Chen, C. C., & Wagner, G., (2004). Vitamin E nanoparticle for beverage applications. *Chemical Engineering Research and Design, 82*, 1432–1437.
19. Chen, J., Xiu, Z., Lowry, G. V., & Alvarez, P. J., (2011). Effect of natural organic matter on toxicity and reactivity of nanoscale zero-valent iron. *Water Research, 45*(5), 1995–2001.
20. Codex Alimentarius Commission, (1987). *General Principles for the Addition of Essential Nutrients to Foods CAC/GL 09–1987 (Amended 1989, 1991)*. Rome, Joint FAO/WHO Food Standards Program, Codex Alimentarius Commission.
21. Coles, D., & Frewer, L. J., (2013). Nanotechnology applied to European food production–a review of ethical and regulatory issues. *Trends in Food Science & Technology, 34*(1), 32–43.
22. De Souza, S. L., Madalena, D. A., Pinheiro, A. C., Teixeira, J. A., Vicente, A. A., & Ramos, O. L., (2017). Micro- and nano bio-based delivery systems for food applications: *In vitro* behavior. *Advances in Colloid and Interface Science, 243*, 23–45.
23. Drakesmit, P., Yates, A. A., Schlicker, S., & Poos, M., (2001). Dietary reference intakes. *Journal of the American Dietetic Association, 101*(3), 294–301.
24. Drakesmith, H., & Prentice, A. M., (2012). Hepcidin and the iron-infection axis. *Science, 338*(6108), 768–772.
25. Dubey, A., Goswami, M., Yadav, K., & Chaudhary, D., (2015). Oxidative stress and nano-toxicity induced by $TiO_2$ and ZnO on WAG cell line. *PLoS One, 10*, e0127493.

26. El-Agamey, A., Lowe, G. M., Mcgarvey, D. J., Mortensen, A., Phillip, D. M., Truscott, T. G., & Young, A. J., (2004). Carotenoid radical chemistry and antioxidant/pro-oxidant properties. *Archives of Biochemistry and Biophysics, 430*, 37–48.
27. El-Shibiny, S., El-Gawad, M., Assem, F. M., & El-Sayed, S. M., (2018). The use of nano-sized eggshell powder for calcium fortification of cow's and buffalo's milk yogurts. *Acta Scientiarum Polonorum. Technologia Alimentaria, 17*(1), 37–49.
28. Erdman, J. W., MacDonald, A. I., & Steven, Z., (2012). *Present Knowledge in Nutrition* ($10^{th}$ edn., Vol. 33, pp. 506–520). New Jersey – USA: Wiley Blackwell.
29. Failloux, N., Bonnet, I., Perrier, E., & Baron, M. H., (2004). Effects of light, oxygen and concentration on vitamin A1. *Journal of Raman Spectroscopy, 35*, 140–147.
30. Faure, B., Sæderup, L. J., Wahlberg, M., Adkins, N., Jackson, P., & Bergstreom, L., (2010). Spray drying of $TiO_2$ Nanoparticles into redispersible granules. *Powder Technology, 203*, 384–388.
31. Fox, T. E., & Fairweather-Tait, S., (1999). Selenium. In: Hurrell, R. F., (ed.), *The Mineral Fortification of Foods* (pp. 112–153). Leatherhead, Surrey – England: Leatherhead Publishing.
32. Gallocchio, F., Belluco, S., & Ricci, A., (2015). Nanotechnology and food: Brief overview of the current scenario. *Procedia Food Science, 5*, 85–88.
33. Gao, J., Liu, Y., Huang, Y., Lin, Z. Q., Banuelos, G. S., & Lam, M. H. W., (2011). Daily selenium intake in a moderate selenium deficiency area of Suzhou, China. *Food Chemistry, 126*, 1088–1093.
34. Gao, L., Zhang, D., & Chen, M., (2008). Drug nanocrystals for the formulation of poorly soluble drugs and its application as a potential drug delivery system. *Journal of Nanoparticle Research, 10*, 845–862.
35. Gill, H., & Walker, G., (2008). Selenium, immune function and resistance to viral infections. *Nutrition and Dietetics, 65*, 41–47.
36. Goodhew, P. J., (2011). General introduction to transmission electron microscopy (TEM). Chapter 1. In: Brydson, R., (ed.), *Aberration-Corrected Analytical Transmission Electron Microscopy* (pp. 1–19). New Jersey – USA: John Wiley & Sons.
37. Guttoff, M., Saberi, A. H., & McClements, D. J., (2015). Formation of vitamin D nanoemulsion-based delivery systems by spontaneous emulsification: Factors Affecting particle size and stability. *Food Chemistry,* 117–122.
38. Hadrup, N., & Lam, H. R., (2014). Oral toxicity of silver ions, silver nanoparticles and colloidal silver: A review. *Regulatory Toxicology and Pharmacology, 68*, 1–7.
39. Hameed, M. A., Nazim, S., Khan, T., & Usman, M. R., (2013). A recent trends in enhancement of solubility and dissolution rate of poorly soluble hydrophobic drugs by using physical and chemical modifications. *Journal of Drug Discovery and Therapeutics, 1*, 13–24.
40. Hategekimana, J., Chamba, M. V. M., Shoemaker, C. F., Majeed, H., & Zhong, F., (2015). Vitamin E nanoemulsions by emulsion phase inversion: Effect of environmental stress and long-term storage on stability and degradation in different carrier oil types. *Colloids and Surfaces A: Physicochemical and Engineering Aspects, 483*, 70–80.
41. Heyneman, C. A., (1996). Zinc deficiency and taste disorders. *The Annals of Pharmacotherapy, 30*, 186–187.
42. Hillyer, J. F., & Albrecht, R. M., (2001). Gastrointestinal persorption and tissue distribution of differently sized colloidal gold nanoparticles. *Journal of Pharmaceutical Sciences, 90*, 1927–1936.

43. Hilty, F. M., Arnold, M., Hilbe, M., Teleki, A., Knijnenburg, J. T. N., Ehrensperger, F., Hurrell, R. F., et al., (2010). Iron from nano compounds containing iron and zinc is highly bioavailable in rats without tissue accumulation. *Nature Nanotechnology, 5*, 374–380.
44. Hilty, F. M., Knijnenburg, J. T. N., Teleki, A., Krumeich, F., Hurrell, R. F., Pratsinis, S. E., & Zimmermann, M. B., (2011). Incorporation of Mg and Ca into nanostructured $Fe_2O_3$ improves Fe solubility in dilute acid and sensory characteristics in foods. *Journal of Food Science, 76*, N2–N10.
45. Hilty, F. M., Teleki, A., Krumeich, F., Buchel, R., Hurrell, R. F., Pratsinis, S. E., & Zimmermann, M. B., (2009). Development and optimization of iron- and zinc-containing nanostructured powders for nutritional applications. *Nanotechnology, 20*, 475101.
46. Hoo, C. M., Starostin, N., West, P., & Mecartney, M. L., (2008). A comparison of atomic force microscopy (AFM) and dynamic light scattering (DLS) methods to characterize nanoparticle size distributions. *Journal of Nanoparticle Research, 10*, 89–96.
47. Horn, D., & Rieger, J., (2001). Organic nanoparticles in the aqueous phase - theory, experiment, and use. *Angewandte Chemie International Edition, 40*, 4331–4361.
48. Huang, S., Chen, J. C., Hsu, C. W., & Chang, W. H., (2009). Effects of nano calcium carbonate and nano calcium citrate on toxicity in ICR mice and on bone mineral density in an ovariectomized mice model. *Nanotechnology, 20*, 375102.
49. Hurrell, R. F., (2002). Fortification: Overcoming technical and practical barriers. *Journal of Nutrition, 132*, 806S–812S.
50. Johnson, E. J., & Russell, R. M., (2010). β-carotene. In: Coates, P. M., Betz, J. M., & Blackman, M. R., (eds.), *Encyclopedia of Dietary Supplements* (2nd edn., pp. 115–120). London -UK: Informa Healthcare.
51. Joosten, E., (2017). Iron deficiency anemia in older adults: A review. *Geriatrics and Gerontology International., 18*(3), 373–379.
52. Kato, H., (2012). Size determination of nanoparticles by dynamic light scattering. Chapter 8. In: Singh, S. C., Zeng, H., Guo, C., & Cai, W., (eds.), *Nanomaterials: Processing and Characterization with Lasers* (pp. 231–265). Weinheim – Germany: Wiley-VCH Verlag & Co.
53. Khan, I., Saeed, K., & Khan, I., (2019). Nanoparticles: Properties, applications and toxicities. *Arabian Journal of Chemistry, 12*, 908–931.
54. Khorasani, S., Danaei, M., & Mozafari, M. R., (2018). Nanoliposome technology for the food and nutraceutical industries. *Trends Food Science Technology, 79*, 106–115.
55. Knijnenburg, J. T. N., Hilty, F. M., Krumeich, F., Zimmermann, M. B., & Pratsinis, S. E., (2013). Multimineral nutritional supplements in a nano-CaO matrix. *Journal of Materials Research, 28*, 1129–1138.
56. Knijnenburg, J. T. N., Seristatidou, E., Hilty, F. M., Krumeich, F., & Deligiannakis, Y., (2014). Proton-promoted iron dissolution from nanoparticles and the influence by the local iron environment. *Journal of Physical Chemistry, 118*, 24072–24080.
57. Koch, C. C., (2003). Top-down synthesis of nanostructured materials: Mechanical and thermal processing methods. *Reviews on Advanced Materials Science, 5*, 91–99.
58. Kumari, A., Yadav, S. K., & Yadav, S. C., (2010). Biodegradable polymeric nanoparticles based drug delivery systems. *Colloids Surfaces B: Biointerfaces, 75*, 1–18.
59. Larue, C., Castillo, M. H., Stein, R. J., Fayard, B., Pouyet, E., Villanova, J., Magnin, V., et al., (2016). Innovative combination of spectroscopic techniques to reveal nanoparticle fate in a crop plant. *Spectrochimica Acta Part B: Atomic Spectroscopy, 119*, 17–24.

60. Lee, J. A., Kim, M. K., Kim, H. M., Lee, J. K., Jeong, J., Kim, Y. R., Oh, J. M., & Choi, S. J., (2015). The fate of calcium carbonate nanoparticles administered by oral route: Absorption and their interaction with biological matrices. *International Journal of Nanomedicine, 10*, 2273–2293.
61. Lequin, R. M., (2005). Enzyme immunoassay (EIA)/enzyme-linked immunosorbent assay (ELISA). *Clinical Chemistry, 51*, 2415–2418.
62. Li, Y. S., & Church, J. S., (2014). Raman spectroscopy in the analysis of food and pharmaceutical nanomaterials. *Journal of Food and Drug Analysis, 22*, 29–48.
63. Li, Z., Wang, Y., Shen, J., Liu, W., & Sun, X., (2014). The measurement system of nanoparticle size distribution from dynamic light scattering data. *Optics and Lasers in Engineering, 56*, 94–98.
64. Lindsay, A., De Benoist, B., Dary, O., & Hurrell, R., (2006). *Guidelines on Food Fortification with Micronutrients*. Geneva, Switzerland: WHO Library. http://www.unscn.org/layout/modules/resources/files/fortification_eng.pdf (accessed on 11 February 2022).
65. Liu, O. R., Molina, R., Wilson, M., & Halpern, B. S., (2018). Global opportunities for mariculture development to promote human nutrition. *Peer J., 6*, e4733.
66. Loveday, S. M., & Singh, H., (2008). Recent advances in technologies for vitamin A protection in foods. *Trends in Food Science and Technology, 19*, 657–668.
67. Lu, L. T., (2011). *Water-dispersible Magnetic NPs for Biomedical Applications: Synthesis and Characterization*. Ph.D. Thesis; University of Liverpool.
68. Mannar, V., & Boy, G. E., (2002). Iron fortification: Country-level experiences and lessons learned. *The Journal of Nutrition, 132*, 856S–858S.
69. Maret, M., & Sandstead, H. H., (2006). Zinc requirements and the risks and benefits of zinc supplementation. *Journal of Trace Elements in Medicine and Biology, 20*, 3–18.
70. Martirosyan, A., & Schneider, Y. J., (2014). Engineered nanomaterials in food: Implications for food safety and consumer health. *International Journal of Environmental Research and Public Health, 11*, 5720–5750.
71. Mason, T. G., Wilking, J. N., Meleson, K., Chang, C. B., & Graves, S. M., (2006). Nanoemulsions: Formation, structure, and physical properties. *Journal of Physics: Condensed Matter, 18*, R635–R666.
72. McClements, D. J., (2017). The future of food colloids: Next-generation nanoparticle delivery systems. *Current Opinion in Colloid & Interface Science,* 28, 7–14.
73. Meschino, J., (2002). Calcium: Requirements, bioavailable forms, physiology and clinical aspects. *Dynamic Chiropractic, 20*(18), 1–12.
74. Minekus, M., Alminger, M., Alvito, P., Ballance, S., Bohn, T., Bourlieu, C., Carrière, F., et al., (2014). A standardized static *in vitro* digestion method suitable for food—An international consensus. *Food and Function, 5*, 1113–1124.
75. Moll, R., & Davis, B., (2017). Iron, vitamin B12 and folate. *Medicine, 45*(4), 198–203.
76. Mourdikoudis, S., Pallares, R. M., & Thanh, N. T. K., (2018). Characterization techniques for nanoparticles: Comparison and complementarity upon studying nanoparticle properties. *Nanoscale, 10*, 12871.
77. Nutan, M. T. H., & Reddy, I. K., (2010). General principles of suspensions. In: Kulshreshtha, A. K., Singh, O. N., & Wall, G. M., (eds.), *Pharmaceutical Suspensions* (pp. 39–65). New York – USA: Springer International Publishers.
78. Pereira, D. I. A., Bruggraber, S. F. A., Faria, N., Poots, L. K., Tagmount, M. A., Aslam, M. F., Frazer, D. M., Vulpe, C. D., et al., (2014). Nanoparticulate iron (III) oxo-hydroxide

delivers safe iron that is well absorbed and utilized in humans. *Nanomedicine: Nanotechnology, Biology and Medicine, 10*, 1877–1886.
79. Perron, N. R., & Brumaghim, J. L., (2009). A review of the antioxidant mechanisms of polyphenol compounds related to iron binding. *Cell Biochemistry and Biophysics, 53*, 75–100.
80. Peters, R. J. B., Bouwmeester, H., Gottardo, S., Amenta, V., Arena, M., Brandhoff, P., Marvin, H. J. P., et al., (2016). Nanomaterials for products and application in agriculture, feed and food. *Trends in Food Science and Technology, 54*, 155–164.
81. Posavec, L., Knijnenburg, J. T. N., Hilty, F. M., Krumeich, F., Pratsinis, S. E., & Zimmermann, M. B., (2016). Dissolution and storage stability of nanostructured calcium carbonates and phosphates for nutrition. *Journal of Nanoparticle Research,* 18.
82. Powell, J. J., Bruggraber, S. F. A., Faria, N., Poots, L. K., Hondow, N., Pennycook, T. J., Latunde-Dada, G. O., et al., (2014). Anano-disperseferritin-coremimeticthatefficiently corrects anemia without luminal iron redox activity. *Nanomedicine: Nanotechnology, Biology and Medicine, 10*, 1529–1538.
83. Prasad, A. S., Beck, F. W., Grabowski, S. M., Kaplan, J., & Mathog, R. H., (1997). Zinc deficiency: Changes in cytokine production and T-cell subpopulations in patients with head and neck cancer and in noncancer subjects. *Proceedings of the Association of American Physicians, 109*, 68–77.
84. Rabinow, B. E., (2004). Nanosuspensions in drug delivery. *Nature Reviews Drug Discovery, 3*, 785–796.
85. Rafferty, K., Walters, G., & Heaney, R. P., (2007). Calcium fortificants: Overview and strategies for improving calcium nutrition of the US population. *Journal of Food Science, 72*, R152–R158.
86. Rohner, F., Ernst, F. O., Arnold, M., Hilbe, M., Biebinger, R., Ehrensperger, F., Pratsinis, S. E., et al., (2007). Synthesis, characterization, and bioavailability in rats of ferric phosphate nanoparticles. *Journal of Nutrition, 137*, 614–619.
87. Ron, N., Zimet, P., Bargarum, J., & Livney, Y. D., (2010). Beta-lactoglobulin-polysaccharide complexes as nanovehicles for hydrophobic nutraceuticals in non-fat foods and clear beverages. *International Dairy Journal, 20*, 686–693.
88. Ross, A., (2006). Vitamin A and carotenoids. In: Shils, M., Shike, M., Ross, A., Caballero, B., & Cousins, R., (eds.), *Modern Nutrition in Health and Disease* (10th edn., pp. 351–375). Baltimore – MD: Lippincott Williams & Wilkins.
89. Ross, C. A., (2010). Vitamin A. In: Coates, P. M., Betz, J. M., & Blackman, M. R., (eds.), *Encyclopedia of Dietary Supplements* (2nd edn., pp. 778–791). London – UK: Informa Healthcare.
90. Rossi, L., Velikov, K. P., & Philipse, A. P., (2014). Colloidal iron (III) pyrophosphate particles. *Food Chemistry, 151*, 243–247.
91. Sabliov, C. M., Fronczek, C., Astete, C. E., Khachaturyan, M., Khachatryan, L., & Leonardi, C., (2009). Effects of temperature and UV light on degradation of A-tocopherol in free and dissolved form. *Journal of the American Oil Chemists' Society, 86*, 895–902.
92. Sadat-Shojai, M., Khorasani, M. T., Dinpanah-Khoshdargi, E., & Jamshidi, A., (2013). Synthesis methods for nanosized hydroxyapatite with diverse structures. *Acta Biomaterialia, 9*, 7591–7621.
93. Santillan-Urquiza, E., Mendez-Rojas, M. A., & Velez-Ruiz, J. F., (2017). Fortification of yogurt with nano and micro-sized calcium, iron and zinc, effect on the physicochemical

and rheological properties. *Lebensmittel-Wissenschaft and-Technology - Food Science and Technology, 80*, 462–469.
94. Sayes, C. M., Fortner, J. D., Guo, W., Lyon, D., Boyd, A. M., Ausman, K. D., & Tao, Y. J., (2004). The differential cytotoxicity of water-soluble fullerenes. *Nano Letters, 4*, 1881–1887.
95. Schmid, G., & Klabunde, K. J., (2001). *Nanoscale Materials in Chemistry* (p. 729). John Wiley & Sons, Inc, New York; USA.
96. Schoepf, J. J., Bi, Y., Kidd, J., Herckes, P., Hristovski, K., & Westerhoff, P., (2017). Detection and dissolution of needle-like hydroxyapatite nanomaterials in infant formula. *Nano impact, 5*, 22–28.
97. Sharifi, F., & Jahangiri, M., (2018). Investigation of the stability of vitamin d in emulsion-based delivery systems. *Chemical Industry and Chemical Engineering Quarterly, 24*, 157–167.
98. Sharma, C. S., Sarkar, S., Periyakaruppan, A., Barr, J., Wise, K., Thomas, R., & Wilson, B. L., (2007). Single-walled carbon nanotubes induces oxidative stress in rat lung epithelial cells. *Journal of Nanoscience and Nanotechnology, 7*, 2466–2472.
99. Shin, G. H., Kim, J. T., & Park, H. J., (2015). Recent developments in nanoformulations of lipophilic functional foods. *Trends in Food Science and Technology, 46*, 144–157.
100. Siegrist, M., Stampfli, N., Kastenholz, H., & Keller, C., (2008). Perceived risks and perceived benefits of different nanotechnology foods and nanotechnology food packaging. *Appetite, 51*, 283–290.
101. Strobel, R., & Pratsinis, S. E., (2007). Flame aerosol synthesis of smart nanostructured materials. *Journal of Materials Chemistry, 17*, 4743–4756.
102. Stuss, M., Michalska-Kasiczak, M., & Sewerynek, E., (2017). The role of selenium in thyroid gland pathophysiology. *Endokrynologia Polska, 68*, 440–454.
103. Sun, L. M., Chow, L. C., Frukhtbeyn, S. A., & Bonevich, J. E., (2010). Preparation and properties of nanoparticles of calcium phosphates with various Ca/P ratios. *Journal of Research of the National Institute of Standards and Technology, 115*, 243–255.
104. Tavakoli, A., Sohrabi, M., & Kargari, A., (2007). A review of methods for synthesis of nanostructured metals with emphasis on iron compounds. *Chemical Papers, 61*, 151–170.
105. Teoh, W. Y., Amal, R., & Madler, L., (2010). Flame spray pyrolysis: An enabling technology for nanoparticles design and fabrication. *Nanoscale, 2*, 1324–1347.
106. Tiede, K., Boxall, A. B. A., Tear, S. P., Lewis, J., David, H., & Hassellov, M., (2008). Detection and characterization of engineered nanoparticles in food and the environment. *Food Additives and Contaminants, 25*, 795–821.
107. Tobelmann, R., (2001). Implementing calcium fortification: An industry case study. *Journal of Food Composition and Analysis, 14*, 241–244.
108. Trumbo, P., Yates, A. A., Schlicker, S., & Poos, M., (2001). Dietary reference intakes. *Journal of the American Dietetic Association, 101*(3), 294–301.
109. Tzitzios, V., Basina, G., Gjoka, G., Alexandrakis, V., Georgakilas, V., Niarchos, D., Boukos, N., & Petridis, D., (2006). Chemical synthesis and characterization of Hcp Ni nanoparticles. *Nanotechnology, 17*, 3750.
110. U.S. Department of Agriculture Agricultural Research Service, (2012). *USDA National Nutrient Database for Standard Reference*, Release 25.
111. U.S. Department of Agriculture, Agricultural Research Service, (2011). *USDA National Nutrient Database for Standard Reference*, Release 24.

112. Van, E. B., Van, D. M. G., & Augustijns, P., (2008). Top-down production of drug nanocrystals: Nanosuspension stabilization, miniaturization and transformation into solid products. *International Journal of Pharmaceutics., 364*(1), 64–75.
113. Van, L. Y. M., Velikov, K. P., & Kegel, W. K., (2012). Stabilization through precipitation in a system of colloidal iron (III) pyrophosphate salts. *Journal of Colloid and Interface Science, 381*, 43–47.
114. Vega-Villa, K. R., Takemoto, J. K., Yáñez, J. A., Remsberg, C. M., Forrest, M. L., & Davies, N. M., (2008). Clinical toxicities of nanocarrier systems. *Advanced Drug Delivery Reviews, 60*, 929–938.
115. Wang, J., Asbach, C., Fissan, H., H€ulser, T., Kuhlbusch, T. A. J., Thompson, D., & Pui, D. Y. H., (2011). How can nanobiotechnology oversight advance science and industry: Examples from environmental, health, and safety studies of nanoparticles (nano-EHS). *Journal of Nanoparticle Research, 13*, 1373–1387.
116. Wegner, K., Schimmoeller, B., Thiebaut, B., Fernandez, C., & Rao, T. N., (2011). Pilot plants for industrial nanoparticle production by fame spray pyrolysis. *KONA Powder and Particle Journal*, 251–265.
117. Whittaker, P., Tufaro, P. R., & Rader, J. I., (2001). Iron and folate in fortified cereals. *Journal of the American College of Nutrition, 20*, 247–254.
118. Winkler, H. C., Suter, M., & Naegeli, H., (2016). Critical review of the safety assessment of nanostructured silica additives in food. *Journal of Nanobiotechnology*, 14.
119. Wojciechowska-Durczynska, K., & Lewinski, A., (2017). Search for relevant indications for selenium supplementation in thyroid diseases. *Neuroendocrinology Letters, 38*, 237–241.
120. World Health Organization (WHO), (1992). *Indicators for Assessing Vitamin A Deficiency and their Application in Monitoring and Evaluating Intervention Programs*. Geneva. Report of a WHO/UNICEF Consultation, Geneva.
121. World Health Organization (WHO), (2002). *World Health Report-Reducing Risks, Promoting Healthy Life: Overview*. Geneva: World Health Organization, https://www.who.int/whr/2002/en/summary_riskfactors_chp4.pdf?ua=1 (accessed on 1 Match 2022).
122. Wu, J., Coradin, T., & Aimé, C., (2013). Reversible bioresponsive aptamer-based nanocomposites: ATP binding and removal from DNA-grafted silica nanoparticles. *Journal of Materials Chemistry B, 1*(39), 5353–5359.
123. Zheng, J., & He, L., (2014). Surface-enhanced Raman spectroscopy for the chemical analysis of food. *Comprehensive Reviews in Food Science and Food Safety, 13*, 317–328.
124. Zimmermann, M. B., & Hilty, F. M., (2011). Nano compounds of iron and zinc: Their potential in nutrition. *Nanoscale, 3*, 2390–2398.

# PART III

# NANOTECHNOLOGY APPLICATIONS IN SUSTAINABLE AGRICULTURE

## CHAPTER 9

# SCOPE OF PLANT-BASED NANOPARTICLES AS ANTIBACTERIAL AND ANTIBIOFILM AGENTS

MUZAMIL AHMAD RATHER, KULDEEP GUPTA, and MANABENDRA MANDAL


## ABSTRACT

Biofilm, a complex community of microbes protected by self-producing extracellular polymeric substances, imparts pathogenicity and extensive antibiotic resistance (ABR) to the microbes. The recurrent use of antibiotics and the subsequent rise of multidrug-resistant microbes demand a new approach to counter the negative impact on human health, animal husbandry, and food industry. Currently, nanotechnology, an interdisciplinary area of modern research, has an advantage as an efficient field for the development of different types of nanomaterials (NMs) with promising antimicrobial and antibiofilm applications. Among various physical, chemical, and biological methods for synthesizing nanoparticles, the plant-based biological approach is gaining interest. The easy availability, biocompatibility, strong antibacterial and antibiofilm potency, non-toxic behavior to the living cells and eco-friendly nature of the plant-based nanoparticles are the attributes that make them potent antibacterial and antibiofilm agents.


## 9.1 INTRODUCTION

Nanotechnology is an evolving area of science for synthesis, development, and utilization of nanoparticles. In 1950, nanotechnology was presented in a

Nanotechnology Horizons in Food Process Engineering, Volume 1. Megh R. Goyal, PhD, PE, Junaid Ahmad Malik, PhD, Satish Kumar, PhD & Ritesh B. Watharkar, PhD (Eds.)

famous lecture, "There is a Plenty of Room at the Bottom" by Nobel laureate Richard P. Feynman [35]. Since decades large developments have been made in the nanotechnology sector in terms of production of different types of materials at the nanoscale. Generally, the range of materials with particle size in-between 1 nm to 100 nm at least in one dimension are known as nanoparticles (NPs) [55]. Although nanomaterials (NMs) have been defined in different aspects by different organization, yet a universal definition for NPs is not available. Nanoscale has been defined as a length range between 1 to 100 nm [58].

Transmittance electron microscopy (TEM) and scanning electron microscopy (SEM) have revealed out the complex nature of NPs. NPs are comprised of three layers *viz.* (i) central part called as core, (ii) middle part called as shell layer with different composition as that of core, and (iii) outer surface layer which could be activated by different molecules [34]. The different types of NPs or NMs have been reported that can be broadly classified as: (a) Carbon-based nanoparticles, (b) Inorganic-based nanoparticles, (c) Organic-based nanoparticles, and (d) Composite-based nanoparticles [36].

Various approaches have been demonstrated for the synthesis of NPs. Iravani [57] classified the synthesis methods in two main categories, (i) simpler substance-based bottom-up synthesis *viz.* spinning, template support synthesis, plasma synthesis, laser pyrolysis, molecular condensation, chemical vapor deposition (CVD), biological synthesis using bacteria, fungi, yeast, algae, plants, etc.; and (ii) larger molecule-based top-down synthesis *viz.* mechanical milling, sputtering, electro-explosion, etc. In order to characterize the formation and physio-chemical properties of NPs different qualitative and quantitative methods have been used. The qualitative analysis includes Fourier transform infrared spectroscopy (FTIR), SEM, energy dispersive x-ray spectroscopy (EDS), UV-Vis spectrophotometry, X-ray diffraction (XRD), while as quantitative approaches include TEM, laser-induced breakdown spectroscopy (LIBS), and intracranial pressure (ICP) [56].

NPs have advanced applications over their bulk counterparts due to their unique physio-chemical properties attributed by nanoscale size and large surface area [61]. Owing to their unique characteristics, NPs have a wide range of applications in various fields, including drugs and medications, manufacturing materials, environment cleansing, electronics, energy harvesting, micro-wiring, and mechanical industries [68]. Due to the high cost and environment degradative nature of NP synthesis using chemicals, the use of microbes [33], enzymes [120] and plants [148] for NP synthesis has been regarded as an alternative environment friendly approach. Therefore, the cost-effectiveness, easy availability, nontoxic, and environment-friendly nature of plant extract mediated NPs synthesis has prompted researchers to focus on the green synthesis of NPs [59].

This chapter describes the history, types, and different methods of synthesis of nanoparticles. The chapter also discusses biosynthesis of NPs, and their scope as antimicrobial and antibiofilm agents (Figure 9.1).

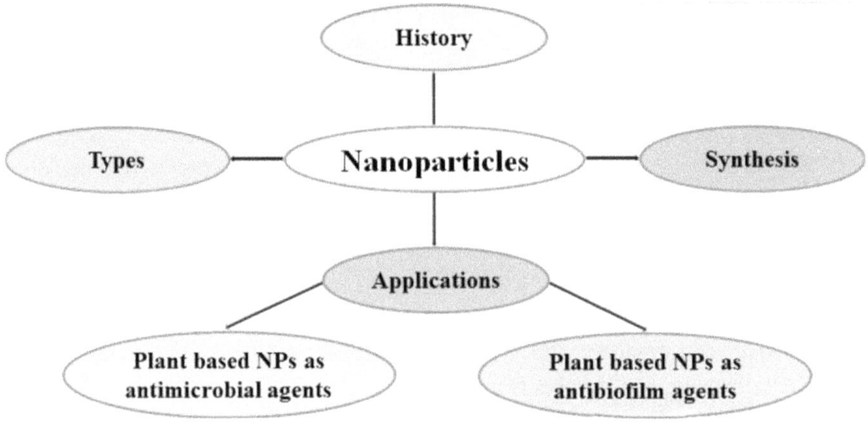


**FIGURE 9.1** Overview of the chapter.

## 9.2 HISTORY

The field of NPs has only developed recently, but people have been employing nanotechnology in their day-to-day life since thousands of years. For example, it has been reported that humans supplemented natural asbestos nanofibers to existing ceramic matrixes earlier than 4,500 years [50]. Around 3$^{rd}$ century B.C. Egyptians developed a synthetic pigment 'Egyptian Blue' using nano-sized quartz and glass [62]. 'Lycurgus Cup' is an evidence of humans using NMs in 4$^{th}$ century AD. The cup has a unique optical characteristic feature of appearing green or red in reflection and transmittance, respectively which has been reported due to the presence of metallic NPs [1]. Cu NPs and cuprous oxide were the major constituents found in Celtic red enamels which has been originated around 4$^{th}$ and 1$^{st}$ century BC [19].

It has been demonstrated that Mesopotamians used varnished ceramics for pronounced luster and beautification purposes during the 9$^{th}$ century. TEM images demonstrated the presence of a bi-layer of silver NPs (in-between 5 nm–10 nm) in the outermost layer [50]. The NMs history in the scientific world is believed to be started by Michael Faraday through synthesis of colloidal silver nanoparticle (AgNP) solution in 1857, wherein he found dissimilarities in optical features of silver colloids and their bulk counterpart [29]. In 1990 the first dedicated journal titled 'Nanobiotechnology' was published by IOP Publishing in the United Kingdom to boost the growing field of nanoscience [54, 80].

Samsung and Mercedes-Benz developed Silver Nano™ antibacterial technology in appliances like washing machine, fridge, etc., and NP-based clear coat for paint finishes, respectively in 2003 [56]. Vance et al. [144] reported about 1,814 nanotechnology-based products available in markets in more than 20 countries.

## 9.3 TYPES OF NPS

NPs could be classified on the basis of morphology (shape and flatness), composition, chemistry (suspensions, colloids, dispersed or agglomerated state), size, dimensions (0D, 1D, 2D, 3D), and state [122]. Some types of NPs are described in this section.

### *9.3.1 ORGANIC NANOPARTICLES (ONPS)*

Organic NPs or polymeric NPs are present in nature as well as in industrial products. Micelles, dendrimers, ferritin, vesicles, polymersomes, polymer conjugates, liposomes, capsules, etc., are some of the known organic NPs [117]. Organic NPs are sensitive to electromagnetic and thermal radiations and could take molecules by conjugation, or encapsulation, thus, serve as the ideal choice for drug delivery, cancer therapy, disease diagnosis, bio-imaging, etc. The predominant shape of organic NPs is nanosphere or nano-capsule and are non-toxic as well as biodegradable in nature [32]. Chitosan nanoparticles (CNPs), composite CNPs, and silk fibroin nanoparticles are the potential ONPs used in the dental field [149].

### *9.3.2 INORGANIC NANOPARTICLES*

The Carbon deficient (Carbonless) nanoparticles are known as inorganic nanoparticles. Stability, biocompatibility, hydrophilicity, and non-toxic nature are some of the characteristic features of inorganic NPs that makes them a versatile tool for biomedical applications like drug delivery, molecular sensing, etc. [70]. Metal and metal oxide nanoparticles are two sub-categories of inorganic nanoparticles. Metallic NPs are metals with nano-sized dimensions within the range of 1 nm–100 nm. Bottom-up or/and top-down methods are used for synthesis of metallic NPs from their respective metal precursors [147]. Metallic NPs have unique properties that make them valuable in various fields, including agriculture and medical sciences. Some of the characteristic features include, low size, high surface to volume ratio, density of external charge, variability in structure, shape, and color. Copper (Cu), Silver (Ag), Titanium (Ti), Gold (Au), Aluminum (Al), Platinum (Pt), Magnesium (Mg), Cadmium (Cd), Iron (Fe), and Zinc (Zn) are few of the commonly used metals for the production of metallic NPs. Metal NPs of noble metals including Pt, Au, and Ag have shown promised role in drug delivery, diagnostic tests, antimicrobial activity, gene delivery, anticancer, and radiotherapy enhancement [152].

Metal oxides are obtained by addition of oxidizing agent to the pure metal. The metal oxide NPs have been designed to enhance the reactivity, durability, and efficiency than their corresponding metal counterparts. Silver(I) oxide ($Ag_2O$), Manganese dioxide ($MnO_2$), Cupric oxide (CuO), MgO or magnesia (MgO), titanium dioxide ($TiO_2$), Ferrous oxide (FeO), Calcium oxide (CaO), Aluminum oxide ($Al_2O_3$), Bismuth(III) oxide ($Bi_2O_3$), and Zinc oxide (ZnO) are some of the metal oxide NPs with potent antibacterial activity [115]. Similarly, metal-based chalcogenides (Metal Sulfide-based NPs) have been regarded as the best semiconductors due to high fluorescence, magnetic, thermal, and structural stability [146]. Nanoceramics or nonmetallic solids are oxides, phosphates, carbides, and carbonates of metals and metalloids. They are being widely used for drug delivery in a number of bacterial infections, and their broad range of applications is due to the high resistance to heat and inertness to the chemicals [20].

## 9.4 SYNTHESIS

Different methods have been employed for the synthesis of NPs. The methods could be broadly classified as: i) bottom-up methods; and (ii) top-down methods and could be further classified into sub-methods on the basis of protocol used, operation time and reaction conditions (Figure 9.2) [57].

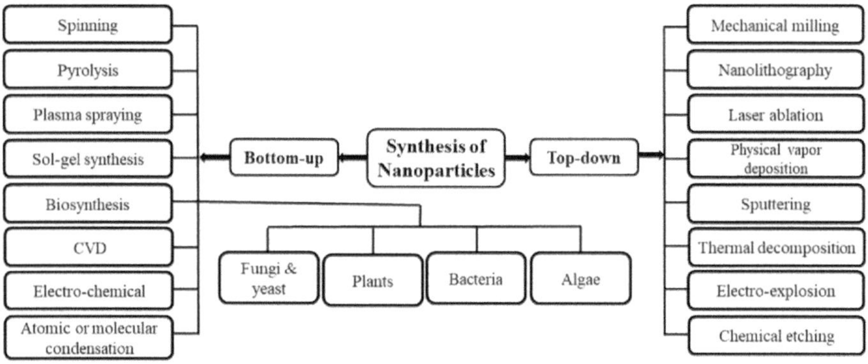


**FIGURE 9.2** Different methods of nanoparticle synthesis.

### *9.4.1 BOTTOM-UP METHODS*

Also, referred as building-up or constructive method because NPs are synthesized from smaller substances. This approach is comprised of spinning, pyrolysis, sol-gel, biological synthesis, etc. [114]. One of the commonly used methods for the synthesis of NPs at large scale is pyrolysis. At high pressure, the liquid or gaseous precursor is passed through a hole and subsequently burned to obtain NPs [154]. Nowadays, the requirement of high flame has been replaced by laser or plasma for a convenient evaporation. Flame spray pyrolysis, laser pyrolysis, ultrasonic spray pyrolysis, and low temperature catalytic pyrolysis are some of the modified forms of pyrolysis employed for the synthesis of NPs [88]. Pyrolysis is an efficient, cost-effective, simple, and high yielding continuous process [38]. The sol-gel process or chemical solution deposition is used mostly in ceramic engineering and material science. In this technique, chemical solutions like metal oxide and chloride are used as precursors [26]. The precursor is dispersed in the liquid by stirring, shaking, and sonication followed by filtration and centrifugation for separating liquid and solid phases to obtain NPs. The step-by-step process includes stable sol solution formation, gelation, gel aging, dehydration, densification, and decomposition [103].

CVD, as the name suggests, is the deposition of a film of gaseous reactants on the substrate for the production of 2D NMs. In this method, the substrate is interacted with volatile precursors (gas or vapor) at high temperature and vacuum leading to the production of desired thin film by reacting or/and decomposing on substrate surfaces [67]. The 2D NMs obtained from CVD are widely used in solar cell devices, and optoelectronics. Limited number of defects on the substrates, high crystal quality and enhanced purity are the characteristic features of 2D NPs obtained by CVD [110]. A spinning disc reactor (SDR) is used for the controlled synthesis of NPs. The rotating disc present within SDR is used to control physical parameters like temperature and inert gases which are used for inhibition of chemical reaction and removal of $O_2$ [87].

Spinning of the precursors for fusion, followed by precipitation, collection, and drying are some of the important steps involved in the synthesis of NPs by spinning method [138]. To replace the chemical synthesis of NPs different biological sources have been used to synthesize NPs at a large scale without affecting environmental health. Actinomycetes, bacteria, fungi, yeast, plants, etc., are used for the production of NPs in the presence of precursors. Biosynthesis of NPs is adventitious over conventional methods in terms of energy efficiency, production cost, safe synthesis, waste production and environmental safety [65].

### 9.4.2 TOP-DOWN APPROACH

In this method, a larger molecule is decomposed into smaller molecules which are later transformed into NPs. It is regarded as the destructive method of NP synthesis since it involves the breakdown of bulk precursors into subsequent smaller particles [39]. Mechanical milling or grinding, nanolithography, laser ablation, physical vapor deposition, sputtering, and thermal decomposition are few of the commonly used destructive methods [55]. Thermal decomposition involves the lysis of chemical bonds due to external heat as the process is endothermal in nature.

Unni et al. [143]; and Maity et al. [83] fabricated iron oxide and monodispersed magnetite ($Fe_3O_4$) NPs respectively using thermal decomposition method. Sputtering is the bombardment of ion energy, usually of noble gases, on solid substances to initiate the process of vaporization. DC diode, RF, DC triode, magnetron, and reactive sputtering are some of the known types of sputtering currently employed [13, 98]. Ball-milling or mechanical alloying is the simplest, cost-effective, fast, and environment friendly mechanical method of NP synthesis in an inert atmosphere by attrition [153]. Ball mill is usually a rotating hollow cylinder fitted with balls made up of rubber, stainless steel, or ceramic. It is one of the potential methods for the synthesis of cellulose nanofibers and nanocrystals [108]. The production of NPs from different solvents could be obtained by laser ablation synthesis in solution (LASiS).

Laser ablation is a common tool for nanofabrication of nanoparticles without addition of a surfactant or a chemical [28]. It has become a common method for the formation of metal NPs, thus, provides an alternative to conventional chemical methods. The metal nanofabrication depends upon repetition rate of laser, ablation time, energy, wavelength, and absorption of a solution [119]. Lithography is an inexpensive, and high-throughput nanofabrication technique for the synthesis of NPs. Some of the sub-types of lithography include photo-lithography, soft lithography, focused ion lithography, dip-pen lithography, electron beam lithography, and nano-imprint lithography [107].

## 9.5 BIOSYNTHESIS OF NANOPARTICLES

The physio-chemical methods for the fabrication of NPs are associated with a number of drawbacks like requirement of large space, increase in environmental temperature due to continuous release of heat, contamination of air, water, and soil due to generation of chemicals as by-products and imperfection of surface structure [74]. In this regard the green approach for synthesis of NPs using biological sources has already replaced existing chemical and physical methods [81, 96]. This eco-friendly approach of NP synthesis is gaining interest as the end products are degradable and non-toxic to the environment, thus, qualifies as a potent eco-friendly NP synthesis approach [105].

The biological sources used for biosynthesis of NPs include microorganisms (fungi, yeast, bacteria, and algae), different types of enzymes, and plant extracts. In green synthesis, the expensive chemical reducing agents are replaced by plant extracts or enzymes for synthesis of metal and oxides of metal NPs [82]. Therefore, sustainability, low cost, contamination-free, easy to synthesize, mass productivity and eco-friendly nature are some of the characteristic features of biosynthetic mode of NP synthesis over other physio-chemical mode of synthesis [53].

### *9.5.1 MICROORGANISMS*

High concentration of ions is toxic and induces mortality to bacterial cells. In order to resist high concentration of ions in the environment, the cellular machinery of microorganisms have adapted to convert reactive species to stable ones [72]. The stable atoms are the NPs of their respective ions. The feature of bacteria to convert ions into respective stable forms is employed for the biosynthesis of NPs [52]. Gold NPs of the size of about 8 nm were synthesized by extremophilic actinomycete *Thermomonospora* sp. [4]. Similarly, CuO NPs were synthesized employing actinomycetes by Nabila and Kannabiran [89]. The average particle size of 61.7 nm depicted by TEM showed potential antibacterial activity against human pathogens as compared to the cell-free supernatant. Fungus is currently being utilized as a natural source for the reduction and stabilization of metal NPs as they are highly tolerant to metal toxicity and are easy to operate [43].

Ahluwalia et al. [3] developed silver NPs from *Trichoderma harzianum* that showed potent antibacterial activity against *Staphylococcus aureus* and *K. pneumoniae.* Algae extracts are being used for the synthesis of NPs through an advanced branch of nanoscience known as phyconanotechnology [100]. Algae provides an important source for the synthesis of NPs as they grow at normal temperature (low temperature), easy to handle, does not require special ingredients for growth and are non-toxic in nature [97].

Hamouda et al. [46] synthesized silver NPs using cyanobacteria *Oscillatoria limnetica* extract. The synthesized NPs were of the size in the range of 3.30–17.97 nm and quasi-spherical shape as revealed by TEM. The *Oscillatoria*-silver nanoparticles (O-AgNPs) demonstrated potent antimicrobial activity against *Bacillus cereus* and *Escherichia coli*. The extract also showed cytotoxic effect against human breast (MCF-7) cell line.

### 9.5.2 PLANTS

Synthesizing NPs using plant extracts is one of the commonly implemented green and eco-friendly approach. The abundance, availability, and medicinal value of plants attracted the researchers for green synthesis of NPs [69]. The parameters which decide the approval of synthesized NPs include the reducing potential, nature of the solvent, and the stabilizing material. Therefore, preparation of NPs using water as a solvent provides the highest stability to the synthesized NPs. Since, phytochemicals are known for stabilization of metal NPs, therefore, act as noble and eco-friendly source for NP synthesis [59]. Extracts from different parts of the plant like seeds, leaves, roots, fruits, bark, and stem have been widely used for the synthesis of NPs due to the abundance of plant metabolites acting as reducing agents or stabilizers during the fabrication of NPs [94]. The kinetics of plant-based NP synthesis is higher than any of the biosynthetic model and is comparable to the preparation of NPs by chemical methods [57]. Some of the metal NPs that have been synthesized using green approach are discussed.

Zinc Oxide NPs (ZnO NPs) have been reported as the nontoxic, safe, and compatible type of metal NPs. It has been widely used in optoelectronics, magnetic sensors, diagnosis, biolabeling, medicinal industry, environmental protection, etc., due to its piezoelectric features, high binding energy, and large bandwidth [25]. Selim et al. [125] synthesized ZnO NPs using aerial parts of *Deverra tortuosa* and the synthesis was verified employing FTIR, UV-Vis, XRD, and HR-TEM. The NPs showed potent cytotoxicity activity against human lung adenocarcinoma 'A549' and human colon adenocarcinoma 'Caco-2.' The phytoconstituents like methylxanthines, terpenoids, tannins, alkaloids, phenolic acids, flavonoids, and saponins have strong reducing power which is harnessed for the synthesis of Zn NPs from zinc precursors [10].

Since we know that phytochemicals act as capping or stabilizing agents, therefore, copper (Cu) and oxides of copper NPs (Cu/CuO-NPs) have been fabricated by using various plant extracts. Cu/CuO-NPs have been used as degradative agent against dyes, enhanced agricultural productivity, bacteria, fungus, and free radicals [129]. Thiruvengadam et al. [142] synthesized about 23 nm spherical Cu-NPs by using flower extract of *Millettia pinnata* as capping or reducing agent. NP synthesis was confirmed by UV-Vis, SEM, TEM, XRD, FTIR, and selected area electron diffraction (SAED). Other Cu-NPs were fabricated employing *Magnolia kobus* [79], copper sulfate and *Eucalyptus* sp. leaf extract [71], *Azadirachta indica* extract [91], *Lawsonia inermis* leaf extract [73], *Citrus medica* juice [127], *Euphorbia esula* leaf extract [95], *Punica granatum* peel extract, etc. [66]. Similarly, different plant extracts were used for synthesis of CuO NPs like gum karaya [102], *Aloe vera* leaf extract [76], *Ixoro coccinea* leaf extract [150], *Psidium guajava* leaf extract [130], *Caesalpinia bonducella* seed extract [134], extract of *Gloriosa superba* plant [92], *Lantana camara* flower extract [24], etc.

Titanium oxide ($TiO_2$) NPs are being synthesized on a large scale for their pronounced antimicrobial, antiparasitic, and photolytic activities. In the medical field, $TiO_2$ NPs has been used widely in diseases diagnostics, production of surgical tools, imaging, sensing, and agriculture. The phytoconstituents and other secondary metabolites from different biological sources have been used for bio-reduction and capping of precursor metal or/and metal salts into respective nanoparticles [90]. In terms of safety, cost-effectiveness, and availability, plants have been regarded as suitable source of NP synthesis including $TiO_2$ NP. *Cicer arietinum* L. bean extract [64], *Cassia fistula* extract [136], *Azadirachta indica* leaf extract [123] are some of the studies performed for the green synthesis of $TiO_2$ NP.

Larue et al. [78] reported that about 500 tons of silver NPs (AgNPs) are synthesized per year in the world. The sustainability and eco-friendly nature of biosynthesis of AgNPs has replaced the chemical synthesis as the latter is associated with environment degradative potential. The potent anti-angiogenesis, antibacterial, antifungal, and anti-inflammatory activity of AgNPs have already been demonstrated [86]. Different extracts of various plants have been employed for the production of AgNPs like *Acalypha indica* leaf extract [73], *Tagetes erecta* flower extract [101], *Acorous calamus* rhizome extract (NP size – 31.83 nm) [93], tea leaf extract (NP size between 20 nm to 90 nm) [135], inflorescence of *Cocous nucifera* (NP size – 20 nm) [85] and many more. The fabrication of AgNPs by using *Salvia spinose* plant extract was first of all demonstrated by Pirtarighat et al. [109] and the same was confirmed by FTIR, XRD, and FESEM studies. Thus, using plant extracts for the fabrication of AgNPs is an energy-efficient, cost-effective, economical, and environmentally friendly approach.

Gold particles are inert in natural form but the Gold NPs (Au NPs) demonstrate optical, catalytic, chemical, and/or thermal properties. Different types of Au-based NMs like Au Nanocages, Au Nanospheres and Au Nanorods show anticancer and antibacterial activities [141]. The wide range of applications of Au NPs in electrical conductivity, catalysis, and antimicrobial activity have increased the focus towards green synthesis of Au NPs. Au NPs were synthesized by using *Mimosa tenuiflora* bark extract. The synthesized Au NPs were verified by TEM, XRD, FTIR, and X-ray photoelectron spectrometry [116]. Similarly, *Stevia rebaudiana* leaf extract [118], Mentha, and *Pelargonium* plant extracts [60] were used for the synthesis of Au NPs from gold ions.

Various NMs like nanocatalysts, nanobiocides, and nanoadsorbents are being used widely for environmental bioremediation. It has been demonstrated that iron NPs (Fe NPs) are being used for water remediation and wastewater management and plants are the richest source for iron and iron oxide NPs fabrication among other biological sources [121]. Devi et al. [27] synthesized $Fe_2O_3$ NPs by employing *Platanus orientalis* leaf extract (tree of Hippocrates in Kashmir). They demonstrated that phytochemicals of the leaf performed the reduction as well as capping during $Fe_2O_3$ NPs synthesis. The fabrication of $Fe_2O_3$ NPs was depicted by XRD, SEM, TEM, IR, and UV spectroscopy, and dynamic light scattering (DLS) techniques.

$Fe_2O_3$ NPs were fabricated by using *Punica granatum* seed extract in the range of 25–55 nm. The $Fe_2O_3$ NPs showed photocatalytic activity for the degradation of dye [17]. Some other plant extracts that have been used for $Fe_2O_3$ NPs synthesis include *Ruellia tuberosa* leaf aqueous extract [145], *Lawsonia inermis* (Henna) extract [22], *Carica papaya* leaf extract [16] and many more.

Palladium NPs (Pd NPs) are produced on a large scale and has attracted scientists because of its catalysis nature in various types of reactions including oxidation, reduction, and carbon-carbon cross-coupling. Although different physio-chemical approaches are available for synthesis of Pd NPs yet biosynthetic approach, especially plants, is being focused for the production because of their biocompatibility and environmental favorable nature [15]. Hazarika et al. [49] evaluated the fabrication of Pd NPs employing *Garcinia pedunculata* leaf extract and was demonstrated by XRD, SEM-EDX, FTIR, TEM, and X-ray photoelectron spectroscopy (XPS). Similarly, *Phoenix dactylifera* leaf extract [137] and lentinan plant extract (2.35–3.32 nm) [47] have already been employed for the production of Pd NPs.

The potential physicochemical properties and superior biomedical applications have favored the synthesis of Platinum nanoparticles (Pt NPs). Plant metabolites have been used for stabilization and bio-reduction of Pt metal precursors and have been regarded as the potent and eco-friendly approach among other living sources (yeast, fungi, algae, etc.) [31]. A single-step synthesis of Pt NPs using *Terminalia chebula* fruit extract was performed by Kumar et al. [75]. The plant was selected because of the abundance of polyphenols which reduced Pt (IV) ions to Pt NPs at 100°C. Al-Radadi, [7] synthesized the best size and shape of Pt NPs by using different ratios of Ajwa and Barni date extracts and metal salts. The shape and size were confirmed by FTIR, XRD, EDX, TGA, and TEM.

Cerium oxide NPs (nanoceria) have been widely used as antioxidant, anticancer, antibacterial, and catalase agent. Arumugam et al. [12]; and Priya et al. [112] synthesized $CeO_2$ NPs employing *Gloriosa superba* L. leaf extract and *Aloe barbadensis* miller plant extract, respectively. Prasad et al. [111] studied the synthesis of Cadmium sulfide NPs (CdS) using *Asparagus racemosus* leaf extract. *Cocos nucifera* extract has been used for the synthesis of 47 nm lead NPs (Pb NPs) which showed photolytic activity [30]. Similarly, ruthenium NPs (25–90 nm) were synthesized by bio-reduction of *Gloriosa superba* leaf extract [41].

## 9.6 APPLICATIONS OF PLANT-BASED NANOPARTICLES

### 9.6.1 PLANT-BASED NANOPARTICLES AS ANTIMICROBIAL AGENTS

Due to increased resistance to commercially used antibiotics and the unavailability of novel antibiotics, the spread of pathogenic microorganisms has created a havoc throughout the world. Treating microbial infections with already known antibiotics has become a challenge and threat to the organisms as the pathogenic microorganisms tend to evolve and develop resistance against antibiotics [132]. It has already been reported that in the developing countries, antibiotic resistance (ABR) has attained an unacceptable level and is increasing at an alarming pace in pathogenic microorganisms. U.S. Centers for Disease Control and Prevention described resistance to antibiotics as "one of the world's most pressing public health problems." Bacteria from both clinical and non-clinical sources especially methicillin-resistant *Staphylococcus aureus* and vancomycin-resistant *Enterococcus* sp. have become resistant to conventional antibiotics [77].

The World Health Organization (WHO) reported that "a post-antibiotic era—in which common infections and minor injuries can kill—far from being an apocalyptic fantasy, is instead a very real possibility for the 21st century" [151]. Antimicrobial resistance (AMR) or ABR poses a threat to global health and a burden on the economy [128]. The Organization for Economic Co-operation and Development (OECD) have reported that it could cost US $3,500 million per year to treat patients infected by antibiotic-resistant microorganisms. It was also discussed that in the coming 30 years, about 2.4 million people from Australia, North America and Europe would die from infections by drug-resistant microorganisms [51]. It was estimated by CDC that annual cost of treating AMR is about US $55,000 million per year in the US alone [21]. In this context, new non-antibiotic approaches have been designed to counter recalcitrant microbial infections and drug resistance like phage therapy, microbiome modulation, biofilm inhibition, inhibition of toxin and secretary systems [48].

NPs are being used as drug delivery systems, cancer therapy, vaccinations, and many other therapeutic applications throughout the world. Furthermore, the use of NPs in microbial implications is intensifying throughout the world. NPs are known to overcome the microbial resistance and to act as an alternative to novel antibiotics [104]. So, the predominance of drug resistance, and unavailability of potent antibiotics have shifted the paradigm towards the use of plant-based NPs as green and potent antimicrobial agents. Such NPs are non-toxic, does not induce microbial resistance, and are eco-friendly in nature [132].

Silver NPs are known as potent antibacterial agents against pathogenic microorganisms. Patra and Baek [106] demonstrated the synthesis of AgNPs employing extract of *Zea mays*. The synthesized NPs showed moderate antibacterial activity against foodborne pathogens *viz. B. cereus, Staphylococcus aureus, Escherichia coli, Listeria monocytogenes, and S. typhimurium*. The antifungal activity of NPs was demonstrated against *C. albicans, Candida glabrata,* and *C. saitoana* as well. Similarly, 20 to 60 nm spherical AgNPs were produced using extract of *Syzygium cumini* and the synthesis was confirmed by AFM, SEM, and UV-visible spectrometer. The AgNPs showed bactericidal activity against Gram-negative and Gram-positive bacteria by agar well diffusion assay [63].

FTIR, Zeta potential, UV-Vis, HRTEM, and XRD techniques demonstrated the synthesis of 7–23 nm using 4-N-methyl benzoic acid, a phenolic derivative from *Memecylon umbellatum* Burm F. The NPs showed antibacterial activity against human pathogens *Acinetobacter baumannii, Salmonella paratyphi,* and *Bacillus subtilis* [8]. Alyousef et al. [11] described the biosynthesis of AgNPs using *Myrtus communis* extract. The synthesis and average size of NPs was established by FTIR and other diffraction techniques. The NPs exhibited antibacterial activity against *E. coli* and MRSA. Platinum NPs using Saudi's Dates extract, ZnO NPs using *Passiflora caerulea* leaf extract, Copper NPs using *Punica granatum* peel extract, ZnO nanoparticles using *Tectona grandis* leaf extract, Silver NPs using fruit extract of *Phyllanthus emblica,* Ruthenium NPs using *Gloriosa superba* leaf extract, Copper NPs using *Magnolia kobus* leaf extract, Silver NPs employing *Acalypha indica* leaf extract, Iron oxide NPs *Carica papaya* extract, Palladium NPs using *Garcinia pedunculata* leaf extract, Palladium NPs using *Phoenix dactylifera* leaf extract, AgNPs *Ocimum canum* leaf extract, showed antibacterial activity against, *E. coli, B. subtilis* [7], *Klebsiella pneumoniae*, *Streptococcus* sp., *Serratia* sp., *B. subtilis*, and *E. coli* [124], *M. luteus*, *P. aeruginosa*, and *E. aerogenes* [66], *S. aureus*, *B. subtilis*, *E. coli* and *S. paratyphi* [126], rice pathogen *Acidovorax oryzae* [86], *B. subtilis*, *S. aureus*, *S. pneumoniae*, *E. coli*, *P. aeruginosa* [41], *E. coli* [79], *E. coli and Vibrio cholera* [73], *E. coli*, *Klebsiella sp.*, *Pseudomonas* sp., and *S. aureus* [16], *Cronobacter sakazakii* [49], *P. aeruginosa* [137], *E. coli* [139], respectively.

### 9.6.2 PLANT-BASED NANOPARTICLES AS ANTIBIOFILM AGENTS

Biofilm has become a problem for human health, food industry, and animal husbandry. Some of the prominent biofilm-forming microorganisms include *P. aeruginosa, Streptococcus gordonii, Staphylococcus epidermidis, Corynebacterium* sp., *Streptococcus oralis, Streptococcus mutans, S. aureus, Enterococcus* sp., *Salmonella enterica, E. coli, Streptococcus sanguis, Actinomyces* sp., *Proteus mirabilis, Streptococcus* sp., *Candida* sp., *Treponema denticola, Actinomyces israelii, Selenomonas flueggei, Haemophilus parainfluenzae,* etc. [40, 42, 113].

The pronounced resistance of biofilm-forming microorganisms against known antibiotics is because of the exopolymeric substances (EPS) production which acts as a barrier against them. Furthermore, the limited penetration of antibiotics, horizontal gene transfer, reduced growth rates, presence of persister cells and escape from the host immune system are some of the properties of biofilm-forming microorganisms which help them to induce resistance against known antibiotics [37]. It has been reported by the Centers for Disease Control, in 2007, that there was an economic burden of US $11,000 million due to 1.7 million hospital-acquired infections caused by biofilm-forming microorganisms in the USA alone. The estimated annual cost of biofilm-associated infections is US $94,000 million and is accountable for about half a million deaths [18]. In this context, nanotechnology has been used as a promising tool to counter the menace of pathogenic microbial biofilms. It has been reported that NP-coated surfaces inhibit biofilm formation and, thus, promises prevention of multidrug-resistant microbial biofilms [140].

EPS acts as a barrier against penetration of antibiotics but NPs being small in size penetrate both biofilm surfaces as well as microbial cell walls inducing damage to DNA and cell membranes. Additionally, due to the large surface to volume ratio and prolonged plasma half-life, the NPs have become a promising approach for drug delivery and targeting [133].

Gold and silver NPs (13–17 nm and 15–30 nm, respectively) were synthesized using rhizome extracts of *Rhodiola rosea*. It was demonstrated that minimum inhibitory concentrations (MIC) and minimum bactericidal concentrations (MBC) of Au NPs against *P. aeruginosa* was 50 and 100 μg/m respectively. Similarly, MIC and MBC of AgNPs against *E. coli* was 100 and 200 μg/ml respectively. The authors also studied the antibiofilm activity of Au NPs and AgNPs and found AgNPs as potent antibiofilm agent both *P. aeruginosa* and *E. coli* [131].

Oliver et al. [99] demonstrated synthesis and enhancement of antibacterial and antibiofilm activity of bio-synthesized AgNPs using polyphenol-rich plant extracts. It was observed that polycat@Ag NPs demonstrated enhanced antibiofilm activity against *P. aeruginosa* and *Acinetobacter baumanni* at sub-MIC. Similarly, the antibiofilm activity of AgNPs using *Oscillatoria* sp. extract against *S. aureus, E. coli, P. aeruginosa, B. cereus, Salmonella typhi* and *Citrobacter* sp. was demonstrated by Adebayo-Tayo et al. [2]. A pronounced antibacterial and antibiofilm activity of AgNPs using extracts of different medicinal plants of India *viz. Semecarpus anacardium, Bridelia retusa* and *Glochidion lanceolarium* was reported against *E. coli, P. aeruginosa,* and *S. aureus.* The fabrication of yellow pepper extract and quercetin-based silver NPs were studied by Ahmed et al. [5]. Both of the biosynthetic NPs showed pronounced antibacterial and antibiofilm activity against MSSA, MRSA, and *P. aeruginosa.* The fluorescent probes of isolates indicate a restricted EPS synthesis due to enhanced production of ROS, superoxide radicals, increase in membrane permeability and DNA leakage.

Gupta et al. [44] fabricated AgNPs employing *Syzygium cumini* which exhibited enhanced antibiofilm activity against *C. albicans, S. aureus,* and *P. aeruginosa.* Ali et al. [6] demonstrated synthesis of microwave-assisted fabrication of AgNPs employing *Eucalyptus globulus* leaf extract as reducing and capping agent. The NPs demonstrated a significant biofilm inhibition in *P. aeruginosa* (84 ± 5%) and *S. aureus* (83 ± 3%). The fabrication of zinc NPs was performed by using *Ochradenus baccatus* leaf extract and the synthesis was confirmed by TEM, XRD, and FTIR. The Zn NPs demonstrated antibiofilm activity against *Chromobacterium violaceum, K. pneumoniae, Serratia marcescens,* and *Listeria monocytogenes* [9].

The fabrication and subsequent characterization of gold NPs (Au NPs) using *Nigella sativa* essential oil (NsEO-AuNPs) was demonstrated by Manju et al. [84]. NsEO-AuNPs showed bactericidal phenomenon against *Vibrio harveyi* and *S. aureus.* It was also observed that biofilm formation of both the strains was inhibited by NsEO-AuNPs due to reduction in hydrophobicity index. Gupta et al. [45] demonstrated the fabrication of AgNPs employing *Psidium guajava* extract and evaluation of antibiofilm activity against *C. albicans, E. coli* and *S. aureus.* Inhibition of synthesis of Quorum sensing (QS) molecules, deep penetration of NPs in EPS, damage to bacterial ETS enzymes, lipid peroxidation of EPS, generation of ROS, cell wall and membrane damages, and rupturing of cell membrane integrity are some of the mechanisms employed by biosynthesized NPs to inhibit biofilm formation [14].

## 9.7 SUMMARY

Increase in ABR, recalcitrant infections, associated deaths and financial distress on the healthcare system is demanding the development of novel antimicrobial approaches. Green synthesis of nanoparticles employing plant extracts is an advanced field of nanotechnology. It is one of the focused areas of researchers because of its non-toxic, cost-effective, and eco-friendly nature as compared to other physio-chemical conventional methods. Biosynthesized NPs serve as an important approach for antimicrobial and antibiofilm activities against infectious human pathogens and foodborne pathogens owing to their unique size, shape, penetrative power, and other physiochemical properties. The increased ABR of microorganisms in biofilm mode and unavailability of novel antibiotics against them have favored the use of green synthesis of NPs for inhibition and eradication of pathogenic biofilms.

Although extensive *in vitro* studies have been demonstrated to check the potency of plant-based NPs as antibacterial and antibiofilm approach, carrying out of *in vivo* studies is the need of the hour to validate and authenticate the results.

## ACKNOWLEDGMENTS

The authors are thankful to Tezpur University for providing us financial support via memo No. DoRD/RIG/10-73/1362-A dated 19/02/2019 and DoRD/RIG/10-73/1592-A dated 07/01/2021.

## KEYWORDS

- **antibacterial agent**
- **antibiofilm agent**
- **Fourier transform infrared spectroscopy**
- **microorganisms**
- **nanomaterial**
- **nanoparticles**

## REFERENCES


1. Abbasi, E., Milani, M., Fekri, A. S., Kouhi, M., Akbarzadeh, A., Tayefi, N. H., & Samiei, M., (2016). Silver nanoparticles: Synthesis methods, bio-applications and properties. *Critical Reviews in Microbiology, 42*(2), 173–180.
2. Adebayo-Tayo, B., Salaam, A., & Ajibade, A., (2019). Green synthesis of silver nanoparticle using *Oscillatoria* Sp. Extract, its antibacterial, antibiofilm potential and cytotoxicity activity. *Heliyon, 5*(10), E02502. doi: 10.1016/j.heliyon.2019.e02502.
3. Ahluwalia, V., Kumar, J., Sisodia, R., Shakil, N. A., & Walia, S., (2014). Green synthesis of silver nanoparticles by *Trichoderma Harzianum* and their bio-efficacy evaluation against *Staphylococcus aureus* and *Klebsiella pneumoniae*. *Industrial Crops and Products, 55*, 202–206.
4. Ahmad, A., Senapati, S., Khan, M. I., Kumar, R., & Sastry, M., (2003). Extracellular biosynthesis of monodisperse gold nanoparticles by a novel extremophilic actinomycete, *Thermomonospora* Sp. *langmuir, 19*(8), 3550–3553.
5. Ahmed, B., Hashmi, A., Khan, M. S., & Musarrat, J., (2018). ROS mediated destruction of cell membrane, growth and biofilms of human bacterial pathogens by stable metallic AgNPs functionalized from bell pepper extract and quercetin. *Advanced Powder Technology, 29*(7), 1601–1616.
6. Ali, K., Ahmed, B., Dwivedi, S., Saquib, Q., Al-Khedhairy, A. A., & Musarrat, J., (2015). Microwave accelerated green synthesis of stable silver nanoparticles with *Eucalyptus globulus* leaf extract and their antibacterial and antibiofilm activity on clinical isolates. *Plos One, 10*(7), E0131178. doi: 10.1371/journal.pone.0131178.
7. Al-Radadi, N. S., (2019). Green synthesis of platinum nanoparticles using Saudi's dates extract and their usage on the cancer cell treatment. *Arabian Journal of Chemistry, 12*(3), 330–349.
8. Alsalhi, M. S., Elangovan, K., Ranjitsingh, A. J. A., Murali, P., & Devanesan, S., (2019). Synthesis of silver nanoparticles using plant-derived 4-N-methyl benzoic acid and evaluation of antimicrobial, antioxidant and antitumor activity. *Saudi Journal of Biological Sciences, 26*(5), 970–978.
9. Al-Shabib, N. A., Husain, F. M., Hassan, I., Khan, M. S., Ahmed, F., Qais, F. A., & Hussain, A., (2018). Biofabrication of zinc oxide nanoparticle from *Ochradenus Baccatus* leaves: Broad-spectrum antibiofilm activity, protein binding studies, and *in vivo* toxicity and stress studies. *Journal of Nanomaterials, 2018*, 8612158. doi: 10.1155/2018/8612158.
10. Altemimi, A., Lakhssassi, N., Baharlouei, A., Watson, D. G., & Lightfoot, D. A., (2017). Phytochemicals: Extraction, isolation, and identification of bioactive compounds from plant extracts. *Plants, 6*(4), 42. doi: 10.3390/plants6040042.
11. Alyousef, A. A., Arshad, M., Alakeel, R., & Alqasim, A., (2019). Biogenic silver nanoparticles by *Myrtus communis* plant extract: Biosynthesis, characterization and antibacterial activity. *Biotechnology & Biotechnological Equipment, 33*(1), 931–936.
12. Arumugam, A., Karthikeyan, C., Hameed, A. S. H., Gopinath, K., Gowri, S., & Karthika, V., (2015). Synthesis of cerium oxide nanoparticles using *Gloriosa superba* L. leaf extract and their structural, optical, and antibacterial properties. *Materials Science and Engineering: C, 49*, 408–415.
13. Asanithi, P., Chaiyakun, S., & Limsuwan, P., (2012). Growth of silver nanoparticles by DC magnetron sputtering. *Journal ofNanomaterials, 2012*, 8. doi: 10.1155/2012/963609.

14. Banerjee, D., Shivapriya, P. M., Gautam, P. K., Misra, K., Sahoo, A. K., & Samanta, S. K., (2020). A review on basic biology of bacterial biofilm infections and their treatments by nanotechnology-based approaches. *Proceedings of the National Academy of Sciences, India Section B: Biological Sciences, 90*, 243–259.
15. Basavegowda, N., Mishra, K., & Lee, Y. R., (2015). Ultrasonic-assisted green synthesis of palladium nanoparticles and their nanocatalytic application in multicomponent reaction. *New Journal of Chemistry, 39*(2), 972–977.
16. Bhuiyan, M. S. H., Miah, M. Y., Paul, S. C., Aka, T. D., Saha, O., Rahaman, M. M., & Ashaduzzaman, M., (2020). Green synthesis of iron oxide nanoparticle using *Carica papaya* leaf extract: Application for photocatalytic degradation of Remazol yellow RR dye and antibacterial activity. *Heliyon, 6*(8), E04603. doi: 10.1016/j.heliyon.2020.e04603.
17. Bibi, I., Nazar, N., Ata, S., Sultan, M., Ali, A., Abbas, A., & Jalal, F., (2019). Green synthesis of iron oxide nanoparticles using pomegranate seeds extract and photocatalytic activity evaluation for the degradation of textile dye. *Journal of Materials Research and Technology, 8*(6), 6115–6124.
18. Brinkman, C. L., Schmidt-Malan, S. M., Karau, M. J., Greenwood-Quaintance, K., Hassett, D. J., Mandrekar, J. N., & Patel, R., (2016). Exposure of bacterial biofilms to electrical current leads to cell death mediated in part by reactive oxygen species. *Plos One, 11*(12), E0168595. doi: 10.1371/journal.pone.0168595.
19. Brun, N., Mazerolles, L., & Pernot, M., (1991). Microstructure of opaque red glass containing copper. *Journal of Materials Science Letters, 10*(23), 1418–1420.
20. Thomas, S. C., Kumar, M. P., & Talegaonkar, S., (2015). Ceramic nanoparticles: Fabrication methods and applications in drug delivery. *Current Pharmaceutical Design, 21*(42), 6165–6188.
21. Centers for Disease Control and Prevention, (2013). *Antibiotic Resistance Threats in the United States* (p. 144). https://www.cdc.gov/drugresistance/pdf/ar-threats-2013-508.pdf (accessed on 11 February 2022).
22. Chauhan, S., & Upadhyay, L. S. B., (2019). Biosynthesis of iron oxide nanoparticles using plant derivatives of *Lawsonia inermis* (Henna) and its surface modification for biomedical application. *Nanotechnology for Environmental Engineering, 4*(1), 8. doi: 10.1007/s41204-019-0055-5.
23. Cheirmadurai, K., Biswas, S., Murali, R., & Thanikaivelan, P., (2014). Green synthesis of copper nanoparticles and conducting nanobiocomposites using plant and animal sources. *RSC Advances, 4*(37), 19507–19511.
24. Chowdhury, R., Khan, A., & Rashid, M. H., (2020). Green synthesis of CuO nanoparticles using *Lantana camara* flower extract and their potential catalytic activity towards the aza-Michael reaction. *RSC Advances, 10*(24), 14374–14385.
25. Dagdeviren, C., Hwang, S. W., Su, Y., Kim, S., Cheng, H., Gur, O., & Rogers, J. A., (2013). Transient, biocompatible electronics and energy harvesters based on ZnO. *Small, 9*(20), 3398–3404.
26. Debecker, D. P., (2018). Innovative sol-gel routes for the bottom-up preparation of heterogeneous catalysts. *The Chemical Record, 18*(7, 8), 662–675.
27. Devi, H. S., Boda, M. A., Shah, M. A., Parveen, S., & Wani, A. H., (2019). Green synthesis of iron oxide nanoparticles using *Platanus orientalis* leaf extract for antifungal activity. *Green Processing and Synthesis, 8*(1), 38–45.

28. Ealias, A. M., & Saravanakumar, M. P., (2017). A review on the classification, characterization, synthesis of nanoparticles and their application. *IOP Conference Series Materials Science and Engineering, 263*(3), 032019. doi: 10.1088/1757-899X/263/3/032019.
29. Edwards, P. P., & Thomas, J. M., (2007). Gold in a metallic divided state - from faraday to present-day nanoscience. *Angewandte Chemie International Edition, 46*(29), 5480–5486.
30. Elango, G., & Roopan, S. M., (2015). Green synthesis, spectroscopic investigation and photocatalytic activity of lead nanoparticles. *Spectrochimica Acta Part A: Molecular and Biomolecular Spectroscopy, 139*, 367–373.
31. Fahmy, S. A., Preis, E., Bakowsky, U., & Azzazy, H. M. E. S., (2020). Platinum nanoparticles: Green synthesis and biomedical applications. *Molecules, 25*(21), 4981. doi: 10.3390/molecules25214981.
32. Fang, F., Li, M., Zhang, J., & Lee, C. S., (2020). Different strategies for organic nanoparticle preparation in biomedicine. *ACS Materials Letters, 2*(5), 531–549.
33. Fang, X., Wang, Y., Wang, Z., Jiang, Z., & Dong, M., (2019). Microorganism assisted synthesized nanoparticles for catalytic applications. *Energies, 12*(1), 190. doi: 10.3390/en12010190.
34. Ferrer, D., Torres-Castro, A., Gao, X., Sepulveda-Guzman, S., Ortiz-Mendez, U., & Jose-Yacaman, M., (2007). Three-layer core/shell structure in Au-Pd bimetallic nanoparticles. *Nano Letters, 7*(6), 1701–1705.
35. Feynman, R. P., (1960). There's plenty of room at the bottom. *Engineering and Science Magazine, 23*(5), 22–36.
36. Foong, L. K., Foroughi, M. M., Mirhosseini, A. F., Safaei, M., Jahani, S., Mostafavi, M., & Khatami, M., (2020). Applications of nanomaterials in diverse dentistry regimes. *RSC Advances, 10*(26), 15430–15460.
37. Gebreyohannes, G., Nyerere, A., Bii, C., & Sbhatu, D. B., (2019). Challenges of intervention, treatment, and antibiotic resistance of biofilm-forming microorganisms. *Heliyon, 5*(8), e02192. doi: 10.1016/j.heliyon.2019.e02192.
38. Ghaffarian, H. R., Saiedi, M., Sayyadnejad, M. A., & Rashidi, A. M., (2011). Synthesis of ZnO nanoparticles by spray pyrolysis method. *Iranian Journal of Chemistry and Chemical Engineering, 30*(1), 1–6.
39. Ghiuţă, I., Cristea, D., & Munteanu, D., (2017). Synthesis methods of metallic nanoparticles-an overview. *Bulletin of the Transylvania University of Brasov. Engineering Sciences; Series I, 10*(2), 133–140.
40. Giaouris, E., Simões, M., & Dubois-Brissonnet, F., (2020). The role of biofilms in the development and dissemination of microbial resistance within the food industry. *Foods, 9*(6), 816. doi: 10.3390/foods9060816.
41. Gopinath, K., Karthika, V., Gowri, S., Senthilkumar, V., Kumaresan, S., & Arumugam, A., (2014). Antibacterial activity of ruthenium nanoparticles synthesized using *Gloriosa superba* L. leaf extract. *Journal of Nanostructure in Chemistry, 4*(1), 83. doi: 10.1007/s40097-014-0083-4.
42. Grigore-Gurgu, L., Bucur, F. I., Borda, D., Alexa, E. A., Neagu, C., & Nicolau, A. I., (2019). Biofilms formed by pathogens in food and food processing environments. In: Dincer, D., Oezdenefe, M. S., & Arkut, A., (eds.), *Bacterial Biofilms* (pp. 1–32). London – UK: IntechOpen.

43. Guilger-Casagrande, M., & De Lima, R., (2019). Synthesis of silver nanoparticles mediated by fungi: A review. *Frontiers in Bioengineering and Biotechnology, 7*, 287. doi: 10.3389/fbioe.2019.00287.
44. Gupta, K., Barua, S., Hazarika, S. N., Manhar, A. K., Nath, D., Karak, N., & Mandal, M., (2014). Green silver nanoparticles: Enhanced antimicrobial and antibiofilm activity with effects on DNA replication and cell cytotoxicity. *RSC Advances, 4*(95), 52845–52855.
45. Gupta, K., Hazarika, S. N., Saikia, D., Namsa, N. D., & Mandal, M., (2014). One-step green synthesis and antimicrobial and anti-biofilm properties of *Psidium Guajava* L. leaf extract-mediated silver nanoparticles. *Materials Letters, 125*, 67–70.
46. Hamouda, R. A., Hussein, M. H., Abo-Elmagd, R. A., & Bawazir, S. S., (2019). Synthesis and biological characterization of silver nanoparticles derived from the cyanobacterium *Oscillatoria limnetica. Scientific Reports, 9*(1), 1–17.
47. Han, Z., Dong, L., Zhang, J., Cui, T., Chen, S., Ma, G., & Wang, L., (2019). Green synthesis of palladium nanoparticles using lentinan for catalytic activity and biological applications. *RSC Advances, 9*(65), 38265–38270.
48. Hauser, A. R., Mecsas, J., & Moir, D. T., (2016). Beyond antibiotics: New therapeutic approaches for bacterial infections. *Clinical Infectious Diseases, 63*(1), 89–95.
49. Hazarika, M., Borah, D., Bora, P., Silva, A. R., & Das, P., (2017). Biogenic synthesis of palladium nanoparticles and their applications as catalyst and antimicrobial agent. *Plos One, 12*(9), e0184936. doi: 10.1371/journal.pone.0184936.
50. Heiligtag, F. J., & Niederberger, M., (2013). The fascinating world of nanoparticle research. *Materials Today, 16*(7, 8), 262–271.
51. Hofer, U., (2019). The cost of antimicrobial resistance. *Nature Reviews Microbiology, 17*(1), 3–3.
52. Hulkoti, N. I., & Taranath, T. C., (2014). Biosynthesis of nanoparticles using microbes: A review. *Colloids and Surfaces B: Biointerfaces, 121*, 474–483.
53. Hussain, I., Singh, N. B., Singh, A., Singh, H., & Singh, S. C., (2016). Green synthesis of nanoparticles and its potential application. *Biotechnology Letters, 38*(4), 545–560.
54. Iijima, S., (1991). Helical microtubules of graphitic carbon. *Nature, 354,* 56–58.
55. Ijaz, I., Gilani, E., Nazir, A., & Bukhari, A., (2020). Detail review on chemical, physical and green synthesis, classification, characterizations and applications of nanoparticles. *Green Chemistry Letters and Reviews, 13*(3), 223–245.
56. Iqbal, T., Ejaz, A., Abrar, M., Afsheen, S., Batool, S. S., Fahad, M., & Tahir, M. B., (2019). Qualitative and quantitative analysis of nanoparticles using laser-induced breakdown spectroscopy (LIBS) and energy-dispersive x-ray spectroscopy (EDS). *Laser Physics, 29*(11), 116001. doi: 10.1088/1555-6611/ab3fa1.
57. Iravani, S., (2011). Green synthesis of metal nanoparticles using plants. *Green Chemistry, 13*(10), 2638–2650.
58. ISO (International Organization for Standardization). (2010). *ISO/TS 80004-1: Nanotechnology - Vocabulary - Part 1: Core Terms and Definitions.* International Organization for Standardization: Geneva, Switzerland., https://www.iso.org/standard/68058.html (accessed on 11 February 2022).
59. Jadoun, S., Arif, R., Jangid, N. K., & Meena, R. K., (2020). Green synthesis of nanoparticles using plant extracts: A review. *Environmental Chemistry Letters, 19*, 1–20.

60. Jafarizad, A., Safaee, K., Gharibian, S., Omidi, Y., & Ekinci, D., (2015). Biosynthesis and in-vitro study of gold nanoparticles using Mentha and pelargonium extracts. *Procedia Materials Science, 11*, 224–230.
61. Jeevanandam, J., Barhoum, A., Chan, Y. S., Dufresne, A., & Danquah, M. K., (2018). Review on nanoparticles and nanostructured materials: History, sources, toxicity and regulations. *Beilstein Journal of Nanotechnology, 9*(1), 1050–1074.
62. Johnson-Mcdaniel, D., Barrett, C. A., Sharafi, A., & Salguero, T. T., (2013). Nanoscience of an ancient pigment. *Journal of the American Chemical Society, 135*(5), 1677–1679.
63. Kajbafvala, A., Li, M., Bahmanpour, H., Maneshian, M. H., & Kauffmann, A., (2013). Nano/microstructured materials: Rapid, low-cost, and eco-friendly synthesis methods. *Journal of Nanoparticles, 2013*, 530170. doi: 10.1155/2013/530170.
64. Kashale, A. A., Gattu, K. P., Ghule, K., Ingole, V. H., Dhanayat, S., Sharma, R., & Ghule, A. V., (2016). Biomediated green synthesis of $TiO_2$ nanoparticles for lithium-ion battery application. *Composites Part B: Engineering, 99*, 297–304.
65. Kaur, P., (2018). Biosynthesis of nanoparticles using eco-friendly factories and their role in plant pathogenicity: A review. *Biotechnology Research and Innovation, 2*(1), 63–73.
66. Kaur, P., Thakur, R., & Chaudhury, A., (2016). Biogenesis of copper nanoparticles using peel extract of *Punica granatum* and their antimicrobial activity against opportunistic pathogens. *Green Chemistry Letters and Reviews, 9*(1), 33–38.
67. Kern, W., & Schuegraf, K. K., (2001). Deposition technologies and applications: Introduction and overview. In: Krisna, S. K., (ed.), *Handbook of Thin Film Deposition Processes and Techniques* (2nd edn., pp. 11–43). New York – UK: William Andrew Publishing.
68. Khan, I., Saeed, K., & Khan, I., (2019). Nanoparticles: Properties, applications and toxicities. *Arabian Journal of Chemistry, 12*(7), 908–931.
69. Khan, T., Ullah, N., Khan, M. A., & Nadhman, A., (2019). Plant-based gold nanoparticles; a comprehensive review of the decade-long research on synthesis, mechanistic aspects and diverse applications. *Advances in Colloid and Interface Science, 272*, 102017. doi: 10.1016/j.cis.2019.102017.
70. Kim, D., Kim, J., Park, Y. I., Lee, N., & Hyeon, T., (2018). Recent development of inorganic nanoparticles for biomedical imaging. *ACS Central Science, 4*(3), 324–336.
71. Kolekar, R., Bhade, S., Kumar, R. A. J. I. V., Reddy, P., Singh, R. A. J. V. I. R., & Pradeepkumar, K., (2015). Biosynthesis of copper nanoparticles using aqueous extract of *eucalyptus* sp. plant leaves. *Current Science, 109*(2), 255–257.
72. Konings, W. N., (2006). Microbial transport: Adaptations to natural environments. *Antonie Van Leeuwenhoek, 90*(4), 325–342.
73. Krishnaraj, C., Jagan, E. G., Rajasekar, S., Selvakumar, P., Kalaichelvan, P. T., & Mohan, N. J. C. S. B. B., (2010). Synthesis of silver nanoparticles using *Acalypha indica* leaf extracts and its antibacterial activity against water borne pathogens. *Colloids and Surfaces B: Biointerfaces, 76*(1), 50–56.
74. Kulkarni, N., & Muddapur, U., (2014). Biosynthesis of metal nanoparticles: A review. *Journal of Nanotechnology, 2014*, 510246. doi: 10.1155/2014/510246.
75. Kumar, K. M., Mandal, B. K., & Tammina, S. K., (2013). Green synthesis of nano platinum using naturally occurring polyphenols. *RSC Advances, 3*(12), 4033–4039.
76. Kumar, P. V., Shameem, U., Kollu, P., Kalyani, R. L., & Pammi, S. V. N., (2015). Green synthesis of copper oxide nanoparticles using *Aloe vera* leaf extract and its antibacterial activity against fish bacterial pathogens. *Bionanoscience, 5*(3), 135–139.

77. Kumarasamy, K. K., Toleman, M. A., Walsh, T. R., Bagaria, J., Butt, F., Balakrishnan, R., & Krishnan, P., (2010). Emergence of a new antibiotic resistance mechanism In India, Pakistan, and the UK: A molecular, biological, and epidemiological study. *The Lancet Infectious Diseases, 10*(9), 597–602.
78. Larue, C., Castillo-Michel, H., Sobanska, S., Cécillon, L., Bureau, S., Barthès, V., & Sarret, G., (2014). Foliar exposure of the crop *Lactuca sativa* to silver nanoparticles: Evidence for internalization and changes in ag speciation. *Journal of Hazardous Materials, 264*, 98–106.
79. Lee, H. J., Song, J. Y., & Kim, B. S., (2013). Biological synthesis of copper nanoparticles using *Magnolia Kobus* leaf extract and their antibacterial activity. *Journal of Chemical Technology & Biotechnology, 88*(11), 1971–1977.
80. Leon, L., Chung, E. J., & Rinaldi, C., (2020). A brief history of nanotechnology and introduction to nanoparticles for biomedical applications. Chapter 1. In: Chung, E. J., Leon, L., & Rinaldi, C., (eds.), *Nanoparticles for Biomedical Applications* (pp. 1–4). Cambridge – USA: Elsevier Inc.
81. Li, X., Xu, H., Chen, Z. S., & Chen, G., (2011). Biosynthesis of nanoparticles by microorganisms and their applications. *Journal of Nanomaterials, 2011*, 270974. doi: 10.1155/2011/270974.
82. Luo, C. H., Shanmugam, V., & Yeh, C. S., (2015). Nanoparticle biosynthesis using unicellular and subcellular supports. *NPG Asia Materials, 7*(8), e209.
83. Maity, D., Ding, J. U. N., & Xue, J. M., (2008). Synthesis of magnetite nanoparticles by thermal decomposition: Time, temperature, surfactant and solvent effects. *Functional Materials Letters, 1*(03), 189–193.
84. Manju, S., Malaikozhundan, B., Vijayakumar, S., Shanthi, S., Jaishabanu, A., Ekambaram, P., & Vaseeharan, B., (2016). Antibacterial, antibiofilm and cytotoxic effects of *nigella sativa* essential oil coated gold nanoparticles. *Microbial Pathogenesis, 91*, 129–135.
85. Mariselvam, R., Ranjitsingh, A. J. A., Nanthini, A. U. R., Kalirajan, K., Padmalatha, C., & Selvakumar, P. M., (2014). Green synthesis of silver nanoparticles from the extract of the inflorescence of *Cocos nucifera* (Family: Arecaceae) for enhanced antibacterial activity. *Spectrochimica Acta Part A: Molecular and Biomolecular Spectroscopy, 129*, 537–541.
86. Masum, M., Islam, M., Siddiqa, M., Ali, K. A., Zhang, Y., Abdallah, Y., & Li, B., (2019). Biogenic synthesis of silver nanoparticles using *Phyllanthus emblica* fruit extract and its inhibitory action against the pathogen *Acidovorax oryzae* strain RS-2 of rice bacterial brown stripe. *Frontiers in Microbiology, 10*, 820. doi: 10.3389/fmicb.2019.00820.
87. Mohammadi, S., Harvey, A., & Boodhoo, K. V., (2014). Synthesis of $TiO_2$ nanoparticles in a spinning disc reactor. *Chemical Engineering Journal, 258*, 171–184.
88. Mueller, R., Mädler, L., & Pratsinis, S. E., (2003). Nanoparticle synthesis at high production rates by flame spray pyrolysis. *Chemical Engineering Science, 58*(10), 1969–1976.
89. Nabila, M. I., & Kannabiran, K., (2018). Biosynthesis, characterization and antibacterial activity of copper oxide nanoparticles (CuO NPs) from actinomycetes. *Biocatalysis and Agricultural Biotechnology, 15*, 56–62.
90. Nadeem, M., Tungmunnithum, D., Hano, C., Abbasi, B. H., Hashmi, S. S., Ahmad, W., & Zahir, A., (2018). The current trends in the green syntheses of titanium oxide

nanoparticles and their applications. *Green Chemistry Letters and Reviews, 11*(4), 492–502.

91. Nagar, N., & Devra, V., (2018). Green synthesis and characterization of copper nanoparticles using *Azadirachta indica* leaves. *Materials Chemistry and Physics, 213*, 44–51.
92. Naika, H. R., Lingaraju, K., Manjunath, K., Kumar, D., Nagaraju, G., Suresh, D., & Nagabhushana, H., (2015). Green synthesis of CuO nanoparticles using *Gloriosa superba* L. extract and their antibacterial activity. *Journal of Taibah University for Science, 9*(1), 7–12.
93. Nakkala, J. R., Mata, R., Gupta, A. K., & Sadras, S. R., (2014). Biological activities of green silver nanoparticles synthesized with *Acorous calamus* rhizome extract. *European Journal of Medicinal Chemistry, 85*, 784–794.
94. Narayanan, K. B., & Sakthivel, N., (2019). Green synthesis of biogenic metal nanoparticles by terrestrial and aquatic phototrophic and heterotrophic eukaryotes and biocompatible agents. *Advances in Colloid and Interface Science, 169*(2), 59–79.
95. Nasrollahzadeh, M., Sajadi, S. M., & Khalaj, M., (2014). Green synthesis of copper nanoparticles using aqueous extract of the leaves of *Euphorbia esula* L and their catalytic activity for ligand-free Ullmann-coupling reaction and reduction of 4-nitrophenol. *RSC Advances, 4*(88), 47313–47318.
96. Nath, D., Manhar, A. K., Gupta, K., Saikia, D., Das, S. K., & Mandal, M., (2015). Phytosynthesized iron nanoparticles: Effects on fermentative hydrogen production by *Enterobacter cloacae* DH-89. *Bulletin of Materials Science, 38*(6), 1533–1538.
97. Negi, S., & Singh, V., (2018). Algae: A potential source for nanoparticle synthesis. *Journal of Applied and Natural Science, 10*(4), 1134–1140.
98. Nguyen, M. T., & Yonezawa, T., (2018). Sputtering onto a liquid: Interesting physical preparation method for multi-metallic nanoparticles. *Science and Technology of Advanced Materials, 19*(1), 883–898.
99. Oliver, S., Wagh, H., Liang, Y., Yang, S., & Boyer, C., (2018). Enhancing the antimicrobial and antibiofilm effectiveness of silver nanoparticles prepared by green synthesis. *Journal of Materials Chemistry B, 6*(24), 4124–4138.
100. Omran, B. A., (2020). Current trends in algae-mediated synthesis of metal and metal oxide nanoparticles (phyconanotechnology). In: Omran, B. A., (ed.), *Nanobiotechnology: A Multidisciplinary Field of Science* (pp. 111–143). Cham – Switzerland: Springer.
101. Padalia, H., Moteriya, P., & Chanda, S., (2015). Green synthesis of silver nanoparticles from marigold flower and its synergistic antimicrobial potential. *Arabian Journal of Chemistry, 8*(5), 732–741.
102. Padil, V. V. T., & Černík, M., (2013). Green synthesis of copper oxide nanoparticles using gum karaya as a biotemplate and their antibacterial application. *International Journal of Nanomedicine, 8*, 889–898.
103. Parashar, M., Shukla, V. K., & Singh, R., (2020). Metal oxides nanoparticles via sol-gel method: A review on synthesis, characterization and applications. *Journal of Materials Science: Materials in Electronics*, 3729–3749.
104. Park, Y., (2014). A new paradigm shift for the green synthesis of antibacterial silver nanoparticles utilizing plant extracts. *Toxicological Research, 30*(3), 169–178.
105. Patil, M. P., & Kim, G. D., (2017). Eco-friendly approach for nanoparticles synthesis and mechanism behind antibacterial activity of silver and anticancer activity of gold nanoparticles. *Applied Microbiology and Biotechnology, 101*(1), 79–92.

106. Patra, J. K., & Baek, K. H., (2017). Antibacterial activity and synergistic antibacterial potential of biosynthesized silver nanoparticles against foodborne pathogenic bacteria along with its anticandidal and antioxidant effects. *Frontiers in Microbiology, 8*, 167. doi: 10.3389/fmicb.2017.00167.
107. Pimpin, A., & Srituravanich, W., (2012). Review on micro and nanolithography techniques and their applications. *Engineering Journal, 16*(1), 37–56.
108. Piras, C. C., Fernández-Prieto, S., & De Borggraeve, W. M., (2019). Ball milling: A green technology for the preparation and functionalization of nanocellulose derivatives. *Nanoscale Advances, 1*(3), 937–947.
109. Pirtarighat, S., Ghannadnia, M., & Baghshahi, S., (2019). Green synthesis of silver nanoparticles using the plant extract of *Salvia spinosa* grown *in vitro* and their antibacterial activity assessment. *Journal of Nanostructure in Chemistry, 9*(1), 1–9.
110. Pottathara, Y. B., Grohens, Y., Kokol, V., Kalarikkal, N., & Thomas, S., (2019). Synthesis and processing of emerging two-dimensional nanomaterials. Chapter 1. In: Pottathara, Y. B., Kalarikkal, N., Kokol, V., Thomas, S., & Grohens, Y., (eds.), *Nanomaterials Synthesis* (pp. 1–25). Cambridge – USA: Elsevier Inc.
111. Prasad, K. S., Amin, T., Katuva, S., Kumari, M., & Selvaraj, K., (2017). Synthesis of water-soluble CDS nanoparticles and study of their DNA damage activity. *Arabian Journal of Chemistry, 10*, S3929–S3935.
112. Priya, G. S., Kanneganti, A., Kumar, K. A., Rao, K. V., & Bykkam, S., (2014). Biosynthesis of cerium oxide nanoparticles using *Aloe barbadensis* miller gel. *International Journal of Scientific and Research Publication, 4*(6), 199–224.
113. Rabin, N., Zheng, Y., Opoku-Temeng, C., Du, Y., Bonsu, E., & Sintim, H. O., (2015). Biofilm formation mechanisms and targets for developing antibiofilm agents. *Future Medicinal Chemistry, 7*(4), 493–512.
114. Rane, A. V., Kanny, K., Abitha, V. K., & Thomas, S., (2018). Methods for synthesis of nanoparticles and fabrication of nanocomposites. Chapter 5. In: Bhagyaraj, S. M., Oluwafemi, O. S., Kalarikkal, N., & Thomas, S., (eds.), *Synthesis of Inorganic Nanomaterials* (pp. 121–139). Cambridge – USA: Woodhead Publishing.
115. Rastogi, A., Zivcak, M., Sytar, O., Kalaji, H. M., He, X., Mbarki, S., & Brestic, M., (2017). Impact of metal and metal oxide nanoparticles on plant: A critical review. *Frontiers in Chemistry, 5*, 78. doi: 10.3389/fchem.2017.00078.
116. Rodríguez-León, E., Rodríguez-Vázquez, B. E., Martínez-Higuera, A., Rodríguez-Beas, C., Larios-Rodríguez, E., Navarro, R. E., & Iñiguez-Palomares, R. A., (2019). Synthesis of gold nanoparticles using *Mimosa tenuiflora* extract, assessments of cytotoxicity, cellular uptake, and catalysis. *Nanoscale Research Letters, 14*(1), 1–16.
117. Romero, G., & Moya, S. E., (2012). Synthesis of organic nanoparticles. Chapter 4. In: Jesus, M., De La, F., & Grazu, V., (eds.), *Frontiers of Nanoscience* (Vol. 4, pp. 115–141). Cambridge – USA: Elsevier Inc.
118. Sadeghi, B., Mohammadzadeh, M., & Babakhani, B., (2015). Green synthesis of gold nanoparticles using *Stevia rebaudiana* leaf extracts: Characterization and their stability. *Journal of Photochemistry and Photobiology B: Biology, 148*, 101–106.
119. Sadrolhosseini, A. R., Mahdi, M. A., Alizadeh, F., & Rashid, S. A., (2019). laser ablation technique for synthesis of metal nanoparticles in liquid. Chapter 4. In: Ma, Y., (ed.), *Laser Technology and its Applications* (pp. 63–81). London – UK: IntechOpen.

120. Şahutoğlu, A. S., & Akgül, C., (2020). One-phase synthesis of single enzyme nanoparticles (Sens) of trametes Versicolor laccase by in situ acrylamide polymerization. *Biocatalysis and Biotransformation, 38*(1), 64–74.
121. Saif, S., Tahir, A., & Chen, Y., (2016). Green synthesis of iron nanoparticles and their environmental applications and implications. *Nanomaterials, 6*(11), 1–26.
122. Saleh, T. A., (2020). Nanomaterials: Classification, properties, and environmental toxicities. *Environmental Technology & Innovation, 20*, 101067. doi: 10.1016/j.eti.2020.101067.
123. Sankar, R., Rizwana, K., Shivashangari, K. S., & Ravikumar, V., (2015). Ultra-rapid photocatalytic activity of *Azadirachta indica* engineered colloidal titanium dioxide nanoparticles. *Applied Nanoscience, 5*(6), 731–736.
124. Santhoshkumar, J., Kumar, S. V., & Rajeshkumar, S., (2017). Synthesis of zinc oxide nanoparticles using plant leaf extract against urinary tract infection pathogen. *Resource-Efficient Technologies, 3*(4), 459–465.
125. Selim, Y. A., Azb, M. A., Ragab, I., & Abd El-Azim, M. H., (2020). Green synthesis of zinc oxide nanoparticles using aqueous extract of *Deverra tortuosa* and their cytotoxic activities. *Scientific Reports, 10*(1), 1–9.
126. Senthilkumar, N., Nandhakumar, E., Priya, P., Soni, D., Vimalan, M., & Potheher, I. V., (2017). Synthesis of ZnO nanoparticles using leaf extract of *Tectona grandis* (L.) and their antibacterial, anti-arthritic, anti-oxidant and *in vitro* cytotoxicity activities. *New Journal of Chemistry, 41*(18), 10347–10356.
127. Shende, S., Ingle, A. P., Gade, A., & Rai, M., (2015). Green synthesis of copper nanoparticles by *Citrus medica* Linn. (Idilimbu) juice and its antimicrobial activity. *World Journal of Microbiology and Biotechnology, 31*(6), 865–873.
128. Shrestha, P., Cooper, B. S., Coast, J., Oppong, R., Thuy, N. D. T., Phodha, T., & Lubell, Y., (2018). Enumerating the economic cost of antimicrobial resistance per antibiotic consumed to inform the evaluation of interventions affecting their use. *Antimicrobial Resistance & Infection Control, 7*(1), 98. doi: 10.1186/s13756-018-0384-3.
129. Siddiqi, K. S., & Husen, A., (2020). Current status of plant metabolite-based fabrication of copper/copper oxide nanoparticles and their applications: A review. *Biomaterials Research, 24*(1), 1–15.
130. Singh, J., Kumar, V., Kim, K. H., & Rawat, M., (2019). Biogenic synthesis of copper oxide nanoparticles using plant extract and its prodigious potential for photocatalytic degradation of Dyes. *Environmental Research, 177*, 108569. doi: 10.1016/j.envres.2019.108569.
131. Singh, P., Pandit, S., Beshay, M., Mokkapati, V. R. S. S., Garnaes, J., Olsson, M. E., & Daugaard, A. E., (2018). Anti-biofilm effects of gold and silver nanoparticles synthesized by the *Rhodiola rosea* rhizome extracts. *Artificial Cells, Nanomedicine, and Biotechnology, 46*, S886-S899.
132. Stevanović, M. M., Škapin, S. D., Bračko, I., Milenković, M., Petković, J., Filipič, M., & Uskoković, D. P., (2012). Poly (lactide-co-glycolide)/silver nanoparticles: Synthesis, characterization, antimicrobial activity, cytotoxicity assessment and ROS-inducing potential. *Polymer, 53*(14), 2818–2828.
133. Suci, P. A., Berglund, D. L., Liepold, L., Brumfield, S., Pitts, B., Davison, W., & Young, M., (2007). High-density targeting of a viral multifunctional nanoplatform to a pathogenic, biofilm-forming bacterium. *Chemistry & Biology, 14*(4), 387–398.

134. Sukumar, S., Rudrasenan, A., & Padmanabhan, N. D., (2020). Green-synthesized rice-shaped copper oxide nanoparticles using *Caesalpinia bonducella* seed extract and their applications. *ACS Omega, 5*(2), 1040–1051.
135. Sun, Q., Cai, X., Li, J., Zheng, M., Chen, Z., & Yu, C. P., (2014). Green synthesis of silver nanoparticles using tea leaf extract and evaluation of their stability and antibacterial activity. *Colloids and Surfaces A: Physicochemical and Engineering Aspects, 444*, 226–231.
136. Swathi, N., Sandhiya, D., Rajeshkumar, S., & Lakshmi, T., (2019). Green synthesis of titanium dioxide nanoparticles using *Cassia fistula* and its antibacterial activity. *International Journal of Research in Pharmaceutical Sciences, 10*(2), 856–860.
137. Tahir, K., Nazir, S., Ahmad, A., Li, B., Shah, S. A. A., Khan, A. U., & Khan, F. U., (2016). Biodirected synthesis of palladium nanoparticles using *phoenix dactylifera* leaves extract and their size-dependent biomedical and catalytic applications. *RSC Advances, 6*(89), 85903–85916.
138. Tai, C. Y., Wang, Y. H., Tai, C. T., & Liu, H. S., (2009). Preparation of silver nanoparticles using a spinning disk reactor in a continuous mode. *Industrial & Engineering Chemistry Research, 48*(22), 10104–10109.
139. Tailor, G., Yadav, B. L., Chaudhary, J., Joshi, M., & Suvalka, C., (2020). Green synthesis of silver nanoparticles using *Ocimum canum* and their antibacterial activity. *Biochemistry and Biophysics Reports, 24*, 100848. doi: 10.1016/j.bbrep.2020.100848.
140. Taylor, E., & Webster, T. J., (2011). Reducing infections through nanotechnology and nanoparticles. *International Journal of Nanomedicine, 6*, 1463–1473.
141. Teimuri-Mofrad, R., Hadi, R., Tahmasebi, B., Farhoudian, S., Mehravar, M., & Nasiri, R., (2017). Green synthesis of gold nanoparticles using plant extract: Mini-review. *Nanochemistry Research, 2*(1), 8–19.
142. Thiruvengadam, M., Chung, I. M., Gomathi, T., Ansari, M. A., Khanna, V. G., Babu, V., & Rajakumar, G., (2019). Synthesis, characterization and pharmacological potential of green synthesized copper nanoparticles. *Bioprocess and Biosystems Engineering, 42*(11), 1769–1777.
143. Unni, M., Uhl, A. M., Savliwala, S., Savitzky, B. H., Dhavalikar, R., Garraud, N., & Rinaldi, C., (2017). Thermal decomposition synthesis of iron oxide nanoparticles with diminished magnetic dead layer by controlled addition of oxygen. *ACS Nano, 11*(2), 2284–2303.
144. Vance, M. E., Kuiken, T., Vejerano, E. P., Mcginnis, S. P., Hochella, Jr. M. F., Rejeski, D., & Hull, M. S., (2015). Nanotechnology in the real world: Redeveloping the nanomaterial consumer products inventory. *Beilstein Journal of Nanotechnology, 6*(1), 1769–1780.
145. Vasantharaj, S., Sathiyavimal, S., Senthilkumar, P., Lewisoscar, F., & Pugazhendhi, A., (2019). Biosynthesis of iron oxide nanoparticles using leaf extract of *Ruellia tuberosa*: Antimicrobial properties and their applications in photocatalytic degradation. *Journal of Photochemistry and Photobiology B: Biology, 192*, 74–82.
146. Vena, M. P., Jobbágy, M., & Bilmes, S. A., (2016). Microorganism mediated biosynthesis of metal chalcogenides; a powerful tool to transform toxic effluents into functional nanomaterials. *Science of the Total Environment, 565*, 804–810.
147. Venkatesh, N., Bhowmik, H., & Kuila, A., (2018). Metallic nanoparticle: A review. *Biomedical Journal of Scientific & Technical Research, 4*(2), 3765–3775.

148. Villaseñor-Basulto, D. L., Pedavoah, M. M., & Bandala, E. R., (2019). Plant materials for the synthesis of nanomaterials: Greener sources. *Journal: Handbook of Ecomaterials,* 105–121. doi: 10.1007/978-3-319-48281-1_88-1.
149. Virlan, M. J. R., Miricescu, D., Radulescu, R., Sabliov, C. M., Totan, A., Calenic, B., & Greabu, M., (2016). Organic nanomaterials and their applications in the treatment of oral diseases. *Molecules, 21*(2), 207. doi: 10.3390/molecules21020207.
150. Vishveshvar, K., Krishnan, M. A., Haribabu, K., & Vishnuprasad, S., (2018). Green synthesis of copper oxide nanoparticles using *Ixiro coccinea* plant leaves and its characterization. *Bionanoscience, 8*(2), 554–558.
151. World Health Organization., (2014). *Antimicrobial Resistance: Global Report on Surveillance* (p. 257). https://www.who.int/publications/i/item/9789241564748 (accessed on 1 March 2022).
152. Yaqoob, A. A., Ahmad, H., Parveen, T., Ahmad, A., Oves, M., Ismail, I. M., & Mohamad, I. M. N., (2020). Recent advances in metal decorated nanomaterials and their various biological applications: A review. *Frontiers in Chemistry, 8*, 341. doi: 10.3389/fchem.2020.00341.
153. Yi, H., Lin, L., Ling, M., Lv, Z., Li, R., Fu, Q., & Li, X., (2019). Scalable and economic synthesis of high-performance $Na_3V_2(PO_4)2F_3$ by a solvothermal-ball-milling method. *ACS Energy Letters, 4*(7), 1565–1571.
154. Zahid, M. U., Pervaiz, E., Hussain, A., Shahzad, M. I., & Niazi, M. B. K., (2018). Synthesis of carbon nanomaterials from different pyrolysis techniques: A review. *Materials Research Express, 5*(5), 052002. doi: 10.1088/2053-1591/aac05b.

## CHAPTER 10

# ANTIBACTERIAL ACTIVITY OF NEEM NANOPARTICLES ON FABRICS

SUNEETA HARLAPUR and SUJATA F. HARLAPUR


## ABSTRACT

Neem (*Azadirachta indica*) nanoparticles (NNPs) for clinical applications have been reported since before times. Progression in current meds brings about an expansion in usage of NNPs for clinical purposes because of their antibacterial, antiviral, antifungal, calming, and antiangiogenic properties. In this part, neem nanoparticles (NNPs) are concentrated widely as amazing nanoweapons for the annihilation of microorganisms. Late exploration gives proof that NNPs are compelling in supporting antimicrobial exercises. The artificially and laser-removed silver, gold, and copper nanoparticles displayed improved antibacterial movement than recently announced. The antibacterial component was revealed portion and size-subordinate and more significant for Gram-negative bacterium when contrasted with Gram-positive ones. For Gram-positive and Gram-negative microbes, the section figures of least inhibitory focus with nanoparticles have been calculated. The most extreme zone of restraint by circle dispersion was additionally tested against different microorganisms. These nanoparticles display magnificent execution genuinely, chemically, and chemically.


## 10.1 INTRODUCTION

Nowadays, naturally extracted nanoparticle usage is increased in the emerging applications of science. These nanomaterials are used in the fabrication,

Nanotechnology Horizons in Food Process Engineering, Volume 1. Megh R. Goyal, PhD, PE, Junaid Ahmad Malik, PhD, Satish Kumar, PhD & Ritesh B. Watharkar, PhD (Eds.)

especially neem nanoparticles (NNPs), due to their attractive properties like size, area, and with extraordinary properties like catalytic, electrical, optical, and antibacterial properties.

The neem NPs are utilized in broad uses of makeup, clinical, and material industry. Physical and compound are the two strategies utilized for a combination of NNPs of wanted shape and size. Yet, these techniques are not effectively attainable and climate agreeable. Green combination is the best technique for amalgamation of NNPs due to its compatibility, reduced hurtfulness and eco-accommodating environment. Numerous microbe and plant life have ended up being a wellspring of encouragement for nanomaterial combination.

In nanotechnology research, silver is one of the most commonly used metals. Its dispersion of size, dependability, morphology, and surface charge/change each assumes an essential part in numerous logical fields [32]. Its fundamental highlights give improvement in photography, microelectronics, photonics, photograph catalysis, lithography, and antimicrobial and antifungal movements [1, 34, 45].

Gold NPs are additionally popular contender for clinical treatment, malignancy treatment, quality treatment, diagnostics, drug conveyance, and natural purposes [15, 19, 26]. Its principal advantage is its straight forward arrangement by compound decrease and displaying low poisonousness. To improve the ability of gold NPs, various procedures have been utilized that give different elements of NPs and their functionalization [5, 51, 52]. Because of latent attributes of gold NPs, it is notable for biocompatibility, yet its cytotoxicity generally relies upon nanoparticle's size having exact focus [7]. Albeit gold is expensive when analogized with silver and copper, the manufacture of its NPs become more unrestricted in late exploration [42].

The subsequent unmistakable metal is copper that has comparative optical, electrical, and warm attributes like silver and gold. The main impediment is its oxidation state during the blend cycle; subsequently, it is the significant test for any researcher. Copper NPs give an ideal trade-off between its novel properties and cost, along these lines demonstrating a significant material for industry. Indeed, its tremendous field of uses is attractive media for capacity gadgets, sun-oriented energy changes, hardware, and catalysis, and moreover, it has uncovered a promising antimicrobial activity [12].

Nanomaterials can be achieved from a piece of plants like stem, leaves, roots, and so on. Usage of plants over microbe is nothing but the disposal of the detailed cycle of cubicle ethos. Nanoparticles blended using natural methods are more helpful for medical applications. Whereas nanoparticles generated from synthetic and actual techniques were more destructive.

Nowadays, expanding anti-toxin opposition by organisms is forcing genuine danger to the wellbeing area. Nanoparticles have being end up an imaginable possibility for antimicrobial specialist due to their huge superficial to capacity proportion guarantees a wide scope of assault on microbial superficial. The most encouraging nanoparticle which goes about as an exceptionally successful antimicrobial specialist is silver. Different examinations on silver nanoparticles have been done to contemplate its antimicrobial action.

AgNPs are more useful in the antibacterial activity. Neem plants are richly available in India. It has a place with the Meliaceae family and is known for its different applications, particularly its therapeutic property. Neem leaves extract is utilized in the combination of different nanoparticles like gold, zinc oxide (ZnO), silver, and so on. Many phytochemicals, including terpenoids and flavanones, are naturally present in neem, which aid in the production of nanoparticles. AgNps are developed with the help of neem leaf extract and silver salt. This chapter discovers the application of neem nanoparticles as an elective antibacterial nanobiotics.

## 10.2 NEEM EXTRACT

Plenty of Neem trees are available freely in the world (Image of Neem tree leaves, as shown in Figure 10.1). Neem is famous for its treating characteristic for antimicrobial and antibacterial infections. Neem tree does not require much care for its growth. Its leaves paste is used for many insect bites and infections. Neem leaves are available with rich resources like alkaloids, resins, gum, sulfur, oils, etc. Its seed contains lots of fatty acids and proteins, but cannot be used because of its bitterness. Neem tree provides a cooling effect to the environment helps to maintain temperature in the atmosphere.

## 10.3 USES OF NEEM

Mainly each and every part of the neem tree is valuable with its significant antibacterial and antifungal properties (Figure 10.2). Neem is generally used in shampoos to cure dandruff and in cleansers or creams for skin conditions like skin inflammation, psoriasis, and competitor's foot. It is additionally a part in certain sorts of toothpaste and mouthwashes, particularly in the

Indian subcontinent the youthful twigs are utilized straightforwardly as unrefined toothbrushes. Sometimes, neem leaves are employed as a conservative management for diabetes and some of the clinical proofs recommending that it might help in controlling glucose levels.

**FIGURE 10.1** Neem leaves and fruits.

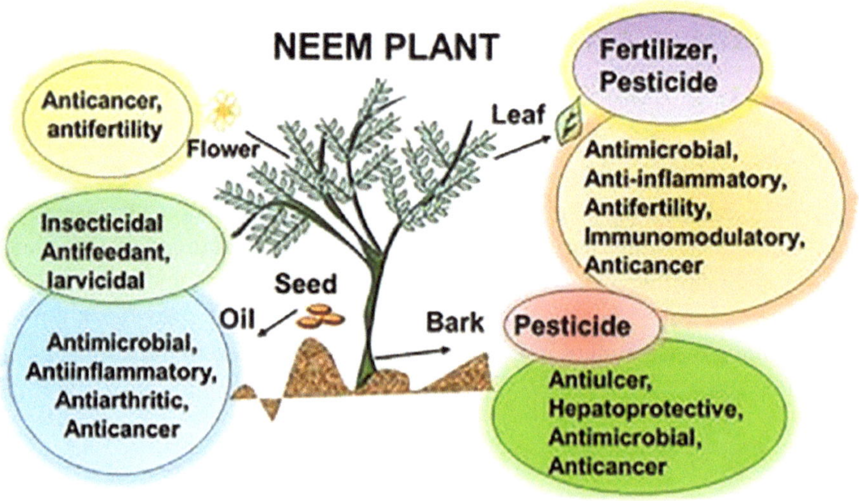


**FIGURE 10.2** Potential of *Azadirachta indica* as biopesticidal and therapeutic agents.

Oil Removed from the seeds can be utilized straightforwardly both pesticide and fungicide. It works on arthropod pests like hornworms, corn earworm, aphids, and whiteflies and besides as a creepy-crawly, bug repellent, bug spray, and fungicide. As a fungicide it controls common fungi, which grow on vegetable plants. Also, it is the wellspring of numerous business pesticide items including tidies, granules, and concentrates.

The essential dynamic insecticidal fixing, azadirachtin, attempts to disturb the hormones associated with creepy-crawly shedding, forestalling hatchlings from forming appropriately into grown-ups and acts as caretaker, an inhibitor. Neem oil can slaughter delicate bodied bugs on contact and diminishes mating and conceptive practices, decreasing irritation fruitfulness. As a fungicide, neem oil is utilized to control rust, dark spot, build up, scab, anthracnose, and curse. Given that neem oil separates rapidly with disclosure to bright light, rehashed applications are frequently fundamental. Neem-based pesticides by and large have low harmfulness for vertebrates and are normal in natural cultivating applications.

## 10.4 NANOPARTICLES

Nanoparticles are very small particles that can be measured in terms of nanometers (nm; 1 nm = $10^{-9}$ meter). Nanoparticles are available in the regular world and are additionally made because of human exercises. Due to their mini structure and wide surface areas, they got more interesting material qualities. Nanoparticles have more value in the market due to their properties to solve many medical-related issues. Its applications are in many fields like medication, designing, catalysis, natural remediation, and so on.

### *10.4.1 NANOPARTICLE APPLICATIONS IN MEDICINE*

These small-sized nanoparticles are highly useful in the medical field. The nanoparticles can roughly pass through the body. These properties enable better methods of optimizing body image, such as tumors and other unhealthy tissues. They have promoted new techniques for transmitting therapies, for example, by blocking the vasculature for infected tissues and tumors, or payloads of drugs.

Following the spread of malignant disease throughout the lymph hubs, attractive nanoparticles were used instead of radioactive technetium. Attractive reverberation imaging nanoparticles work by manipulating the coordination achieved by small particles of superparamagnetic iron oxide. These are used to kill tumors through hyperthermia, by destroying the warmth and tissue at a nearby level.

The nanoparticles may be intended to improve images from positron discharge tomography. These techniques usually require the nanoparticle to have the option of perceiving a particular disease. In principle, a similar idea can be used to help the delivery of drug accurately to a specific diseased location.

The drug can be conveyed through a nanocapsule or it can be deposited in bonds in a permeable nanosponge structure and later, allowing for the modest arrival of the drug. Improving nanoparticles to deliver medicine to the mind through internal breathing has been widely argued to treat neurological problems, Such as skin infection, allergic disease and numerous sclerosis.'

Nanoparticles and nano-fibers have a significant influence in assembling of unique framework constructions for flesh and bone fix. Nanomaterials utilized here are biocompatible. For instance, nanoparticles of calcium hydroxyapatite, a characteristic part of bone, collagen substitutes could be utilized in upcoming tissue-fix treatments.

Nano-particles additionally have been utilized in the improvement of wellbeing-related items. For instance, a sun-screen known as Optisol, concocted at the University of Oxford during the 1990s, was intended with the goal of building up a protected sun-screen that was straight forward in noticeable light, however held bright hindering activity on the skin. The fixings generally utilized in sun-screens depended on huge particles of either zinc oxide (ZnO)/titanium dioxide ($TiO_2$). Nonetheless, these materials were not palatable: ZnO and $TiO_2$ are exceptionally powerful photocatalysts, and within the sight of water and daylight they create free revolutionaries, which can possibly harm skin cells and DNA.

Researchers continued to build up a nanoparticle type of titanium oxide that contained a modest quantity of manganese. Studies demonstrated that the nanoparticle-based sun-screen was more secure than sun-screen items fabricated by utilizing conventional materials. The improvement in security was ascribed to the presentation of manganese, which changed the semi-conducting properties of the compound from n-type to p-type, accordingly moving its Fermi level, and making the age of free revolutionaries more uncertain.

Medicines and indicative methodologies dependent on the utilization of nanoparticles are relied upon to have significant advantages for medication later on, however the utilization of nanoparticles additionally presents critical difficulties, especially with respect to effects on human wellbeing. For instance, little is thought about the destiny of nanoparticles that are brought into the body. Broad clinical preliminaries are required to completely address worries about the wellbeing and viability of nanoparticles utilized in medication. There likewise are fabricating issues to be survived, for example, the capacity to deliver nanoparticles under clean conditions, which is needed for clinical applications.

## 10.5 PROCESS TO ATTAIN NEEM NANO-PARTICLES (NNPS)

### *10.5.1 COLLECTION OF NEEM LEAF EXTRACT*

Collect the fresh neem leaves from the tree and grind them to prepare a smooth green paste out of it (Figure 10.3). Grinded paste has to undergo the centrifugation process. This procedure aids in the removal of debris. Supernatant has to be taken out on filter paper from the processed sample to produce Neem NPs.

**FIGURE 10.3** Neem leaf extract.

### *10.5.2 SYNTHESIS OF NANOPARTICLES USING NEEM EXTRACT*

Greener NNP synthesis has been achieved using extracts from the neem plant [18, 43, 48]. The key objective is to employ neem leaves since they are abundant in nature and are known to have antibacterial properties [16].

The production of gold NNPs was validated by color changes and described by ultraviolet-visible spectroscopy employing neem leaf extract [39, 44]. Mixing an aqueous extract of neem leaves and $10^{-3}$ M aqueous chloroauric acid produced gold nanoparticles (AuNPs). Isomorphic spherical gold NPs of 15–18 nm in size were synthesized. Gold NPs were synthesized using neem leaf extract and 2% auric chloride solution and were characterized [18].

Platinum NPs were synthesized from chloroplatinic acid employing neem leaf extract. On heating, the color of the solution changes from light yellow to black, suggesting the production of platinum NNPs. The size of the NNPs ranged from 5 to 50 nm. These platinum NNPs might be used in a variety of therapeutic and catalytic purposes in the bioelectronics and chemical sectors [43].

The extract of leaves of the neem plant was used to generate zerovalent iron NPs. The size of iron NNPs ranged between 50 and 100 nm [31]. A basic co-precipitation process was used to make spherical-shaped iron oxide particles with diameter varying from 10 to 20 nm. Biocapping the generated NNPs with aqueous neem leaf extract stabilized them. The antibacterial activity of neem extract encapsulated magnetite NNPs was tested using the agar well diffusion technique against *Bacillus* sp., *Staphylococcus aureus*, *Escherichia coli*, and *Pseudomonas* sp. The encapsulating compounds on the surface of magnetite NNPs were ineffective in stopping bacteria from growing [36].

### *10.5.3 SYNTHESIS OF SILVER NANOPARTICLES (AGNPS) USING NEEM EXTRACT*

Silver, one of the most popular nanomaterials with the creation of 500 tons of AgNPs per year [24] and is estimated to rise in subsequent years. Memorizing a significant part of the field of high impact biomolecular identification, catalysts, biosensors, and pharmaceuticals; It has been recognized to have solid inhibitory and antibacterial effects in addition to antifungal, calming, and angiogenesis exercises [10, 47]. AgNPs have interesting optical, electrical, and warm properties that assume an essential part in medication conveyance, diagnostics, imaging, detecting, quality conveyance, counterfeit embeds, and tissue designing [40].

For the combination of AgNPs, there are a number of options, for example, particle faltering, compound decrease, sol-gel, and so on [4, 13, 33, 41]. Be that as it may, these techniques additionally bring about toxins during the amalgamation systems or in later applications with related restrictions. AgNPs can be delivered at a low grouping of leaf separate without utilizing any extra hurtful substance/actual techniques. This greener blend of AgNPs is one-venture, savvy, climate agreeable and generally reproducible and frequently brings about more steady materials [27].

This eco-accommodating strategy could be a serious option to the ordinary physical/substance strategies utilized for the combination of silver nanoparticle and can possibly be utilized in biomedical applications and will assume a significant part in optoelectronics and clinical gadgets is not so distant future.

AgNPs, because of their antimicrobial properties, have been utilized most generally in the wellbeing business, medication, material coatings, food stockpiling, color decrease, wound dressing, germicide creams and various ecological applications [13]. Since old occasions, natural silver and its mixes have been utilized as antimicrobial specialists and were utilized to store water in silver vessels [9, 30]. Antibacterial properties of silver have been reported since 1000 B.C. at the point when silver vessels were utilized to protect water [6, 37].

Silver particles and their mixes are exceptionally poisonous to microorganisms showing solid biocidal impacts on numerous types of microbes yet have a low harmfulness toward creature cells [35]. Bacterial cells presented to the $Ag^{+}$ particles actuate pressure reaction that prompts the buildup of DNA in the focal point of the cell and this buildup of DNA forestalls cell replication by keeping the DNA from being gotten to by transcriptional catalysts, for example, DNA polymerase. It can likewise cause cell film separation from the cell divider, cell divider harm, and electron thick granules made out of silver outside and, in certain occasions, inside the cell [11].

AgNPs anchor to and enter the cell mass of microbes [28], recommending that the resultant primary change in the cell layer could cause an expansion in cell porousness, prompting an uncontrolled vehicle through the cytoplasmic film, and at last cell demise. Silver inactivates proteins by official to sulfur-containing mixes [22]. It was likewise detailed that getting cells with silver leads cell shrinkage and drying out [17]. It has likewise been suggested that the antibacterial instrument of AgNPs is identified with the arrangement of free extremists and ensuing free revolutionary initiated layer harm [8, 21]. Notwithstanding, it has been accounted that AgNPs are non-harmful to people however best against infection, microbes, and other eukaryotic microorganisms even at low fixations with no results [20].

AgNPs are orchestrated by decreasing silver particle utilizing lessening specialists present in neem separate and are portrayed by UV-noticeable spectrophotometer, molecule size analyzer, SEM, transmission electron microscopy (TEM), energy-dispersive spectroscopy, XRD investigation, nanoparticle following analyzer examination and FTIR examination [2, 3, 14, 29]. Change of silver particles' into AgNPs during presentation to the neem leaf extricate is trailed by shading change from light yellow to dull yellowish earthy colored tone in fluid arrangement [50].

Studies have shown that the decreasing phytochemicals in the neem leaf comprised the most part of terpenoids. It was discovered that these lessening segments additionally filled in as covering and balancing out specialists notwithstanding decrease as uncovered from FTIR contemplates [38].

### *10.5.4 ANTIMICROBIAL ACTIVITY OF NEEM SYNTHESIZED AGNPS*

Surgical masks treated with AgNPs were found to have antibacterial effects across various studies. After incubation, nanoparticle-coated masks reduced the number of viable *E. coli* and *S. aureus* cells by 100%. Furthermore, according to the research, there is no skin irritation [25].

The bactericidal impact of AgNPs produced by aqueous extract of neem leaves and 0.01 M silver nitrate mixture in a 1:4 mixing ratio on cotton fabric against E. coli was investigated using the disc diffusion method. The findings indicated antibacterial action, indicating that cotton cloths infused with biologically produced AgNPs might be used for wound treatment [46].

Newly produced AgNPs reduced the development of gentamycin- and ampicillin-resistant *Klebsiella pneumoniae*, gentamicin-, and piperacillin-resistant *Salmonella typhi*, fluconazole-resistant *Candida albicans*, and various drug-resistant *E. coli*. AgNPs produced from neem leaves extract had an average size of 43 nm [14]. Urinary tract infection bacteria such as *S. aureus*, *Proteus* sp., *Pseudomonas* sp., and *E. coli* were significantly inhibited by neem-mediated biosynthesized NPs with a size of roughly 70–80 nm [49]. Green AgNPs prepared from aqueous preparation of neem and 1 $cm^3$ $AgNO_3$ solution demonstrated antimicrobial property against *S. typhi* and *Klebsiella pneumoniae* [23].

At ambient temperature, AgNPs were formed from aqueous neem plant extracts and $AgNO_3$ mixture. The processed NPs were coated on polyurethane foams. After many washes and dry gives stable PU foam with uniform coating. After 30 minutes of contact, this silver nanoparticle (AgNP)-coated PU was effective for eliminating the whole load of microorganisms from water samples with preliminary loads of $1\times10^5$ and $1\times10^6$ CFU/$cm^3$. On the plates, the water treated with the sample group exhibited significant growth. The AgNPs were shown to exhibit stable binding with PU and did not mix with water following a 30-minute treatment [29].

## 10.6 PROCEDURE TO USE NEEM NANO-PARTICLES (NNPS) IN TEXTILE INDUSTRY

### *10.6.1 COLLECTION, PROCESSING, AND EXTRACTION OF NEEM*

Haveri in Karnataka is surrounded by many Neem trees. The collected leaves were concealed to reduce the moisture content to dry at room temperature. Dried leaves and crumbled. 20 g of natural powder was diluted in 100 $cm^3$ of methanol, and it is currently brewed.

### *10.6.2 PREPARATION OF THE NANO-PARTICLES*

Herbal solenoidal albumin nanoparticles are prepared with a coarsening process, followed by crosslinking to glutaraldehyde. The *Azadirachta indica* (Neem) extract was adjusted to pH 5.5 by 1 M HCl using an appropriate quantity of the protein solution and digital pH meter for 1 hour at room temperature. Later ethanol is added.

Biological diluents are eliminated in controlled pressure. Balls are then adjourned in phosphate buffer and each sample finally was lyophilized with mannitol at –480°C and 28 x $10^{-3}$ M Bar pressure for 24 hours. Albumin Nano-particles controls were also prepared by the same procedure without herbal extracts loading.

### *10.6.3 FABRIC SPECIFICATION AND COATING OF COTTON FABRICS WITH NANO-PARTICLES*

A fine-medium weight 100% cotton woven fabric was used for the application purpose. Herbal Nano-particles and plain herbal extract are coated on cotton using pad-dry-cure method. For 5 minutes, the tiny cotton fabric was submerged in a solution comprising Nanoparticles (2%) and citric acid binder (1%) before being pushed through a padded mangle at a speed of 15 m/min and a compression of 2 kg/$cm^2$ to eliminate surplus solution.

For all of the interventions, a 100% wet pick-up was preserved. After packing, the material was air-dried before being cured for 3 minutes at 140°C, submerged for 5 minutes in 2 g/L sodium lauryl sulfate to eliminate unattached nanoparticles, washed to remove the soap solution, and air-dried. The cloth that had not been treated served as a control.

### 10.6.4 ANTIMICROBIAL ASSESSMENT OF FINISHED COTTON FABRICS

Conferring to AATCC (American Association of Textile Chemists and Colorists) guidelines, the investigation employed *Staphylococcus aureus* and *Escherichia coli* as tested microorganisms. The qualitative antibacterial evaluation was performed using the Parallel Strike method (AATCC test method 147), and quantitative antibacterial evaluation was performed using the percentage reduction test (AATCC 100).

#### 10.6.4.1 PARALLEL STREAK METHOD (AATCC 147)

Test examples (nonsterile) were cut into small pieces (25 mm x 50 mm). Sterile AATCC bacteriostasis agar plates were readied. Utilizing a clean 4 mm immunizing circle, one circle brimming with culture was stacked and moved to the outside of the agar plate by making five equal inoculum streaks dispersed 10 mm covering the focal region of the Petri dish without topping off the circle. The test example was delicately squeezed dynamically, over the five inoculums of streaks to guarantee cozy contact with the agar surface.

The plates were hatched at 37°C for 18–24 hours. The immunized plates were inspected for the interference of development alongside the dashes of inoculum underneath the texture and for an away from of hindrance past the texture edge. The normal width of the zone of restraint around the test example is determined in mm utilizing the accompanying recipe Zone of hindrance (mm) = [(T-I)/2], where T-width of the zone of hindrance and I-width of example.

#### 10.6.4.2 PERCENTAGE REDUCTION TEST (AATCC 100)

The non-sterile test examples were taken, and they were sliced into pieces as indicated by standard size (20 mm span) with round shape suggested by

AATCC and sanitized under UV. Test examples were drenched inside the test tubes containing sterile AATCC bacteriostasis stock and a circle loaded with test living beings was vaccinated. Introductory cell fixation was determined by utilizing reasonable check technique.

The assessment tests were hatched at 370°C for 18 hrs. After 18 hrs. of brooding, last convergence of cell in charge and the test tests were determined utilizing reasonable tally strategy. The plates were hatched at 37°C for 18–24 hours. After brooding period, the quantity of states presents in every weakening plate was checked. The rate in bacterial decrease was determined by utilizing "Bacterial Reduction (%)= [(A-B)/A] × 100," Where, A = Introductory no. of cells; B = last no. of cells.

#### *10.6.4.3 WASH DURABILITY TESTING*

The wash solidness testing of the completed textures was conveyed utilizing an impartial cleanser at 40°C (± 2°C) for 30 minutes, keeping the M:L proportion at 1: 50, trailed by flushing washing and drying. In the wake of drying the test textures and the control were surveyed for antimicrobial action by the strategy as depicted before.

## 10.7 SUMMARY

According to the conversations hung on the texture kicking the bucket cycle utilizing NNPs causes us to reach an overall resolution. Indian neem fills in all kind of soil with various air conditions and the necessary consideration to develop the neem plant is exceptionally negligible. It has been utilized in the meds from the old days. The neem removes are powerful for shielding the skin from parasitic contaminations.

These days it is utilized in the manufacture of textures to give antibacterial, antifungal, and antimicrobial impacts on texture. NNPs falls under regular class and it is liberated from unsafe synthetics. Extraction measure is exceptionally straightforward and easy to use.

The antimicrobial exercises of the nanoparticles treated cotton textures were discovered to be higher than that of the neem remove alone treated textures and untreated textures and the outcomes were affirmed by standard AATCC tests. Since the leaves are richly accessible in huge numbers of the nations, the extent of execution and commercialization of natural concentrate nanoparticles to bestow antibacterial completion in material is high. It is likewise considered as eco-accommodating treatment since it utilizes just spices yet just synthetic buildups are to be stayed away from by intensive flushing and balance. The flushing specialist, for example, acidic corrosive could be suggested extraordinarily it is utilized by the cleanliness material industry. The decrease in antibacterial action is clear for the situation of neem extricate for the two sorts of microorganisms after three washes the bacterial decrease is thoroughly zero, while on account of Nano neem remove it was up to 25 washes where there is as yet bacterial decrease.

## KEYWORDS


- **antibacterial activity**
- ***Azadirachta indica***
- **cotton fabrics**
- **leaf extract**
- **neem nanoparticles**
- **silver nanoparticles**


## REFERENCES


1. Abkhoo, J., & Panjehkeh, N., (2017). Evaluation of antifungal activity of silver nanoparticles on *Fusarium oxysporum*. *International Journal of Infection, 4*(2), e1126. doi: 10.5812/iji-41126.
2. Ahmed, S., Saifullah, Ahmad, M., Swami, B. L., & Ikram, S., (2016). Green synthesis of silver nanoparticles using *Azadirachta indica* aqueous leaf extract. *Journal of Radiation Research and Applied Sciences, 9*(1), 1–7.
3. Banerjee, P., Satapathy, M., Mukhopahayay, A., & Das, P., (2014). Leaf extract mediated green synthesis of silver nanoparticles from widely available Indian plants: Synthesis; characterization; antimicrobial property and toxicity analysis. *Bioresources and Bioprocessing, 1*, 3. doi: 10.1186/s40643-014-0003-y.
4. Bindhu, M. R., & Umadevi, M., (2015). Antibacterial and catalytic activities of green synthesized silver nanoparticles. *Spectrochimica Acta Part A: Molecular and Biomolecular Spectroscopy, 135*, 373–378.
5. Capek, I., (2013). Preparation, functionalization of gold nanoparticles. *Journal of Surface Science and Technology, 29*(3, 4), 1–18.
6. Castellano, J. J., Shafii, S. M., Ko, F., Donate, G., Wright, T. E., Mannari, R. J., Payne, W. G., et al., (2007). Comparative evaluation of silver-containing antimicrobial dressings and drugs. *International Wound Journal, 4*(2), 114–122.
7. Coradeghini, R., Gioria, S., García, C. P., Nativo, P., Franchini, F., Gillil, D., Ponti, J., & Rossi, F., (2013). Size-dependent toxicity and cell interaction mechanisms of gold nanoparticles on mouse fibroblasts. *Toxicology Letter, 217*(3), 205–216.

8. Danilczuk, M., Lund, A., Sadlo, J., Yamada, H., & Michalik, J., (2006). Conduction electron spin resonance of small silver particles. *Spectrochimica Acta Part A: Molecular and Biomolecular Spectroscopy, 63*(1), 189–191.
9. Devi, L. S., & Joshi, S. R., (2015). Ultrastructures of silver nanoparticles biosynthesized using endophytic fungi. *Journal of Microscopy and Ultrastructure, 3*(1), 29–37.
10. El-Chaghaby, G., & Ahmad, A. F., (2011). Biosynthesis of silver nanoparticles using *Pistacia lentiscus* leaves extract and investigation of their antimicrobial effect. *Oriental Journal of Chemistry, 27*(3), 929–936.
11. Feng, Q. L., Wu, J., Chen, G. Q., Cui, F. Z., Kim, T. N., & Kim, J. O., (2000). A mechanistic study of the antibacterial effect of silver ions on *Escherichia Coli* and *Staphylococcus aureus*. *Journal of Biomedical Material Research, 52*, 662–668.
12. Galletti, A. M. R., Antonetti, C., Marracci, M., Piccinelli, F., & Tellini, B., (2013). Novel microwave synthesis of Cu nanoparticles in the absence of any stabilizing agent and their antibacterial and antistatic applications. *Applied Surface Science, 280*, 610–618.
13. Gao, X., Yourick, J. J., Topping, V. D., Black, T., Olejnik, N., Keltner, Z., & Sprando, R. L., (2015). Toxicogenomic study in rat thymus of F1 generation offspring following maternal exposure to silver ion. *Toxicology Reports, 2*, 341–350.
14. Gavhane, A. J., Padmanabhan, P., Kamble, S. P., & Jangle, S. N., (2012). Synthesis of silver nanoparticles using extract of neem leaf and triphala and evaluation of their antimicrobial activities. *International Journal of Pharma BioSciences, 3*(3), 88–100.
15. Giasuddin, A. S. M., Jhuma, K. A., & Mujibul, H. A. M., (2012). Use of gold nanoparticles in diagnostics, surgery and medicine: A review. *Bangladesh Journal of Medicine and Biochemistry, 5*(2), 56–60.
16. Girish, K., & Shankara, B., (2008). Neem - a green treasure. *Electronic Journal of Biology, 4*(3), 102–111.
17. Guggenbichler, J. P., Boswald, M., Lugauer, S., & Krall, T., (1999). A new technology of micro dispersed silver in polyurethane induces antimicrobial activity in central venous catheters. *Infection, 27*, 16–23.
18. Gujral, S. S., (2014). Synthesis and characterization of gold and silver nanoparticles using different leaf extracts, namely *Catharanthus roseus*; *Datura metel* and *Azadirachta indica* and estimation of antimicrobial activity of silver nanoparticles using disc diffusion method. *International Journal of Drug and Development and Research, 6*(4), 222–232.
19. Jain, S., Hirst, D. G., & O'Sullivan, J. M., (2012). Gold nanoparticles as novel agents for cancer therapy. *The British Journal of Radiology, 85*(1010), 101–113.
20. Jeong, S. H., Yeo, S. Y., & Yi, S. C., (2005). The effect of filler particles size on the antibacterial properties of compounded polymer or silver fiber. *Journal of Material Science, 40*(20), 5407–5411.
21. Kim, J. S., Kuk, E., Yu, K. N., Kim, J. H., Park, S. J., Lee, H. J., Kim, S. H., et al., (2007). Antimicrobial effects of silver nanoparticles. *Nanomedicine: Nanotechnology, Biology and Medicine, 3*(1), 95–101.
22. Klueh, U., Wagner, V. E., Kelly, S., Johnson, A., & Bryers, J. D., (2000). Efficacy of silver-coated fabric to prevent bacterial colonization and subsequent device based biofilm formation. *Journal of Biomedical Material Research Part B: Applied Biomaterials, 53*(6), 621–631.
23. Lalitha, A., Subbaiya, R., & Ponmurugan, P., (2013). Green synthesis of silver nanoparticles from leaf extract *Azadirachta indica* and to study its antibacterial and

antioxidant property. *International Journal of Current Microbiology and Applied Sciences, 2* 6), 228–235.

24. Larue, C., Castillo-Michel, H., Sobanska, S., Cécillon, L., Bureau, S., Barthès, V., Ouerdane, L., Carriere, M., & Sarret, G., (2014). Foliar exposure of the crop *Lactuca sativa* to silver nanoparticles: Evidence for internalization and changes in Ag speciation. *Journal of Hazardous Materials, 264*, 98–106.
25. Li, Y., Leung, P., Yao, L., Song, Q. W., & Newton, E., (2006). Antimicrobial effect of surgical masks coated with nanoparticles. *The Journal of Hospital Infection, 62*(1), 58–63.
26. Misbahi, A., (2010). A review on gold nanoparticles radiosensitization effect in radiation therapy of cancer. *Reports of Practical Oncology and Radiotherapy, 15*(6), 176–180.
27. Mittal, J., Batra, A., Singh, A., & Sharma, M. M., (2014). Phytofabrication of nanoparticles through plant as nanofactories. *Advances in Natural Sciences: Nanoscience and Nanotechnology, 5*, 043002. doi: 10.1088/2043-6262/5/4/043002.
28. Morones-Ramirez, J. R., Elechiguerra, J. L., Camacho, A., Holt, K., Kouri, J. B., Tapia, J., & Yacaman, M. J., (2005). The bactericidal effect of silver nanoparticles. *Nanotechnology, 16*(10), 2346–2353.
29. Namratha, N., & Monica, P. V., (2013). Synthesis of silver nanoparticles using *Azadirachta indica* (Neem) extract and usage in water purification. *Asian Journal of Pharmaceutical Technology, 3*(4), 170–174.
30. Padalia, H., Moteriya, P., & Chanda, S., (2015). Green synthesis of silver nanoparticles from marigold flower and its synergistic antimicrobial potential. *Arabian Journal of Chemistry, 8*(5), 732–741.
31. Pattanayak, M., & Nayak, P. L., (2013). Green Synthesis and characterization of zero-valent iron nanoparticles from the leaf extract of *Azadirachta indica* (Neem). *World Journal of Nano Science and Technology, 2*(1), 6–9.
32. Popa, M., Pradell, T., Crespo, D., & Calderon-Moreno, J. M., (2007). Stable silver colloidal dispersions using short-chain polyethylene glycol. *Colloids and Surfaces A: Physicochemical and Engineering Aspects, 303*(3), 184–190.
33. Pourmortazavi, S. M., Taghdiri, M., Makari, V., & Rahimi-Nasrabadi, M., (2015). Procedure optimization for green synthesis of silver nanoparticles by aqueous extract of *Eucalyptus oleosa*. *Spectrochimica Acta Part A: Molecular and Biomolecular Spectroscopy, 136*(1), 1249–1254.
34. Prabhu, S., & Poulose, E. K., (2012). Silver nanoparticles mechanism of antimicrobial action, synthesis, medical applications, and toxicity effects. *International Nano Letters, 2*, 32. doi: 10.1186/2228-5326-2-32.
35. Prema, P., (2011). Chemical mediated synthesis of silver nanoparticles and its potential antibacterial application. Chapter 6. In: Carpi, A., (ed.), *Progress in Molecular and Environmental Bioengineering - From Analysis and Modeling to Technology Applications*. London – UK: IntechOpen Limited., doi: 10.5772/22114.
36. Rajendran, K., & Sen, S., (2015). Effect of capping agent on antimicrobial activity of nanoparticles. *The Pharma Letter, 7*, 37–42.
37. Richard, J. W., Spencer, B. A., McCoy, L. F., Carina, E., Washington, J., & Edgar, P., (2002). Acticoat versus silver ion: The Truth. *Journal of Burns and Surgery Wound Care, 1*, 11–20.
38. Sajesh, K. N. K., Vazhacharickal, P. J., Mathew, J. J., & Joy, J., (2015). Synthesis of silver nanoparticles from neem leaf (*Azadirachta indica*) extract and its antibacterial activity. *CIB Tech Journal of Biotechnology, 4*(2), 20–31.

39. Shankar, S. S., Rai, A., Ahmad, A., & Sastry, M., (2004). Rapid synthesis of Au, Ag, and bimetallic Au core-Ag shell nanoparticles using neem (*Azadirachta indica*) leaf broth. *Journal of Colloid and Interface Science, 275*(2), 496–502.
40. Sharma, S., Kumar, S., Bulchandini, B. D., Taneja, S., & Banyal, S., (2013). Green synthesis of silver nanoparticles and their antimicrobial activity against gram-positive and gram-negative bacteria. *International Journal of Biotechnology and Bioengineering Research, 4*(4), 341–346.
41. Sridevi, A., Sandhya, A., & Suvarnalatha, D. P., (2015). Characterization and antibacterial studies of leaf assisted silver nanoparticles from *Carica papaya*: A green synthetic approach. *International Journal of Pharmacy and Pharmaceutical Sciences, 7*(5), 143–146.
42. Tan, K. S., & Cheong, K. Y., (2013). Advances of Ag, Cu, Ag-Cu alloy nanoparticles synthesized via chemical reduction route. *Journal of Nanoparticle Research, 15*, 1537. doi: 10.1007/ s11051-013-1537-1.
43. Thirumurugan, A., Aswitha, P., Kiruthika, C., Nagarajan, S., & Christy, A. N., (2016). Green synthesis of platinum nanoparticles using *Azadirachta indica*-an eco-friendly approach. *Materials Letters, 170*, 175–178.
44. Thirumurugan, A., Jiflin, G. J., Rajagomathi, G., Tomy, N. A., Ramachandran, S., & Jaiganesh, R., (2010). Biotechnological synthesis of gold nanoparticles of *Azadirachta indica* leaf extract. *International Journal of Biological Technology, 1*(1), 75–77.
45. Tolaymat, T. M., El Badawy, A. M., Genaidy, A., Scheckel, K. G., Luxton, T. P., & Suidan, M., (2010). An evidence-based environmental perspective of manufactured silver nanoparticle in syntheses and applications: A systematic review and critical appraisal of peer-reviewed scientific papers. *Science of the Total Environment, 408*(5), 999–1006.
46. Tripathi, A., Chandrasekaran, N., Raichur, A. M., & Mukherjee, A., (2009). Antibacterial applications of silver nanoparticles synthesized by aqueous extract of *Azadirachta indica* (Neem) leaves. *Journal of Biomedical Nanotechnology, 5*(1), 93–98.
47. Veerasamy, R., Xin, T. Z., Gunasagaran, S., Xiang, T. F. W., Yang, E. F. C., Jeyakumar, N., & Dhanaraj, S. A., (2011). Biosynthesis of silver nanoparticles using mangosteen leaf extract and evaluation of their antimicrobial activities. *Journal of Saudi Chemical Society, 15*(2), 113–120.
48. Verma, A., & Mehata, M. S., (2016). Controllable synthesis of silver nanoparticles using neem leaves and their antimicrobial activity. *Journal of Radiation Research and Applied Sciences, 9*(1), 109–115.
49. Vijayameena, C., & Kavitha, K., (2013). Studies and evaluation of photosynthetic silver nanoparticles against UTI causing pathogens. *International Journal of Current Biotechnology, 1*, 22–24.
50. Vivehananthan, K., Sudha, L. S. Y., De Silva, W. H., & Kumar, S. A., (2015). Silver nanoparticles of *Azadirachta indica*: Biological method of synthesis and characterization. *World Journal of Pharmaceutical Research, 4*(11), 77–86.
51. Yugang, S., & Changhua, A., (2011). Shaped gold and silver nanoparticles. *Frontier of Material Science, 5*, 1–24.
52. Zhang, Z., Ross, R. D., & Roeder, R. K., (2010). Preparation of functionalized gold nanoparticles as a targeted X-ray contrast agent for damaged bone tissue. *Nanoscale, 2*(4), 582–586.

## CHAPTER 11

# ROLE OF NANOMATERIALS FOR REMOVAL OF BIOMATERIALS AND MICROBES FROM WASTEWATER

SHAZIA HASAN and ABDUL RAUF


### ABSTRACT

Industrial wastewater is a leading environmental issue all over the globe. Chemical dyes, toxins, harmful chemicals, and pathogenic bacteria are among the diverse organic and inorganic contaminants found in extreme concentrations in water bodies. Removing these detrimental contaminants from industrial wastewater in an effective method has become a momentous concern. Efficient treatment methods are unavoidable to remove those contaminants before disposal. In this direction, wastewater remediation has become a specific application. Furthermore, nanomaterials (NMs) are exceptionally effective for treating water by employing low-cost nanoadsorbents and nanocatalysts. This chapter documents the implication of metal nanoparticles, metal oxide nanoparticles, metal nanocomposites, and capped metal nanoparticles in the treatment of organic dyes, organic and inorganic molecules, and pathogenic microbes present in industrial wastewater.


### 11.1 INTRODUCTION

Environmental chemists and eco-toxicologists primarily investigated and studied one of the most necessary subjects; the behavior and detrimental effects of synthetic organic chemicals in the general environment and health. "Health" here is not only restricted to humans but involves the well-being

Nanotechnology Horizons in Food Process Engineering, Volume 1. Megh R. Goyal, PhD, PE, Junaid Ahmad Malik, PhD, Satish Kumar, PhD & Ritesh B. Watharkar, PhD (Eds.)

of animal and plant species too, since human health is correlated to the well-being of other species along with the availability of pure air, water, and food [47]. The rapid pace of industrialization during the last 10 decades produces million tons (MT) of industrially and domestically synthetic organic discharges. This leads to an unavoidable enhancement in the occurrence and concentration of organic contaminants in sewage and effluents with different degradation rates. The extent of damage an organism experiences by such type of substance is known as toxicity while their adverse effects on health and environment are dealt with by environmental toxicology. Toxicity includes the effect either on a whole living organism (unicellular/multicellular) or on the subcellular level, such as cytotoxicity and hepatotoxicity.

The toxicity of a material is dose-specific such as high intake of minerals can cause mineral intoxication while even a very little amount of fatal scorpion venom is toxic [106, 123]. Environment especially water bodies are skeptically disturbed by diversified organic pollutants resulting into emerging issues worldwide. On combustion and application of a class of organic pollutants in daily routines, endocrine-disrupting chemicals are released into the environment in the form of aerosols, dusts, and particulates which creates disturbance in the endocrine system. Eventually such interference in hormonal signals leads to reproductive problems (decreased fertility, miscarriages, and other birth effects) and specific types of cancer [9]. Urbanization in both economically developed and developing countries have driven an intensive and still growing use of water bodies resulting in discharge of chemical contaminants and ultimately deterioration of water sources in many regions.

There are enormous varieties of industrially synthesized organic compounds which are classified as industrial wastewater constituents. Trace metal is one such category that effortlessly dissolved and transported in water. Due to this behavior, trace metals are easily taken up by the organisms present in water bodies. Usually as defined above, intake of trace metals in low dose act as nutrients while at an increased concentration they behave as toxic which interfere with the cellular enzymatic system and finally damage the whole organism [116, 123]. The textile industries are the developer of major earth and water pollutants due to the involvement of as many as 8,000 chemicals and abundant quantity of water. Reports proposed that about 1.6 million liters of water are employed by an average sized textile industry each day. Such industries due to the implication of dyeing and finishing treatments on the processed fabrics produce 17–20% of water pollution since they are difficult to remove through waste treatment processes [41, 125].

Dyes are the chemical substances whose crystal structures get temporarily modified on application to a substrate and provide color. These coloring substances are broadly used in the textile, paper, cosmetics, plastics; etc. industries. Dyes are categorized according to their chromophoric groups, applications, and chemical structures. Chromophores contain various functional groups like azo, anthraquinone, nitro, carbonyl, and several others. Additional presence of electron-withdrawing or donating substituents called auxochromes like amine, carboxyl, sulphonate, and hydroxyl escalate the color of chromophores. Dyes on the one side, hinder sunlight from penetrating inside the water bodies which influence aquatic life badly, while on the other side, their well-known degradation products are dangerous to humans by causing skin problems and cancer. These organic contaminants subsequently are posing a serious threat to aquatic organisms along with animals, including humans as consumers of aquatic resources [18, 33].

In order to monitor environmental pollution and wastewater purification, there is an unavoidable requirement of efficient removal and degradation of water pollutants particularly cost-effective, stable, and large-scale materials. Several approaches respective to such demand exist, among which separation based upon adsorption method is the most widely employed due to its uncomplicated nature. Moreover, traditional wastewater treatment technologies fail to provide an abundant amount of safe water among increasing population with changing basic purification criteria with time [35, 77, 120, 123]. Table 11.1 shows some of the conventional methods and their most related limitations in wastewater treatment.

Many existing wastewater-related problems may be effectively addressed using different nanostructures based on the chemical and physical properties of sorbents and pollutants, according to advances in nanoscience and nanotechnology [129]. The NMs are primarily involving materials of size less than 100 nm at least in a dimension in order of $10^{-9}$ m (i.e., 1 billionth of a meter). The difference in physicochemical behavior compared to bulk materials is due to nanomaterial's shape/size or both, and by modifying these two factors, unique characteristics with new capabilities can be developed. Thus, developments in nanotechnology provide economical solutions to wastewater remediation that generally do not require large or centralized systems and are more efficient. The most widely studied nanosized metals and metal oxides include silver, gold, zinc, iron oxides, titanium oxides, etc. [74, 75].

**TABLE 11.1** Conventional Methods and Limitations of Wastewater Treatment

| Conventional Methods | Limitations |
|---|---|
| Carbon filters | • Improper for removing salts like nitrates and fluorides, metals; etc. |
| | • Prone to clogging and fungi, demand cleansing through frequent changing of filters. |
| Distillation | • High demand of energy and water. |
| | • Incapable to remove contaminants with higher boiling point (>100°C). |
| Flocculation | • A complicated and inefficient method. |
| | • Specification of alkaline additives to obtain optimum pH. |
| Reverse osmosis | • Unhealthy as it eliminates important minerals from water. |
| | • Makes treated water acidic. |
| | • Fail to remove volatile organics and other large chemical molecules. |
| Ultrafiltration | • Demands high energy. |
| | • Unable to separate dissolved inorganics. |
| | • Difficult to clean. |
| UV treatment | • Extravagant and sensitive method. |
| | • Impotent to remove heavy metals. |

This chapter focuses on how metal NMs may be used to remove organic compounds from environmentally dangerous chemicals. This chapter briefly overviews employment of different NMs such as metal, metal oxide nanoparticles, metal composites and capped nanostructures for the successful elimination of biologically toxic materials catalytically and photo-catalytically. Eventually, recommendations for eliminating of such toxic materials are made based on the current practices of NMs.

## 11.2 APPLICATIONS FOR THE REMOVAL OF BIOTOXIC MATERIALS

### *11.2.1 BY USING METAL NANOPARTICLES*

The priority of nanoparticles in any area of research is due to the availability of high surface area for chemical reactivity as compared to bulk.

The influencing factors in the synthesis of metal nanoparticles decide their performance in various applications. These factors chiefly include the kind of precursors and reducing agents involved during synthesis, such as silver nitrate is the mostly used precursor for synthesis while the type of reducing agent decides the overall efficiency of the synthesized silver nanoparticles (AgNPs). Biosynthesis employing different plant extracts and microorganisms are also employed for the efficient synthesis of AgNPs by enhanced reduction of metal salts [80, 148].

#### *11.2.1.1 REMOVAL OF DYES FROM WATER*

Dyes constitute a large proportion of chemicals among the different hazardous pollutants of the aquatic system. They are visually noticeable even in a very low concentration, disturbing aquatic life and thus food chain. Iron has long been used to save surface sites from contamination but was ultimately stopped due to its corrosive nature. However, the unique characteristic of nanoscale materials makes nanosized iron particles suitable for water remediation and enhances its reaction ability [76]. The initial implications of iron nanoparticles for dye degradation are presented. The zero-valent iron nanoparticles donate electrons by reducing into ferrous/ferric ions which are accepted by reacting dye molecules. Additional released protons also promote dye reduction and assist in cleavage of auxochrome bonds to decolorize the dyes [30]. The participation of silver in various applications including medicine and water remediation has been known from ancient period. Moreover, introduction of AgNPs has now elaborated its role in a different form [86].

The general mechanism of dye removal with AgNPs involves either adsorption on to the surface of nanoparticles or photocatalytic degradation or both. However, flocculation of nanoparticles during catalytic degradation results into impoverished catalytic efficiency [64]. Biosynthesized AgNPs by using plant extracts have reportedly shown photocatalytic degradation of dyes like methyl orange, methyl violet from 90 to 97% efficiency [12, 90]. The intrinsic non-toxic nature of the gold nanoparticles (AuNPs) respective to environment exhibits an additional advantage over other metal nanoparticles. Its inert characteristic with unique size and shape dependent optical and electronic properties makes it chemical and physical stable [34, 149]. Reduction of auric acid by polyethyleneimine (PEI) produced low-cost cationic AuNPs which has the potential of catalytically degrading both cationic and anionic dyes [26].

Electrostatic attraction along with cation and lone pair interactions are the reason behind such catalytic participation of nanoparticles in the presence of reductant like metal hydride. The metal nanoparticles have an overall charge through which it participates in electron relay mechanism between reductant and dye molecules for degradation (Figure 11.1) [26]. As compared to AgNPs, AuNPs has not been extensively biosynthesized. *Sueda fruciotosa* extracts was used by Khan et al. [73] in developing and using AuNPs in photodegradation of methylene blue (MB) dye.

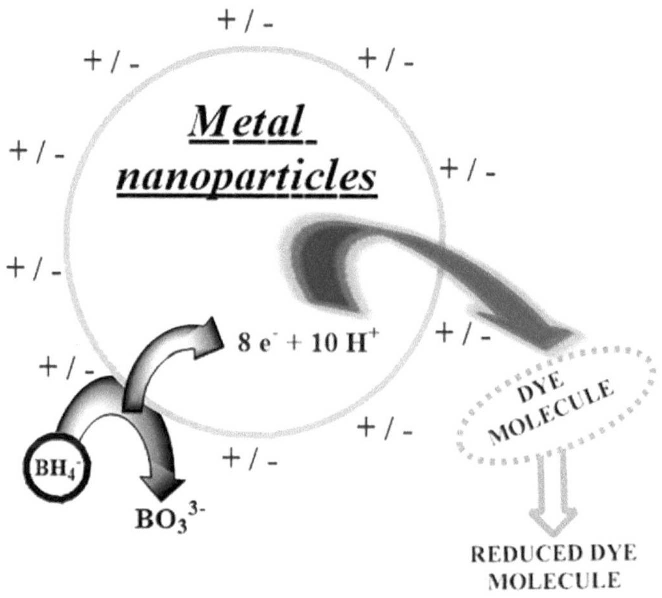


**FIGURE 11.1** The relay effect involved in catalysis of dye degradation by metal nanoparticles and reducing agent.

### *11.2.1.2 REMOVAL OF ORGANIC MOLECULES, HEAVY METALS, AND IONS FROM WATER*

Undoubtedly, the importance of commercial use of organochlorine pesticides uplifted the overall plant growth performance and created food security across the earth. However, their harmful consequences on human health and continuing impacts on the food chain and eventually in environmental contamination make them mandatory to be removed. Similarly, another chemical which saves numerous lives due to malaria is dichlorodiphenyl-trichloroethane (DDT) has also long-term participation in environmental contamination. In addition to these, vulnerability to trace amount of heavy metals is a threat for human beings [69, 117]. Few of the major contaminants and their respective sources with health issues are present in Table 11.2.

**TABLE 11.2** Major Organic and Inorganic Contaminants of Wastewater

| Major Contaminant | Source | Health Effects |
|---|---|---|
| Arsenic | Geological origin, combustion of coal | Hypertension, glucosuria, hyperpigmentation, cancer |
| Fluoride | Geological origin, mineral weathering, coal mining volcanic sources | Dental and skeletal fluorosis, disturbed nervous system |
| Halogenated organics | Chlorination, insecticide, municipal, and industrial wastes | Carcinogenic, lung/liver tumor |
| Lead | Old piping lines and fittings, mineral weathering, paint | Delays in physical/mental development, renal issues, hypertension |
| Mercury | Industrial pollutants, dental filling, aquatic food | Neurotoxicant, body tremors, respiratory, gastrointestinal failures, kidney damage |
| Pesticides | Agriculture, effluents, shampoos | Cancer, liver/kidney problems, cardiovascular/reproductive/neurological disorders |

These compounds are extracted from water using either homogeneous or heterogeneous methodology. Although, homogeneous procedure exploits all the available surface area offered by metal nanoparticles but its presence in processed water due to completely dispersible nature and thus inseparable is a major concern. Heterogeneous chemistry also involves highly dispersed nanoparticles but attached with supports like polymers, hydrocarbons, oxides, etc. This attachment or binding of nanoparticles with support provides enough strength to inhibit leaching of nanoparticles in the medium such as water [14, 117].

The identification of zero-valent iron mediated oxidation of halogenated aliphatics sparked investigations into the interactions of halogenated organics with different transition metals. Carbon tetrachloride, same as air and moisture activates corrosion in the metals resulting in metal chlorides formation on the metal surface along with partially dehalogenated organic compound. This motivated further study of other organic compounds with zero-valent iron and subsequently enhanced degradation of pesticides with iron nanoparticles as reducing agent became famous [117, 160]. By early 2003, detailed investigation on the interaction between halocarbons and noble metal nanoparticles has been reported. The reports claimed that nano-sized noble metals react with halocarbons in a fashion similar to metal's reductive dehalogenation method but with no by-products [108].

Some of the broadly used pesticides like endosulfan, malathion, and chlorpyrifos were degraded in the presence of noble metal nanoparticles [92, 107, 109]. Silver and AuNPs interacted with pesticides differently and exhibited different spectroscopic behavior which might be due to the particle aggregation [92]. Most of the pesticides present in drinking water are either organochlorine or organosulfur or organonitrogen such as simazine, quinalphos, carbofuran, etc. This makes the entire removal of such pesticides by noble metal nanoparticles simple. Similarly, reports on nanomaterial-based degradation of heavy metals like mercury, cadmium, copper, arsenic, lead, etc. AgNPs with high adsorption capacity has the potential to adsorb heavy metals such as $Hg^{2+}$ since silver is a more capable reducing agent and it can exist with mercury in different phases [117]. It is more likely that adsorbed mercury atoms diffused into the inner silver core gradually from the surface. The existence of AuHg, $AuHg_3$ and $Au_3Hg$ clearly displays the affinity of Hg towards Au. The reduction of $Hg^{2+}$ to $Hg^0$ by $NaBH_4$ in the presence of AuNPs supported by aluminum showed much higher removal of mercury by Au. Another study used modified AuNPs by citrate coating to treat $Hg^{2+}$ in water, where citrate ions behaved as weak reducing agent and evaded the use of harsh reducing agent like $NaBH_4$ [154].

#### 11.2.1.3 REMOVAL OF PATHOGENIC MICROBES FROM WATER

With an exponential increase in the anthropogenic activities the microbiological pollution in water sources has been escalated all over the globe. The reports of UNICEF, 1995 (Oxford University Press) mentioned that about 5 million deaths per year happened because of waterborne pathogens. Thus, undeniable requirement of water disinfection is present worldwide [113]. A summary of waterborne pathogens and related diseases has been summarized in Table 11.3.

**TABLE 11.3** Waterborne Pathogens and Related Diseases (Compiled from Multiple Sources)

| Pathogens | Diseases |
|---|---|
| *Ascaris lumbricoides* | Ascariasis |
| *Clostridium botulinum* | Botulism |
| *Entamoeba histolytica* | Amebiasis |
| *Escherichia coli* | Gastroenteritis |
| *Hepatitis viruses* A and E | Hepatitis |
| *Poliovirus* | Polio |
| *Pseudomonas aeruginosa* | Pulmonary disease, skin infection |
| *Salmonella typhi* | Typhoid |

Nanometals such as nano Cu, Ag, Au, Pd, Al, Ni, etc., have proven to show microbial property [49]. Noble metal nanoparticles particularly show an impressive result for the disinfection of *E. coli, Pseudomonas aeruginosa, Vibrio cholera* and *Bacillus subtilis*. The antimicrobial effects of $Ag^0$ and ionic form have been thoroughly investigated and employed as common disinfectant, wound dressing, etc. [144]. The mechanism based upon AgNPs attack on microbial cell can be summarized as follows [117]:

- Due to interaction of silver with proteins, inactivation of protein takes place.
- The cell surface has some charge because of the existence of different functional groups which leads to increased or decreased contact towards the antimicrobial composition due to attraction or repulsion. This is a key role in deciding the effectiveness of anti-bacterial composition.
- The cellular permeability of Gram-negative bacteria is primarily determined by a lipopolysaccharide (LS) layer present in cellular membrane. This LS has an affinity towards cations and on binding with cationic antibiotics, the cellular membrane gets weak. Therefore, small sized nanoparticles can easily permeate across cellular membranes with triangular shaped particles has an advantage over other shaped nanoparticles.
- The complete destructive oxidation inside microorganism is also facilitated by the silver in oxygen charged aqueous medium.

- The DNA inside cell is protected by a defense around the nucleus which on external severe stimuli fails and leads to denaturation through losing replication ability. A similar experience was found in bacterial cells such as *E. coli* and *Staphylococcus aureus* when subjected to AgNPs, resulting in generation of protective layers around DNA and ultimately DNA's condensation.

Biogenic AgNPs have an additional synergistic effect with antibiotics like ampicillin, erythromycin, etc., against both Gram-positive and Gram-negative bacterial strains [37]. AuNPs in solution synthesized through citrate reduction procedure are photomutagenic against *Salmonella typhimurium* strain. The activity of AuNPs relied upon the coexistence of $Au^{3+}$ ions and citrate. Free radicals generate when $Au^{3+}$ gets oxidized and citrate gets decarboxylated in the presence of light irradiation and eventually proteins and DNA gets destroyed [129].

### 11.2.2 BY USING METAL OXIDES NANOPARTICLES

Many NMs are currently being investigated because of their unique results towards obstinate pollutants, for the processing of contaminated groundwater, surface water as well as drinking water. In particular, metal oxide NMs are receiving more attention for environmental monitoring and remediation due to the interesting properties based on rich valence states, huge surfaces, and changeable electronic structures. Also, they exhibit splendid magnetic and optical properties due to surface mobility [44]. Among the accessible metal oxides adsorbents, ferric oxides, manganese oxides, aluminum oxides ($Al_2O_3$), titanium oxides and cerium oxides are categorized as the promising NMs for purifying water bodies from contaminants.

#### 11.2.2.1 REMOVAL OF DYES FROM WATER

Recently, the employment of metal oxide especially transition metal nanoparticles in the remediation of wastewater comprising organic dyes have received immense attention due to their characteristics. Unsystematic implication of colorants is a proof of environmental pollution. Green synthesis of ZnO, $Cr_2O_3$, $Co_3O_4$, CuO, and NiO nanoparticles using sunlight irradiation approach [135]. The results exhibited well-defined with uniformly distributed nanoparticles. The synthesized nanoparticles were used against the mixture of alizarin red S and MB dyes which are hazardous discharge into an aqueous system. The degradation efficiency was highest in $Cr_2O_3$ while others followed the efficiency as ZnO>CuO>NiO and least in $Co_3O_4$. The zeta potential implies the stability of particles, i.e., with increasing zeta potential, the probability of stability of particles also increases because charged particles repel each other and thus overcome the aggregation process [102]. $Cr_2O_3$ has the most negative zeta potential value, i.e., of –25.2 mV signifying a strong repellent force and absence of aggregation and greater surface activity [135].

Laser irradiation as a simple and green technique was implied to synthesize amorphous transitional metal oxides of Fe, Co, and Ni nanoparticles as super dyes adsorbents [85]. The fabricated amorphous metal oxides were employed against the adsorption of MB and reported the highest 10584.6 $mgg^{-1}$ adsorption capacity till date. Amorphous NiO degraded 99% MB in 1 min while raw nanoparticles took 150 min to degrade 90.3% of dye. Nanoscale hematite synthesized by Mansour exhibited photocatalysis degradation against MB solution under UV-light radiation [93]. CuO is one of the widest used nanosized metal oxides for the removal of water contaminants. *Madhuca longifolia* plant extract was used for biosynthesizing cupric oxide nanoparticles in a well dispersed and spherical shape. Depending upon the plant extract used for biosynthesizing, the average particle sizes estimated through particle distribution graphs were 30 and 120 nm. The different particle size was defined by varied band gap values and photoluminescence intensity. The as-prepared nanoparticles were investigated for photocatalytic degradation against MB dye for wastewater treatment [36]. Some of the reported metal oxide nanoparticles are mentioned in Table 11.4.

**TABLE 11.4** Metal Oxide as Photocatalyst in Few Reported Dyes

| Metal Oxide | Radiation | Dye | Degradation Efficiency (%) | References |
|---|---|---|---|---|
| $Ag_2O$ | UV | Methyl orange | ~100 | [62] |
| $CeO_2$ | Sunlight | Acidic black 10B | 97 | [159] |
| $CeO_2$ | Visible light | Acid orange 7 | 95 | [61] |
| CuO | Sunlight | Methylene blue | 93 | [72] |
| $TiO_2$ | UV | Methyl orange | 77.19 | [6] |
| $WO_3$ | Visible | Indigo carmine | 93 | [95] |
| ZnO | Sunlight | Methyl orange, rhodamine 6G | >90 | [65] |

#### *11.2.2.2 REMOVAL OF ORGANIC MOLECULES, HEAVY METALS, AND IONS FROM WATER*

Due to exceptional properties of high removal capability and selectivity, nanometal oxides are employed as promising adsorbent for the removal of contaminated organic molecules and heavy metals from aqueous system. Iron is one of the most abundant elements in the Earth's crust because of which iron oxides nanomaterial contribute widely among various fields. Goethite, hematite, maghemite, magnetite, and hydrous iron oxides are the most studied iron oxides [23, 59]. High adsorption and affordable cost make goethite a better adsorbent against heavy metal. According to the research, nanoscale goethite removed uranium, $Cu^{2+}$ from water more efficiently compared to non-nanoscale goethite [31]. Other heavy metals like $V^{5+}$, $Mn^{2+}$, $Ni^{2+}$, $Zn^{2+}$, etc., are reportedly removed through goethite [45, 83].

Hematite is the most stable iron oxide which is comparable less prone to resistance, and also its nanoparticles is more effective adsorbent against heavy metals [148]. Through the removal of $Cr^{6+}$ from water, impact of different morphology of hematite nanoparticles was also investigated; indicating the significance of morphology on capacity [1]. Maghemite nanoparticles are most extensively used to treat heavy metals due to its large surface area and responsive nature towards magnetic field. Maghemite nanoparticles were utilized to treat heavy metals such as $Cu^{2+}$, $Ni^{2+}$, $Mn^{2+}$, $Cd^{2+}$, $Cr^{2+}$, $Pb^{2+}$, $Cu^{2+}$, etc., in wastewater [5, 150]. Similar to maghemite, magnetite-based NMs has also the capability to be separated by adding a magnetic field. Researchers have used magnetite to treat $Pb^{2+}$, $Hg^{2+}$, $Cd^{2+}$, $Cr^{6+}$ and $Ni^{2+}$ from wastewater [51].

ZnO nanoparticles have large surface area which enhance its removal capacity and thus gained attention as adsorbent in treating heavy metals such as $Cr^{6+}$, $Cu^{2+}$, $Ni^{2+}$ and so on [36]. Mahdavi and group examined the adsorption effects of $TiO_2$, $Al_2O_3$ and MgO nanoparticles against $Cd^{2+}$, $Cu^{2+}$, $Ni^{2+}$ and $Pb^{2+}$. Among which, MgO exhibited the highest efficiency through adsorption and precipitation while $TiO_2$ and $Al_2O_3$ used only adsorption as a removal mechanism [89]. Most of the synthesized metal oxides display large surface areas encompassing partially combined homogeneous nanoparticles with cores and shells. Other studied transition metal oxides with mesoporous structure towards heavy metals in solutions are $ZrO_2$, $HfO_2$, $Nb_2O_5$ and $Ta_2O_5$ [132].

Out of all pesticides, about 10–15% only reaches the target pests while the remaining major proportion dispersed off in environment and declines its quality, since most of them are highly toxic and carcinogenic [25]. The photocatalytic degradation of organophosphate pesticides through $TiO_2$ aqueous suspension under artificial and solar light is not fully discussed yet. Photocatalytic degradation of organochlorine pesticides, including their degraded metabolites, using $TiO_2$ has proven to be an efficacious approach. The rate of degradation might be increased by doping with nitrogen or others for modifying bandgap and thus easier photooxidation and photoreduction [120].

#### *11.2.2.3 REMOVAL OF PATHOGENIC MICROBES FROM WATER*

Currently, most of waterborne microorganisms have developed their resistance to available treatments, which leads to either requirement of new disinfectants or enhanced already introduced disinfectant doses which eventually produce higher amounts of disinfection by-products [101]. Processes like ozonation or UV treatment offer nil protection against reinfection in treated water system which opens the door for consideration of new approaches. Nanotechnology presents novel methods due to improvised properties for wastewater treatment without the production of disinfection by-products [38].

Nanoscale ZnO has attracted intensive research interest in the past years as an antibacterial agent into textiles and fibers, surface coatings and cosmetics in order to prevent microbial growth. ZnO is exceptionally active even in small amount as compared to organic materials and its stability under harsh processing environment makes it safe for humans and animals. ZnO is more durable and heat resistant like most of the other inorganic NMs. A large number of different microbial strains have been examined as antimicrobial agent. The physicochemical properties of ZnO along with the existence of surface defects like edges and corners contribute towards antibacterial activity through disruption of bacterial cell walls. Like other application, properties like particle size, shape, surface area and dispersity of ZnO nanoparticles affect the antibacterial activity [38, 143].

Among many possible mechanistic pathways, the release of $Zn^{2+}$ ions for destroying cell membrane and penetrating the intracellular contents is mostly studied. Investigating physicochemical characterization and antibacterial tests of ZnO nanoparticles against E. coli found that the release of $Zn^{2+}$ ions has contributed primarily [86]. Another waterborne pathogen *Saccharomyces cerevisiae* as exhibited toxicity of ZnO nanoparticles, resulting from the solubility of $Zn^{2+}$ ions in the experimental medium [71]. From the results, it was concluded that nano-sized ZnO as an antibacterial agent was more abrasive and behaved better than micron-sized nanoparticles suggesting particle size influence the outcome [114]. In contrast, there was no clear relationship between the size of ZnO nanoparticles and antibacterial activity against municipal wastewater bacteria and *Pseudokirchneriella subcapitata* respectively [43, 46]. The antimicrobial activity of ZnO was investigated by Narayanan et al. [110] against pathogenic microorganisms present in water bodies like *S. aureus, E. coli, Klebsiella pneumoniae, Enterococcus faecalis* and *Pseudomonas aeruginosa*. The results exhibited a strong antibacterial activity even at lower concentration against both Gram-negative/positive bacteria [110].

Another big issue regarding the need of antibacterial agents is the ineffectiveness of antibacterial therapy against biofilms. Biofilm is a broad category of bacteria that comprises microorganism culture and extracellular polymer matrix (EPM). EPM itself is a biochemical mixture of polysaccharides, proteins, glycopeptides, lipids, and nucleic acids which gives an appearance of slimy 3D biopolymer with structures similar to transport and water channels. Thus, biofilm as a complex 3D biological structure results into higher microbial life organization and protect microorganisms from adverse physical, chemical, and biological environmental factors like temperature, UV radiation, antibacterial agents, etc. The resulting resilience to antibiotics creates the progression of infections to a chronic level [29, 40, 116].

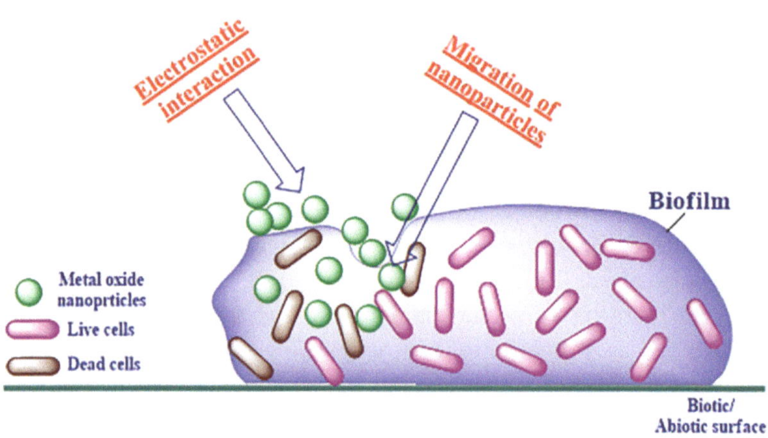


**FIGURE 11.2** The interaction of metal oxide nanoparticles with biofilm.

Among various nanoparticles, the most effective and widely investigated metal oxide nanoparticles are $TiO_2$, $Fe_3O_4$, ZnO, and CuO. Shkodenko et al. [140] performed a comparative analysis of CuO, $Fe_3O_4$, $TiO_2$, ZnO, MgO, and $Al_2O_3$ nanoparticles against biofilms. The interaction of nanoparticles and biofilm follows three-stage procedures, i.e., transfer of nanoparticles in the vicinity of biofilm followed by attachment to the surface of biofilm and ultimately movement inside biofilm (Figure 11.2). CuO and ZnO nanoparticles from the chosen ones exhibited the most substantial antibiofilm properties while $Al_2O_3$ nanoparticles were weakest but can be potentially active in high concentrations [140].

### *11.2.3 BY USING METAL NANOCOMPOSITES*

Composites are heterogeneous/hybrid synthesized or naturally occurring solid materials resulting from combining together of two or more different components with its own particular physical/chemical characteristic. Moreover, the newly formed composite has superior characteristics than the constituent original materials which commonly offer properties like stiffness, strength, resistance to corrosion, etc. [139]. Nanocomposites are especially those multiphasic composites of dimensions within 10–100 nm range that contain at least one phase with nanoscale morphology such as nanoparticles, nanotubes, or another nanostructure. Nanocomposites are excellent alternatives to overcome the limitation of various constructed materials. The observed property of nanocomposites does not necessarily depend upon the respective characteristics of parent constituents but also on their interfacial and morphological characteristics, while sometimes a newly generated property also occurs [133].

#### *11.2.3.1 REMOVAL OF DYES FROM WATER*

The manufacturing industries of textile, leather, dyeing, and cosmetics are important sources of dye pollution. Among various methods, photocatalytic removal of dyes and other effluents has been considered as an adequate approach. The gradual recombination of electron-hole pairs usually reduces the overall photocatalytic efficiency [50]. For removing dyes from aqueous systems, various nanocomposites through photocatalytic method have been employed and reported. $ZrO_2$ and ZnO as an individual photocatalyst have shown enhanced removal of organic dyes from wastewater [3, 68]. ZnO is an affordable semiconductor with a wide bandgap. Additionally, by mixing particulate semiconductors, elevated photocatalytic activity can be observed due to simultaneous excitation of electrons and movement of holes to conduction and valence bands, respectively, in both semiconductors on illumination. This results into improvised electron-hole pair separation which results into better photocatalysis [32]. In, $ZnO/ZrO_2$ nanocomposites were synthesized under ultrasonic irradiation [2].

The nanocomposites were investigated for photocatalytically degradation of Congo red dye in the aqueous system under UV light. The ultrasonic irradiation during the synthesis process generates a unique environment along with elevation of the hydrolysis of Zn and Zr precursors. Moreover, the enhanced surface to volume proportion of small sized nanoparticles increases the photodegradation rate constants of sonicated samples as compared to unsonicated ones. The results also implied that more ratio of ZnO respective to $ZrO_2$ in nanocomposite influenced the degradation of dyes [2]. In recent years, magnetic nanoparticles have developed enormously against various environmental related applications for pollutants adsorption from contaminated aqueous systems [51].

Activated carbon is a well-known sorbent for wastewater remediation with great adsorption capacity but high-cost development, poor mechanical characteristics, and poor recyclability limit its usage. Its porous structure with high surface area (500–2,000 $m^2/g$) makes it a good adsorbent towards several pollutants [91, 125, 134]. As a result, an interaction of activated carbon and magnetic nanocomposite formed by combustion method was used with the aim of developing a new material as an adsorbent. The magnetic nanoparticles were ingrained within a matrix of activated carbon. The synthesized nanocomposite has adsorption efficiency closer to that of activated carbon along with good regeneration and recyclability. The magnetic nanoparticles were selectively used for removal of Nylosan Blue, Chromazurol S and Basic Red 2 industrial dyes from wastewater by implying adsorption technology. The nanocomposite was potentially active even after seven catalytic cycles with 68% removal efficiency [103].

Ni-doped $TiO_2$ nanostructure through chemical bath deposition procedure and investigated its photocatalytic efficiency against Ponceau S dye [4]. The results reveal that the efficiency under UV light irradiation was higher compared to sunlight. Adsorbents based on natural biopolymers are extensively employed to maintain the view of green chemistry. In this direction, xanthan gum can be involved as an organic moiety due to the presence of carboxyl and hydroxyl functional groups to synthesize bionanocomposite and has received enormous attention due to its availability, biodegradability, and affordability [4].

Ecofriendly amino acid, glutathione (GSH) acts as a cross-linker for further enhancement of adsorbability of xanthan. They also used zeolites as inorganic nanofiller due to its porous and crystalline characteristic. Zeolites are aluminosilicates with negative lattice sites that can be balanced by cations existing in the solution [58]. The modification along with cross-linking contributes in large surface area and also provides plenty of adsorption sites for adsorbing pollutants. The xanthan gum-glutathione/zeolite bionano-composite have been used for the removal of anionic dye, Congo red. The same group used the fact that functional groups present on the adsorbent's surface have a keen role in adsorbent effectiveness and capacity. They cross linked two natural and easily available biopolymers; alginate and pectin to synthesize more functionalized nanocomposite [9].

The presence of carboxylic functional group and (1–4) linked α-d-galacturonic acid in alginate and pectin respectively create binding possibility with other inorganic or organic molecules [99]. The resulting nanocomposite was amorphous in nature and porous in morphology with 13–34 nm particle size. Thus, due to the presence of such structure, the synthesized Alginate/Pectin nanocomposite was used for the removal of crystal violet dye from wastewater and proved as an excellent adsorbent [97].

#### *11.2.3.2 REMOVAL OF ORGANIC MOLECULES, HEAVY METALS, AND IONS FROM WATER*

An abundant number of adsorbents has potential to remove heavy metals and other organic molecules from wastewater. The elements with atomic weights under the range 63.5–200.6 amu and specific gravity greater than 5 $kgm^{-3}$ are heavy metals [48, 146]. Adsorption over other methods found to be more superior due to its simplicity of designing and insensitivity to toxic contaminants beyond other advantages [78]. Nanocomposite materials as compared to pure organic and inorganic materials show improved adsorption capabilities, granulometric behavior, thermal, and chemical stability, recyclability along with better selectivity towards heavy metal ions. Adsorbents based polymer nanocomposites display some extraordinary characteristic of providing effective binding sites across the polymer surface with increased surface area and pore volume [136].

From diversified options of polymer nanocomposites, conducting polymer-based adsorbents have gained extensive focus due to the potential of adsorbing various heavy metals. Few of these polymers are polyaniline (PANI), polypyrrole (PPy), PEI, polythiophene (PTh) and their derivatives. By observing the structure, it can be concluded that the presence of N and S atoms in the main framework of conducting polymers is primarily accountable for the contaminant adsorption. Apart from this, N atoms of PPy elevate adsorption capacity more than S atoms of PTh [158]. Few of the reported natural and synthetic polymers for removal of heavy metal ions from wastewater are mentioned in Table 11.5.

**TABLE 11.5** Natural and Synthetic Polymers for Modification of Metal Nanocomposites

| Polymers | Heavy Metals | References |
|---|---|---|
| Cellulose | $Cd^{2+}$, $Pb^{2+}$ | [155] |
| Chitosan | $Cr^{6+}$, $Cd^{2+}$ | [24] |
| Dextran | $Cu^{2+}$ | [84] |
| PAANAF (Poly anthranilic acid/4-nitroaniline/formaldehyde) | $Fe^{3+}$, $Co^{2+}$, $Ni^{2+}$, $Cu^{2+}$, $Zn^{2+}$, $Pb^{2+}$ | [15] |
| PPy | $Cd^{2+}$ | [57] |
| PPy/$Fe_3O_4$ | $F^-$, $Cr^{6+}$ | [21, 20] |
| PSMA (Poly styrene maleic anhydride) | $Zn^{2+}$, $Cu^{2+}$, $Fe^{2+}$ | [98] |
| Starch | $As^{5+}$, $As^{3+}$ | [111] |

Conducting polymers of PANI and its derivatives as mentioned above are good adsorbents to remove several types of heavy metal ions from aqueous systems. Other functional materials like natural polymers, agricultural remnants, inorganic, and organic molecules are mixed with PANI and its derivatives to form PANI-based nanocomposites in order to improve its surface area and also to modify its surface morphology. The incorporation of different functional materials with PANI and its derivatives make the resulting structure as either of PANI and its derivatives/natural/organic/inorganic material nanocomposites. Several forms of affordable materials such as starch, dextrin, chitosan (CS), sawdust, etc., can be used as the incorporating natural material, and with PANI have been intensively investigated as promising bio-nanoadsorbents to remove heavy metal ions [158].

Sawdust (SD), accessed from wood has the potential to be coated with PANI nanoparticles to generate PANI/SD nanocomposites that can bind heavy metal ions through various mechanisms. Ansari et al. [10, 11] showed the considerable potential of PANI/SD nanocomposite for removing up to 100% $Cr^{6+}$ and $Pb^{2+}$ metal ions at room temperature from an aqueous system. The polysaccharide which could be any chemical and biological polymer, in the last decade has been significantly involved in synthesizing of PANI/polysaccharide due to its application in environmental and material science. In a biodegradable PANI/dextrin nanocomposite was employed as a sorbent to remove heavy metals ions of Cu, Pb, and Cd [157]. The result did not show high adsorption of the mentioned cations which could be due to the protonation of active sites present in PANI/dextrin. Polyethylene (PE) glycol (PEG), polystyrene (PS), polyvinyl alcohol (PVA), etc., synthetic materials have been used in several forms with PANI to be applied in wastewater purification.

A study showed the ability of PANI/PS synthetic nanocomposite for the adsorptive removal of $Hg^{2+}$ heavy ions up to 67% from water solution [54]. The nanocomposite was synthesized through solution mixing and micro-emulsion polymerization and reported the maximum adsorption capacity of 19 $mgg^{-1}$ heavy metal ions. PANI/PVA nanocomposite was synthesized using an in-situ polymerization process for the removal of $Cr^{6+}$ with a maximum adsorption capacity of 111.23 mg at pH 4.12 [67].

The high adsorption of heavy metal ions by using nanocomposites fabricated with PANI and various inorganic materials has gained wide attention in water purification recently (Figure 11.3). Some nanocomposites consisting of PANI/iron oxide have been reported for removing pollutants through adsorption like $Pb^{2+}$, $Cr^{+3}$, $As^{5+}$, etc., from water [19]. Magnetic nanoparticles with nanocomposites are selectively utilized for separating pesticides. The magnetic nanoparticles contribute towards large surface area to modify it into an effective adsorbent. The separation of targeted pesticides with the application of a magnet near the beaker containing nanocomposites and pesticides is rapid and simple to use. The successful removal of endrin and dieldrin pesticides from aqueous sample by graphene oxide/$Fe_3O_4$ was performed by Shrivas et al. [141] is such an application.

### *11.2.3.3 REMOVAL OF PATHOGENIC MICROBES FROM WATER*

The antibacterial activity of nanomaterial has numerous medical implications, including implants, medical devices, etc. Nanoproducts made from CS and its derivatives have shown good antimicrobial activity and thus, are used for various medical applications such as CS-carbon dots, CS-based biosensors and biomarkers, scaffolds, etc. [60]. Chitin based nanomaterial has also proposed implications in scaffold and wound healing materials [70]. Silver from pre-historic age is known for antibacterial activity, and this fact was used to prepare CS and polyvinylpyrrolidone film along with an addition of silver oxide nanoparticles [13].

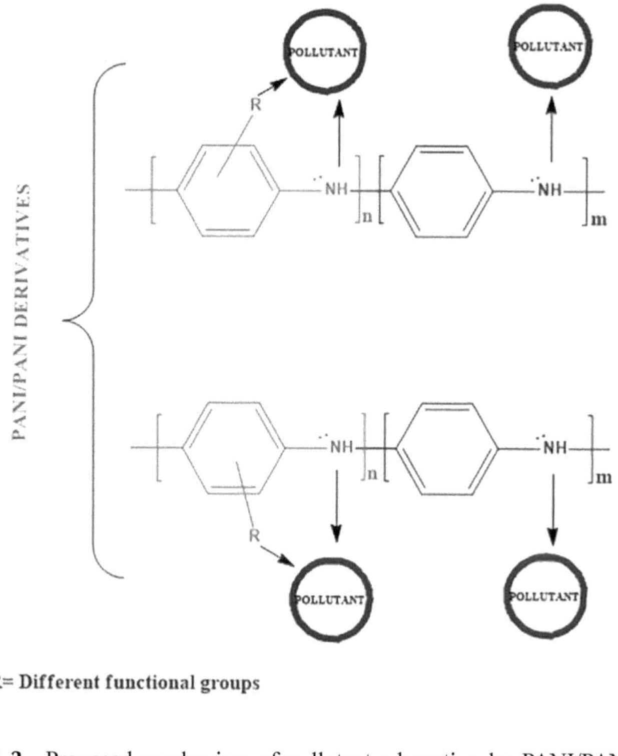


**FIGURE 11.3** Proposed mechanism of pollutant adsorption by PANI/PANI derivatives-based nanocomposite [158].

*Source:* Reprinted permission from: Zare et al. [158]. © 2018 Elsevier.

The nanocomposite exhibited substantial antibacterial activity against *S. aureus* and *E. coli* [137]. Chitin is comprised of multiple units of N-acetyl glucosamine and its wound-healing ability and antibacterial activity can be elevated by the addition of AgNPs. Implementing this fact, synthesized chitin/nanosilver composite scaffold from chitin hydrogel [87]. The scaffolds have shown successful bactericidal effect against *S. aureus* and *E. coli* along with blood clotting and wound healing capability. Titania/polymethyl methacrylate (PMMA) nanocomposite through electrochemical anodization was used for investigation against *E. coli* [55].

Sankar et al. [127] synthesized a group of aluminum oxyhydroxide (AlOOH)-chitosan nanocomposite. The preparation was performed at near room temperature via an aqueous path resulting into granular nanocrystalline materials. The CS is composed of abundant –O and –OH functional groups that result into the crystallinity nature and eventually crystallization of metal oxyhydroxide. The functional groups also assist into strong covalent binding of the nanocomposite surface to the matrix. CS and AlOOH synergistically cooperate in the structural stability of the composite.

Hydrogen bonding present in CS destroys on swelling due to water, a behavior similar to cellulose. In the same way, AlOOH on introduction to water dismantles into finer particles but the AlOOH-chitosan nanocomposite does not undergo any disintegration in water. The cage like structure of AlOOH-chitosan composite shows an important role in stabilizing AgNPs due to heavy metal ion-binding capacity. This unique scheme helps in reactivation of AgNPs for uninterrupted antimicrobial activity in water [127].

The graphene oxide-based nanocomposite combined with AgNPs was developed for adsorbing a variety of water contaminants simultaneously in a short time by carrying out multiple tasks. It acts as an antibacterial agent and hence capable to purify wastewater. The nanocomposite was found to be more thermally stable compared to bare graphene oxide sheets because the AgNPs regulate the thermal motion of graphene sheets. The model organisms to investigate the antimicrobial effect of synthesized nanocomposite were Gram-negative *E. coli* and Gram-positive *S. aureus*, though results indicated that destruction of *E. coli* was more than *S. aureus* strain. The inhibition zone with increasing amount of nanocomposite also increases due to increase in antimicrobial effect [106].

### *11.2.4 BY USING CAPPED METAL NANOPARTICLES*

NMs show exceptional physical and chemical properties, but in some cases, they do not exhibit particular surface properties for distinct applications which subsequently points toward a surface modification. The most obvious method to modify such nanomaterial is to add suitable organic groups with the surface atoms [126]. The surface modification can potentially stabilize nanoparticles by preventing agglomeration and enables their self-organization [112]. By modifying nanoparticles, a possibility of compatibility with another phase is also present such as by attaching appropriate groups can make nanoparticles hydrophilic which also promotes mechanical properties by avoiding homogeneity [53]. Usually, noble metal nanoparticles are modified by adding thiols, disulfides, amines, and carboxylic acids [131, 151].

Organosulfur groups due to the huge affinity of sulfur have the potential to coordinate with several metals like Ag, Cu, Pt, Au, etc. [39, 52]. The interaction between metal and sulfur atoms immobilizes the thiol-groups on metal nanoparticles surface [82]. Attachment of amines with metal nanoparticles is generally performed to stabilize the particles. The weaker interaction between amino groups and metal nanoparticles are stabilizing nanoparticles to a lesser extent when compared to thiols attachment, thus the size of amine-modified nanoparticles are bigger than respective organosulfur-modified nanoparticles [112]. The interaction between ligand with carboxylic acid group and the nanoparticle surface as observed by Fourier transform infrared spectroscopy (FTIR) study occurs primarily via carboxylate chemisorption [22].

#### *11.2.4.1 REMOVAL OF DYES FROM WATER*

The World Bank reported that dyes and textile sectors create approximately 17–20% of water pollution. Among various treatment procedures, catalytic degradation of organic dyes is fast and economically proven method with the simplest experimental set-up [8]. AgNPs are especially and majorly used catalyst for degradation of organic dyes because of easy surface functionalization besides several advantages [27, 152]. The metallic nanoparticles as homogeneous catalyst is receiving more importance due to their compatibility behavior. The capping agent has a profound impact on the functionalized nanoparticles which influence the surface structure. PEI, an extremely charged cationic polymer is a capping agent to stabilize nanoparticles via electrostatic and steric factors. The primary, secondary, and tertiary amines of PEI promote reversible pH-responsive modification in the structure leading to chain conformation with branching of the polymer at the surface of nanoparticles [118, 138].

Karimi et al. [66] investigated the behavior and structural effects of PEI polymer chain on the catalytic effect of PEI capping in PEI-AgNPs during decomposition of MB dye through collision electrochemistry. The results from UV-Vis spectroscopy (UV-Vis) conclude that no degradation of MB by $NABH_4$ was accomplished at pH 12 while gradual 80% degradation at acidic and neutral pH was observed; indicating PEI-AgNPs as a catalyst can efficiently and rapidly degrade MB [66]. Various fatty acids like oleic acid, castor oil, etc., as capping agent of AgNPs to enhance stability by preventing aggregation have also been employed [81, 121]. Fatty acid, a carboxylic acid with a long aliphatic chain, comprises both hydrophilic and hydrophobic group that makes it the most convenient and flexible coating material [81, 156].

In this part, the visible light photocatalyst for the degradation of colored organic dyes by the fatty acid-capped AgNps that can act as an efficient degradation compound has been described. This process was a simple, one-step, and low-cost synthesis method which was developed [56]. The various nanoparticles, i.e., stearic acid capped AgNPs (SNP), oleic acid capped AgNPs (ONP) and 9,10-dibromooctadecanoic acid capped AgNPs (DOAN) were used for degradation of two well-known organic dyes.

They found that photocatalytic efficiency by DOAN was significantly higher as compared to ONP and SNP under visible light. There are primarily three specific novelties of the presented approach. First, SNP, and ONP has been synthesized by simple precursor of silver, i.e., $AgNO_3$, triethylamine (TEA) and respective fatty acid where TEA acts as reducing agent. The nanoscale synthesis of ONP and SNP can be easily implied for larger-scale synthesis. DOAN was easily formed from ONP by using $HBr/H_2O_2$. Second, the plasmonic characteristic of Ag has been employed for the efficient trapping of visible radiation and the photodegradation of organic molecules which are impotent to absorb any radiations. Third, it was found that the different capping agent influences the photocatalytic efficiency (Figure 11.4).

Two well-known organic dyes such as MB and eriochrome black T (EBT) have been chosen for photocatalytic degradation study. These two contaminants were preferred because they are usually found in wastewater with hazardous impacts, and they are incapable of absorbing any radiation. Typically, an aqueous solution of organic dye was mixed with individual capped nanoparticles and stirred in dark to for 30 minutes to attain the equilibrium adsorption-desorption state. Afterward, the mixture solution was irradiated with visible light and a fraction of the solution was used for adsorption investigation.

Figure 11.5 summarizes the results of photocatalytic degradation of MB and EBT under visible light. The presence of catalyst enhances the degradation of dyes under visible light. Most importantly, the degradation was most effective by DOAN as compared to ONP or SNP. By performing trapping experiment under visible radiation, the photocatalytic mechanism was observed due to the generation of electrons and holes in conduction and valence band respectively. The electrons and holes interact with dissolved oxygen and water molecules to generate superoxide and hydroxyl radicals, respectively.

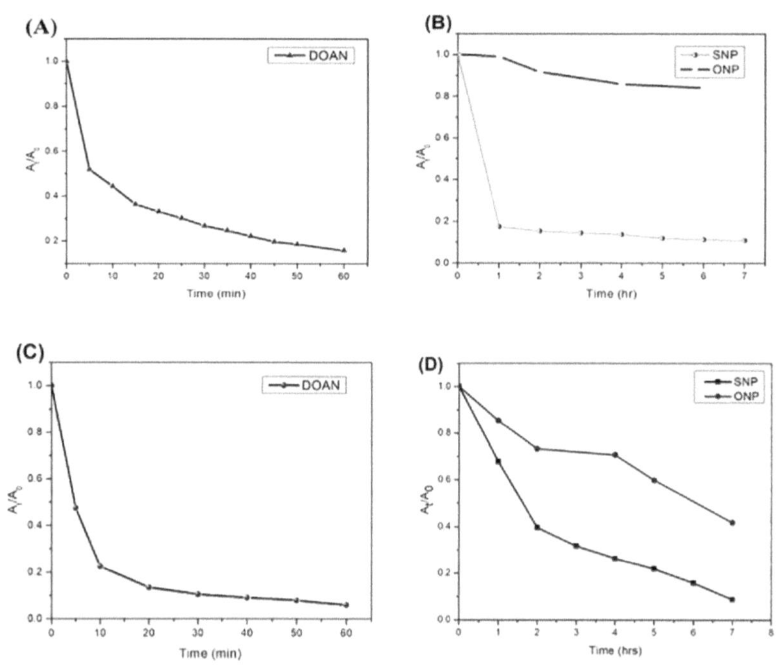


**FIGURE 11.4** Plots of degradation of (A and B) EBT and (C and D) MB by DOAN, SNP, and ONP.

Figure 11.6 shows representative results of the removal efficiency of MB and EBT by using these fatty acid-capped AgNPs. The results conclude that DOAN was more potentially active in removal of MB and EBT with 94 and 84.5% degradation respectively within 60 minutes while SNP also performed well with degradation of 92 and 88% in a time period of 7 hours. The results indicate the reactivity of nanoparticles is more towards cationic dye, i.e., MB compared to anionic dye, EBT.

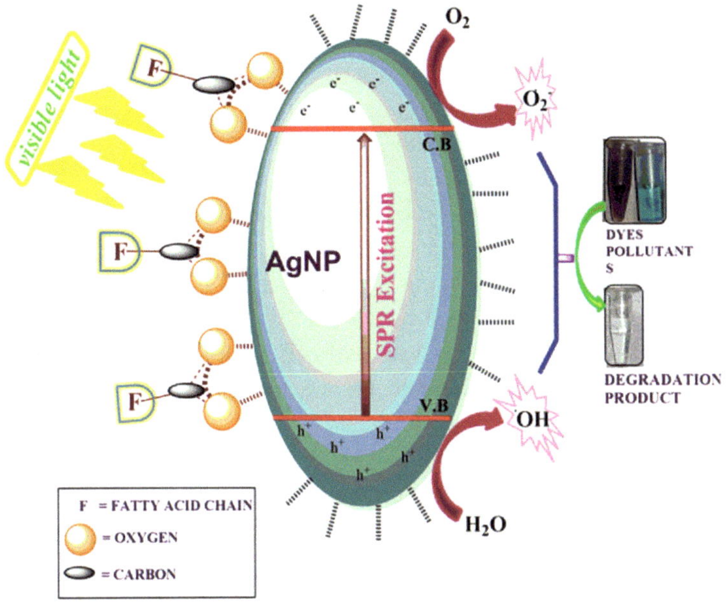


**FIGURE 11.5** Schematic diagram of the photocatalytic mechanism.

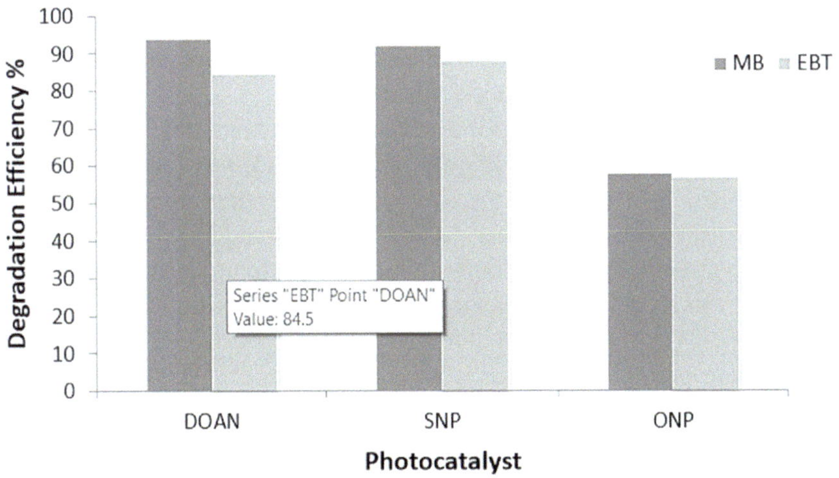


**FIGURE 11.6** Comparative MB and EBT degradation efficiencies over photocatalysts.

#### *11.2.4.2 REMOVAL OF ORGANIC MOLECULES, HEAVY METALS, AND IONS FROM WATER*

Bioaccumulation of heavy metal ions in cells result into irregular function of organs in both humans and animals. Cd (II) ion is one such heavy metal ion which hazardously affects kidneys, lungs, and other organs in humans and eventually to various diseases such as itai-itai disease [16, 88]. Venkateswarlu and Yoon [153] biosynthesized $Fe_3O_4$ nanoparticles and functionalized the surface with diethyl-4-(4-amino-5-mercapto-4h-1,2,4-triazol-3-yl) phenyl phosphonate (DEAMTTP) to remove Cd (II) from wastewater [153]. The functionalization enhanced the removal efficiency up to 96.1% due to the strong affinity of thiol functional group present in DEAMTTP and Cd (II). Ni (II) is another heavy metal ion which can be easily ingested with either water or foodstuff and its high uptake can be very dangerous and also influence genetic transmission. Systemic nickel allergy syndrome (SNAS) caused by Ni (II) is affecting about 15–20% of world's population [17].

Bettini et al. [17] implied the anti-cancer and anti-inflammatory characteristics of curcumin extraction of turmeric with AgNPs as substrate for removal of Ni (II) from aqueous system. Curcuminoids has the potential to chelate at alkaline pH and remove Ni (II) from water via the generation of nickel square planar complexes with two curcumin molecules. The proposed method was effective to remove about 70% Ni (II) from supernatant [123]. Cellulose shows a great surface area along with high affinity towards metals. It is the most abundant and one of the eco-friendly biopolymers with many (β-1,4) linked b-D-glucopyranose units [145].

The designed metallic nano-traps of cellulose by binding zinc and AgNPs improve adsorption capacity of cellulose [7]. Also, the cellulose presence minimizes the release of nanoparticles in treated water. The nano-trap was employed for the removal of $Hg^{2+}$, $Cr^{3+}$, $Co^{2+}$, $Ni^{2+}$ and $Pb^{2+}$ heavy ions from water. Silver-cellulose nano-trap exhibited more removal of heavy metals and came out as a versatile heavy metal adsorbent within a short contact time.

Pesticide presence in water is another component which increases the toxicity of water and is a great threat to the inhabitants of earth. Chlorpyrifos (O, O-diethyl-O-(3,5,6-trichloro-2-pyridyl) phosphorothioate) and Malathion (S-1, 2-bis (ethoxycarbonyl) ethyl O, O-dimethyl phosphorodithioate) are very common and highly involved ingredients in pesticides worldwide, causing several human defects [42, 96].

Singhal and Lind [142] studied cysteine-capped AgNPs to find their affinity towards above mentioned two pesticides. The modified AgNPs was more active towards chlorpyrifos because of the presence of aromatic amine group compared to malathion [142]. Glyphosate (N-phosphono-methylglycine) is one of the most utilized herbicides and is well known for its harmful impacts on humans like Alzheimer and Parkinson's disease, depression including genotoxic effects. Its stability and presence in aqueous and groundwater system makes its removal crucial [63]. Fluorescence technique to determine and remove glyphosate from drinking water through polyvinylpyrrolidone (PVP) capped silver nanocrystals [130]. The nanocrystals were synthesized by chemical reduction method, and through photoluminescence study, they confirmed that PVP capped nanocrystals successfully removed the glyphosate, suggesting the fluorescence-based procedure is gradual and facile method to remove the particular herbicide from water.

### *11.2.4.3 REMOVAL OF PATHOGENIC MICROBES FROM WATER*

Infectious diseases caused by pathogenic bacteria are the result of antibiotic resistance (ABR) in pathogens and host's immune system. The other types of pathogens are opportunistic pathogens which use their ability to form biofilms to adapt in adverse conditions. Lambadi et al. [79] performed an experiment to increase the antibacterial activity of AgNPs through functionalization. They selected polymyxin B, an antimicrobial peptide, which is a well-developed cationic and nonribosomal peptide with an ability to exhibit antibacterial activity via interaction with lipid A [100]. Lipid is a component of LS which is present in the outer membrane of Gram-negative bacteria [104].

The investigation of antibacterial and antibiofilm activity was performed against multidrug-resistant strain of *Vibrio fluvialis* and *Pseudomonas aeruginosa* PAO1. The removal efficiency of endotoxin from solution was also examined by functionalized nanoparticles. The results revealed that functionalized nanoparticles had a 3-fold higher antibacterial effect than the citrate-capped nanoparticles and also the nanoparticles were 97% efficient in endotoxin removal; from solution [79].

Chaudhary et al. [28] investigated the ability of $CeO_2$ nanoparticles to treat wastewater [28]. The surface of bare $CeO_2$ is anionic in nature and modification was done by utilizing cationic surfactants, i.e., hexadecyltrimethylammonium bromide (CTAB), hexadecyltrimethylammonium chloride (CTAC) and cetylpyridinum chloride (CPC). The nanoparticles were examined for their antimicrobial property against *S. aureus* and *E. coli* bacterial strain while antifungal property against *Echinodontium taxodii* fungi. The results confirmed that compared to *E. coli*, *S. aureus* exhibited higher zone of inhibition due to more affinity towards generated reactive oxygen species (ROS). In the case of fungi, bare $CeO_2$ nanoparticles displayed enhanced antifungal activity than surfactant coated nanoparticles. Thus, $CeO_2$ nanoparticles can solve numerous waters related issues and can behave as disinfectant [28]. Fatty acid coated AgNPs as antibacterial agent against water borne bacteria, Gram-negative *E. coli* and *S. marcescens* strains with 12- and 12.5-mm zone of inhibition (Figure 11.7) [56].

Gram-negative bacteria have a cell wall with two layers of LS for protecting cell from outer risks and maintenance of homeostasis. The presence of excess negative charges on LS promotes interaction of nanoparticles through electrostatic attraction. This leads to the morphological modification in the membrane which disturbs membrane permeability. Thus, compared to untreated cell, AgNPs-treated cell has disrupted cell walls [56]. The various pathways of AgNPs causing leakage, shrinkage, and death of cell are shown in Figure 11.8.

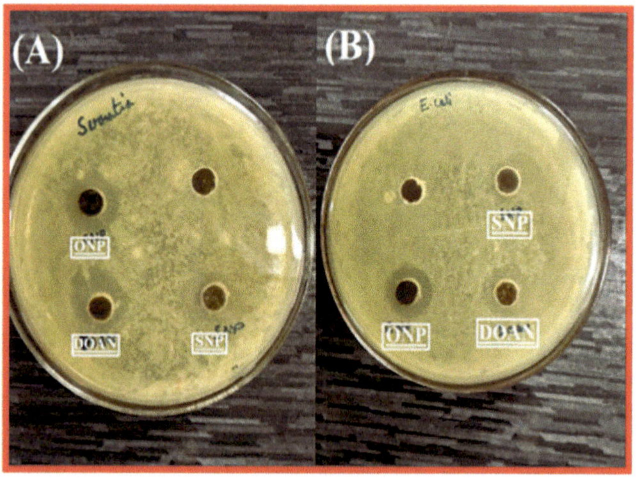


**FIGURE 11.7** The antibacterial activity of nanoparticles against: (A) *S. marcescens*; and (B) *E. coli*.

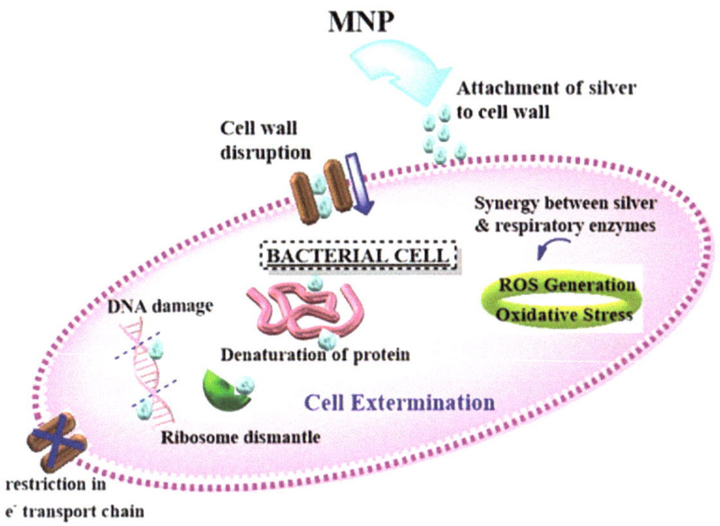


**FIGURE 11.8** An illustrative mechanism of antibacterial activity inside a gram-negative cell.

## 11.3 SUMMARY

Metal NMs owe a variety of unique physicochemical features which make them attractive for wastewater remediation. Some of these properties like larger surface area, capacity of being functionalized with diversified chemical groups and high selectivity towards detrimental elements or ions or microbes present in affluent, broad the regions of their applications. Various chemical modifications such as capping of nanoparticles lead to encouraging stability and results for the specific removal of organic dyes and heavy metal ions. This shows great achievement in wastewater management, owing to their antimicrobial characters, in an attempt to improve their lifetime and elimination potency. However, most of the nanomaterial applications for large-scale industrial wastewater treatment possess some major limitations and hazardous effects on the environment along with living creatures, thus restricting the development in sustainable removal of environmental pollutants. Hence, new research initiatives are desirable to counter these threats and conquer these hurdles.

## KEYWORDS


- **antimicrobial agents**
- **capping**
- **extracellular polymer matrix**
- **heavy-metal ions**
- **nanomaterial**
- **organic dyes**
- **pesticides**
- **wastewater**


## REFERENCES


1. Adegoke, H. I., Amooadekola, F., Fatoki, O. S., & Ximba, B. J., (2014). Adsorption of Cr (VI) on synthetic hematite (A-$Fe_2O_3$) nanoparticles of different morphologies. *Korean Journal of Chemical Engineering, 31*(1), 142–154.
2. Aghabeygi, S., & Khademi-Shamami, M., (2018). ZnO/$ZrO_2$ nanocomposite: Sonosynthesis, characterization and its application for wastewater treatment. *Ultrasonics Sonochemistry, 41*, 458–465.
3. Ahmad, M., Ahmed, E., Hong, Z. L., Ahmed, W., Elhissi, A., & Khalid, N. R., (2014). Photocatalytic, sonocatalytic and sonophotocatalytic degradation of rhodamine B using ZnO/CNTs composites photocatalysts. *Ultrasonics Sonochemistry, 21*(2), 761–773.
4. Ahmad, R., & Mirza, A., (2018). Adsorptive removal of heavy metals and anionic dye from aqueous solution using novel xanthan gum-glutathione/zeolite bionanocomposite. *Groundwater for Sustainable Development, 7*, 305–312.
5. Akhbarizadeh, R., Shayestefar, M. R., & Darezereshki, E., (2014). Competitive removal of metals from wastewater by maghemite nanoparticles: A Comparison between simulated wastewater and AMD. *Mine Water and the Environment, 33*, 89–96.
6. Alammar, T., Birkner, A., Shekhah, O., & Mudring, A. V., (2010). Sonochemical preparation of $TiO_2$ nanoparticles in the ionic liquid 1-(3-Hydroxypropyl)-3-methylimidazolium-bis (trifluoromethylsulfonyl) amide. *Materials Chemistry and Physics, 120*(1), 109–113.
7. Ali, A., Mannan, A., Hussain, I., Hussain, I., & Zia, M., (2018). Effective removal of metal ions from aqueous solution by silver and zinc nanoparticles functionalized cellulose: Isotherm, Kinetics and statistical supposition of process. *Environmental Nanotechnology, Monitoring & Management, 9*, 1–11.
8. Anantha, M. S., Olivera, S., Hu, C., Jayanna, B. K., Reddy, N., Venkatesh, K., Muralidhara, H. B., & Naidu, R., (2020). Comparison of the photocatalytic, adsorption and electrochemical methods for the removal of cationic dyes from aqueous solutions. *Environmental Technology & Innovation, 17*, 100612. doi: 10.1016/j.eti.2020.100612.

9. Annamalai, J., & Namasivayam, V., (2015). Endocrine-disrupting chemicals in the atmosphere: Their effects on humans and wildlife. *Environment International, 76*, 78–97.
10. Ansari, R., (2006). Application of polyaniline and its composites for adsorption/recovery of chromium (VI) from aqueous solutions. *Acta Chimica Slovenica, 53*(1), 88–94.
11. Ansari, R., & Raofie, F., (2006). Removal of lead ion from aqueous solutions using sawdust coated by polyaniline. *E-Journal of Chemistry, 3*, 1–11.
12. Anu, Y., & Supriya, K., (2016). Photochemical decolorization of methyl violet dye using *Azadirachta indica* (neem) mediated synthesized silver nanoparticles. *Der Pharmacia Lettre, 8*(7), 119–128.
13. Archana, D., Singh, B. K., Dutta, J., & Dutta, P. K., (2015). Chitosan-PVP-nano silver oxide wound dressing: *In vitro* and *in vivo* evaluation. *International Journal of Biological Macromolecules, 73*, 49–57.
14. Astruc, D., (**2020**). Introduction: Nanoparticles in catalysis. *Chemical Reviews, 120(2)*, 461–463.
15. Azarudeen, R. S., Riswan, A. M. A., Thirumarimurugan, M., Prabu, N., & Jeyakumar, D., (2016). Synthetic functionalized terpolymeric resin for the removal of hazardous metal ions: Synthesis, characterization and batch separation analysis. *Polymers for Advanced Technologies, 27*(2), 235–244.
16. Benoff, S., Jacob, A., & Hurley, I. R., (2000). Male infertility and environmental exposure to lead and cadmium. *Human Reproduction Update, 6*(2), 107–121.
17. Bettini, S., Pagano, R., Valli, L., & Giancane, G., (2014). Drastic nickel ion removal from aqueous solution by curcumin-capped Ag nanoparticles. *Nanoscale, 6*(17), 10113–10117.
18. Bhatia, D., Sharma, N. R., Singh, J., & Kanwar, R. S., (2017). Biological methods for textile dye removal from wastewater: A review. *Critical Reviews in Environmental Science and Technology, 47*(19), 1836–1876.
19. Bhaumik, M., Choi, H. J., Mccrindle, R. I., & Maity, A., (2014). Composite nanofibers prepared from metallic iron nanoparticles and polyaniline: High performance for water treatment applications. *Journal of Colloid and Interface Science, 425*, 75–82.
20. Bhaumik, M., Leswifi, T. Y., Maity, A., Srinivasu, V. V., & Onyango, M. S., (2011). Removal of fluoride from aqueous solution by polypyrrole/$Fe_3O_4$ magnetic nanocomposite. *Journal of Hazardous Materials, 186*(1), 150–159.
21. Bhaumik, M., Maity, A., Srinivasu, V. V., & Onyango, M. S., (2011). Enhanced removal of Cr (VI) from aqueous solution using polypyrrole/$Fe_3O_4$ magnetic nanocomposite. *Journal of Hazardous Materials, 190*(1–3), 381–390.
22. Bieri, M., & Bürgi, T., (2006). D-penicillamine adsorption on gold: An in-situ ATR-IR spectroscopic and QCM study. *Langmuir, 22*(20), 8379–8386.
23. Birch, F., (1964). Density and composition of mantle and core. *Journal of Geophysical Research, 69*(20), 4377–4388.
24. Borsagli, F. G. M., Mansur, A. A., Chagas, P., Oliveira, L. C., & Mansur, H. S., (2015). O-carboxymethyl functionalization of chitosan: Complexation and adsorption of Cd (II) and Cr (VI) as heavy metal pollutant ions. *Reactive and Functional Polymers, 97*, 37–47.
25. Cahill, M. G., Caprioli, G., Stack, M., Vittori, S., & James, K. J., (2011). Semi-automated liquid chromatography-mass spectrometry (LC-MS/MS) method for basic pesticides in wastewater effluents. *Analytical and Bioanalytical Chemistry, 400*(2), 587–594.

26. Cavuslar, O., Nakay, E., Kazakoglu, U., Abkenar, S. K., Ow-Yang, C. W., & Acar, H. Y., (2020). Synthesis of stable gold nanoparticles using linear polyethyleneimines and catalysis of both anionic and cationic azo dye degradation. *Materials Advances, 1*(7), 2407–2417.
27. Chandhru, M., Rani, S. K., & Vasimalai, N., (2020). Reductive degradation of toxic six dyes in industrial wastewater using diaminobenzoic acid capped silver nanoparticles. *Journal of Environmental Chemical Engineering, 8*(5), 104225. doi: 10.1016/j.jece.2020.104225.
28. Chaudhary, S., Sharma, P., Singh, D., Umar, A., & Kumar, R., (2017). Chemical and pathogenic cleanup of wastewater using surface-functionalized $CeO_2$ nanoparticles. *ACS Sustainable Chemistry & Engineering, 5*(8), 6803–6816.
29. Chen, C. P., Chen, C. T., & Tsai, T., (2012). Chitosan nanoparticles for antimicrobial photodynamic inactivation: Characterization and *in vitro* investigation. *Photochemistry and Photobiology, 88*(3), 570–576.
30. Chen, S. Y., Li, Y. F., & Zhang, J., (2008). Vision processing for real-time 3-D data acquisition based on coded structured light. *IEEE Transactions on Image Processing, 17*(2), 167–176.
31. Chen, Y. H., & Li, F. A., (2010). Kinetic study on removal of copper (II) using goethite and hematite nano-photocatalysts. *Journal of Colloid and Interface Science, 347*(2), 277–281.
32. Chen, Y. L., Zhang, C. E., Deng, C., Fei, P., Zhong, M., & Su, B. T., (2013). Preparation of ZnO/GO composite material with highly photocatalytic performance via an improved two-step method. *Chinese Chemical Letters, 24*(6), 518–520.
33. Chequer, F. M. D., De Oliveira, G. A. R., Ferraz, E. R. A., Cardoso, J. C., Zanoni, M. V. B., & De Oliveira, D. P., (2013). Textile dyes: Dyeing process and environmental impact. Chapter 6. In: Günay, M., (ed.), *Eco-Friendly Textile Dyeing and Finishing* (pp. 151–176). London – UK: IntechOpen Limited.
34. Chung, J., Ko, S., Bieri, N. R., Grigoropoulos, C. P., & Poulikakos, D., (2004). Conductor microstructures by laser curing of printed gold nanoparticle ink. *Applied Physics Letter, 84*, 801–803.
35. Crini, G., & Lichtfouse, E., (2019). Advantages and disadvantages of techniques used for wastewater treatment. *Environmental Chemistry Letters, 17*, 145–155.
36. Das, P., Ghosh, S., Ghosh, R., Dam, S., & Baskey, M., (2018). *Madhuca longifolia* plant-mediated green synthesis of cupric oxide nanoparticles: A promising environmentally sustainable material for wastewater treatment and efficient antibacterial agent. *Journal of Photochemistry and Photobiology B: Biology, 189*, 66–73.
37. Devi, L. S., & Joshi, S. R., (2012). Antimicrobial and synergistic effects of silver nanoparticles synthesized using: Soil fungi of high altitudes of Eastern Himalaya. *Mycobiology, 40*, 27–34.
38. Dimapilis, E. A. S., Hsu, C. S., Mendoza, R. M. O., & Lu, M. C., (2018). zinc oxide nanoparticles for water disinfection. *Sustainable Environment Research, 28*(2), 47–56.
39. Dong, T. Y., Wu, H. H., & Lin, M. C., (2006). Superlattice of octanethiol-protected copper nanoparticles. *Langmuir, 22*(16), 6754–6756.
40. Donlan, R. M., & Costerton, J. W., (2002). Biofilms: survival mechanisms of clinically relevant microorganisms. *Clinical Microbiology Reviews, 15*(2), 167–193.

41. Dsikowitzky, L., & Schwarzbauer, J., (2014). Industrial organic contaminants: Identification, toxicity and fate in the environment. *Environmental Chemistry Letters, 12*(3), 371–386.
42. Dyer, S. M., Cattani, M., Pisaniello, D. L., Williams, F. M., & Edwards, J. W., (2001). Peripheral cholinesterase inhibition by occupational chlorpyrifos exposure in Australian termiticide applicators. *Toxicology, 169*(3), 177–185.
43. Elmi, F., Alinezhad, H., Moulana, Z., Salehian, F., Mohseni, T. S., Asgharpour, F., Fallah, H., & Elmi, M. M., (2014). The use of antibacterial activity of ZnO nanoparticles on the treatment of municipal wastewater. *Water Science and Technology, 70*(5), 763–770.
44. Fei, J., & Li, J., (2010). Metal oxide nanomaterials for water treatment. Chapter 9. In: Kumar, C. S. S. R., (ed.), *Nanotechnologies for the Life Sciences* (pp. 287–314). Weinheim – Germany: Wiley Interscience.
45. Forbes, E. A., Posner, A. M., & Quirk, J. P., (1976). The specific adsorption of divalent Cd, Co, Cu, Pb, and Zn on goethite. *Journal of Soil Science, 27*(2), 154–166.
46. Franklin, N. M., Rogers, N. J., Apte, S. C., Batley, G. E., Gadd, G. E., & Casey, P. S., (2007). Comparative toxicity of nanoparticulate ZnO, Bulk ZnO, And $ZnCl_2$ to a freshwater microalga (*Pseudokirchneriella subcapitata*): The importance of particle solubility. *Environmental Science & Technology, 41*(24), 8484–8490.
47. Frumkin, H., (2001). Beyond toxicity: Human health and the natural environment. *American Journal of Preventive Medicine, 20*(3), 234–240.
48. Fu, F., & Wang, Q., (2011). Removal of heavy metal ions from wastewaters: A review. *Journal of Environmental Management, 92*(3), 407–418.
49. García, A., Delgado, L., Torà, J. A., Casals, E., González, E., Puntes, V., Font, X., et al., (2012). Effect of cerium dioxide, titanium dioxide, silver, and gold nanoparticles on the activity of microbial communities intended in wastewater treatment. *Journal of Hazardous Materials, 199*, 64–72.
50. Gaya, U. I., & Abdullah, A. H., (2008). Heterogeneous photocatalytic degradation of organic contaminants over titanium dioxide: A review of fundamentals, progress and problems. *Journal of Photochemistry and Photobiology C: Photochemistry Reviews, 9*(1), 1–12.
51. Giraldo, L., Erto, A., & Moreno-Piraján, J. C., (2013). Magnetite nanoparticles for removal of heavy metals from aqueous solutions: Synthesis and characterization. *Adsorption, 19*(2), 465–474.
52. Gooding, J. J., Mearns, F., Yang, W., & Liu, J., (2003). Self-assembled monolayers into the $21^{st}$ century: Recent advances and applications. *Electroanalysis: An International Journal Devoted to Fundamental and Practical Aspects of Electroanalysis, 15*(2), 81–96.
53. Grancharov, S. G., Zeng, H., Sun, S., Wang, S. X., O'Brien, S., Murray, C. B., Kirtley, J. R., & Held, G. A., (2005). Bio-functionalization of monodisperse magnetic nanoparticles and their use as biomolecular labels in a magnetic tunnel junction based sensor. *The Journal of Physical Chemistry B, 109*(26), 13030–13035.
54. Gupta, R. K., Singh, R. A., & Dubey, S. S., (2004). Removal of mercury ions from aqueous solutions by composite of polyaniline with polystyrene. *Separation and Purification Technology, 38*(3), 225–232.
55. Han, C., Lalley, J., Namboodiri, D., Cromer, K., & Nadagouda, M. N., (2016). Titanium dioxide-based antibacterial surfaces for water treatment. *Current Opinion in Chemical Engineering, 11*, 46–51.

56. Hasan, S., Rauf, A., & Saleem, S., (2020). Nanoformulation of C-18 long fatty acid-capped silver nanoparticles with exploration of photocatalytic and antibacterial activities. *New Journal of Chemistry, 44*(33), 13979–13989.
57. Hasani, T., & Eisazadeh, H., (2013). Removal of Cd (II) by using polypyrrole and its nanocomposites. *Synthetic Metals, 175*, 15–20.
58. Hernández-Montoya, V., Pérez-Cruz, M. A., Mendoza-Castillo, D. I., Moreno-Virgen, M. R., & Bonilla-Petriciolet, A., (2013). Competitive adsorption of dyes and heavy metals on zeolitic structures. *Journal of Environmental Management, 116*, 213–221.
59. Hua, M., Zhang, S., Pan, B., Zhang, W., Lv, L., & Zhang, Q., (2012). Heavy metal removal from water/wastewater by nanosized metal oxides: A review. *Journal of Hazardous Materials, 211*, 317–331.
60. Jain, T., Kumar, S., & Dutta, P. K., (2015). Chitosan in the light of nanobiotechnology: A mini-review. *Journal of Biomedical Technology and Research, 1*(1), 101–107.
61. Ji, P., Zhang, J., Chen, F., & Anpo, M., (2008). Ordered mesoporous $CeO_2$ synthesized by nanocasting from cubic Ia3d mesoporous MCM-48 silica: Formation, characterization and photocatalytic activity. *The Journal of Physical Chemistry C, 112*(46), 17809–17813.
62. Jiang, W., Wang, X., Wu, Z., Yue, X., Yuan, S., Lu, H., & Liang, B., (2015). Silver oxide as superb and stable photocatalyst under visible and near-infrared light irradiation and its photocatalytic mechanism. *Industrial & Engineering Chemistry Research, 54*(3), 832–841.
63. Jiang, X., Ouyang, Z., Zhang, Z., Yang, C., Li, X., Dang, Z., & Wu, P., (2018). Mechanism of glyphosate removal by biochar supported nano-zero-valent iron in aqueous solutions. *Colloids and Surfaces A: Physicochemical and Engineering Aspects, 547*, 64–72.
64. Jiang, Z. J., Liu, C. Y., & Sun, L. W., (2005). Catalytic properties of silver nanoparticles supported on silica spheres. *Journal of Physical Chemistry B, 109*(5), 1730–1735.
65. Kansal, S. K., Singh, M., & Sud, D., (2007). Studies on photodegradation of two commercial dyes in aqueous phase using different photocatalysts. *Journal of Hazardous Materials, 141*(3), 581–590.
66. Karimi, A., Kirk, K. A., & Andreescu, S., (2017). Electrochemical investigation of pH-dependent activity of polyethylenimine-capped silver nanoparticles. *Chemelectrochem, 4*(11), 2801–2806.
67. Karthik, R., & Meenakshi, S., (2015). Adsorption study on removal of Cr (VI) ions by polyaniline composite. *Desalination and Water Treatment, 54*(11), 3083–3093.
68. Karunakaran, C., Dhanalakshmi, R., & Gomathisankar, P., (2012). Phenol-photodegradation on $ZrO_2$. enhancement by semiconductors. *Spectrochimica Acta Part A: Molecular and Biomolecular Spectroscopy, 92*, 201–206.
69. Karunakaran, C. O., (1958). The Kerala food poisoning. *Journal of the Indian Medical Association, 31*(5), 204–207.
70. Karwowska, E., (2017). Antibacterial potential of nanocomposite-based materials-a short review. *Nanotechnology Reviews, 6*(2), 243–254.
71. Kasemets, K., Ivask, A., Dubourguier, H. C., & Kahru, A., (2009). Toxicity of nanoparticles of ZnO, CuO and $TiO_2$ to yeast saccharomyces cerevisiae. *Toxicology In Vitro, 23*(6), 1116–1122.
72. Katwal, R., Kaur, H., Sharma, G., Naushad, M., & Pathania, D., (2015). Electrochemical synthesized copper oxide nanoparticles for enhanced photocatalytic and antimicrobial activity. *Journal of Industrial and Engineering Chemistry, 31*, 173–184.

73. Khan, Z. U., Khan, A., Chen, Y. M., Khan, A. U., Shah, N. S., Muhammad, N., Murtaza, B., et al., (2017). Photocatalytic applications of gold nanoparticles synthesized by green route and electrochemical degradation of phenolic azo dyes using AuNPs/GC as modified paste electrode. *Journal of Alloys Compound, 725*, 869–876.
74. Khin, M. M., Nair, A. S., Babu, V. J., Murugan, R., & Ramakrishna, S., (2012). A review on nanomaterials for environmental remediation. *Energy & Environmental Science, 5*(8), 8075–8109.
75. Kolahalam, L. A., Viswanath, I. K., Diwakar, B. S., Govindh, B., Reddy, V., & Murthy, Y. L. N., (2019). Review on nanomaterials: Synthesis and applications. *Materials Today: Proceedings, 18*, 2182–2190.
76. Krishnan, S. K., Subbiah, K., Kandasamy, S., & Subramaniam, K., (2020). Application of metal nanoparticles for textile dye remediation. Chapter 17. In: Sivasubramanian, V., Pugazhendhi, A., & Moorthy, I. G., (eds.), *Sustainable Development in Energy and Environment* (pp. 217–223). The Gateway East – Singapore: Springer International Publishers.
77. Kunduru, K. R., Nazarkovsky, M., Farah, S., Pawar, R. P., Basu, A., & Domb, A. J., (2017). Nanotechnology for water purification: Applications of nanotechnology methods in wastewater treatment. Chapter 2. In: Grumezescu, A. M., (ed.), *Water Purification* (pp. 33–74). New York – USA: Academic Press.
78. Lakouraj, M. M., Hasanzadeh, F., & Zare, E. N., (2014). Nanogel and superparamagnetic nanocomposite of thiacalix [4] arene functionalized chitosan: Synthesis, characterization, and heavy metal sorption. *Iranian Polymer Journal, 23*(12), 933–945.
79. Lambadi, P. R., Sharma, T. K., Kumar, P., Vasnani, P., Thalluri, S. M., Bisht, N., Pathania, R., & Navani, N. K., (2015). Facile biofunctionalization of silver nanoparticles for enhanced antibacterial properties, endotoxin removal, and biofilm control. *International Journal of Nanomedicine, 10*, 2155–2171.
80. Lee, K. J., Lee, Y. I., Shim, I. K., Jun, B. H., Cho, H. J., & Joung, J. W., (2007). Large-scale synthesis of polymer-stabilized silver nanoparticles. *Solid State Phenomena, 124*, 1189–1192.
81. Lee, S. Y., & Harris, M. T., (2006). Surface modification of magnetic nanoparticles capped by oleic acids: Characterization and colloidal stability in polar solvents. *Journal of Colloid and Interface Science, 293*(2), 401–408.
82. Leff, D. V., Ohara, P. C., Heath, J. R., & Gelbart, W. M., (1995). Thermodynamic control of gold nanocrystal size: Experiment and theory. *The Journal of Physical Chemistry, 99*(18), 7036–7041.
83. Leiviskä, T., Khalid, M. K., Sarpola, A., & Tanskanen, J., (2017). Removal of vanadium from industrial wastewater using iron sorbents in batch and continuous flow pilot systems. *Journal of Environmental Management, 190*, 231–242.
84. Li, K., Satoh, M., Tagaya, M., & Kobayashi, T., (2015). Coprecipitated removal of $Cu^{2+}$ using dextran in cationic porphyrin aqueous solution. *Sensors and Materials, 27*(10), 899–906.
85. Li, L. H., Xiao, J., Liu, P., & Yang, G. W., (2015). Super adsorption capability from amorphization of metal oxide nanoparticles for dye removal. *Scientific Reports, 5*, 9028. doi: 10.1038/srep09028.
86. Li, Q., Mahendra, S., Lyon, D. Y., Brunet, L., Liga, M. V., Li, D., & Alvarez, P. J. J., (2008). Antimicrobial nanomaterials for water disinfection and microbial control: Potential applications and implications. *Water Research, 42*(18), 4591–4602.

87. Madhumathi, K., Kumar, P. S., Abhilash, S., Sreeja, V., Tamura, H., Manzoor, K., Nair, S. V., & Jayakumar, R., (2010). Development of novel chitin/nanosilver composite scaffolds for wound dressing applications. *Journal of Materials Science: Materials in Medicine, 21*(2), 807–813.
88. Mahalik, M. P., Hitner, H. W., & Prozialeck, W. C., (1995). Teratogenic effects and distribution of cadmium ($Cd^{2+}$) administered via osmotic minipumps to gravid CF-1 mice. *Toxicology Letters, 76*(3), 195–202.
89. Mahdavi, S., Jalali, M., & Afkhami, A., (2013). Heavy metals removal from aqueous solutions using $TiO_2$, MgO, and $Al_2O_3$ nanoparticles. *Chemical Engineering Communications, 200*(3), 448–470.
90. Malaikozhundan, B., Vijayakumar, S., Vaseeharan, B., Jenifer, A. A., Chitra, P., Prabhu, N. M., & Kannapiran, E., (2017). Two potential uses for silver nanoparticles coated with *Solanum Nigrum* unripe fruit extract: Biofilm inhibition and photodegradation of dye effluent. *Microbial Pathogenesis, 111*, 316–324.
91. Malik, P. K., (2004). Dye removal from wastewater using activated carbon developed from sawdust: Adsorption equilibrium and kinetics. *Journal of Hazardous Materials, 113*(1–3), 81–88.
92. Manimegalai, G., Shanthakumar, S., & Sharma, C., (2014). Silver nanoparticles: Synthesis and application in mineralization of pesticides using membrane support. *International Nano Letters, 4*(2), 105. doi: 10.1007/s40089-014-0105-8.
93. Mansour, A. M., (2015). Photocatalytic degradation of methylene blue with hematite nanoparticles synthesized by thermal decomposition of fluoroquinolones oxalato-iron (III) complexes. *RSC Advances, 5*(76), 62052–62061.
94. Marathe, S. D., & Shrivastava, V. S., (2015). Photocatalytic removal of hazardous ponceau S dye using nanostructured Ni-doped $TiO_2$ thin film prepared by chemical method. *Applied Nanoscience, 5*(2), 229–234.
95. Martínez-De La, C. A., Martínez, D. S., & Cuéllar, E. L., (2010). Synthesis and characterization of $WO_3$ nanoparticles prepared by the precipitation method: Evaluation of photocatalytic activity under vis-irradiation. *Solid-State Sciences, 12*(1), 88–94.
96. Mehdipour, M., Ansari, M., Pournamdari, M., Zeidabadinejad, L., & Kazemipour, M., (2020). Selective extraction of malathion from biological fluids by molecularly imprinted polymer coated on spinel $ZnFe_2O_4$ magnetic nanoparticles based on green synthesis. *Separation Science and Technology*, 1–11. doi: 10.1080/01496395.2020.1803912.
97. Mirza, A., & Ahmad, R., (2020). An efficient sequestration of toxic crystal violet dye from aqueous solution by alginate/pectin nanocomposite: A novel and eco-friendly adsorbent. *Groundwater for Sustainable Development, 11*, 100373. doi: 10.1016/j.gsd.2020.100373.
98. Moghadam, P. N., Hasanzadeh, R., & Khalafy, J., (2013). Preparation of SMA functionalized sulfanilic acid hydrogels and investigation of their metal ions adsorption behavior. *Iranian Polymer Journal, 22*(2), 133–142.
99. Monier, M., Abdel-Latif, D. A., & Mohammed, H. A., (2015). Synthesis and characterization of uranyl ion-imprinted microspheres based on amidoximated modified alginate. *International Journal of Biological Macromolecules, 75*, 354–363.
100. Morrison, D. C., & Jacobs, D. M., (1976). Binding of polymyxin B to the lipid a portion of bacterial lipopolysaccharides. *Immunochemistry, 13*(10), 813–818.
101. Motshekga, S. C., Ray, S. S., Onyango, M. S., & Momba, M. N., (2015). Preparation and antibacterial activity of chitosan-based nanocomposites containing bentonite-supported

Silver and zinc oxide nanoparticles for water disinfection. *Applied Clay Science, 114*, 330–339.
102. Mu, L., & Feng, S. S., (2001). Fabrication, characterization and *in vitro* release of paclitaxel (Taxol®) loaded poly (lactic-co-glycolic acid) microspheres prepared by spray drying technique with lipid/cholesterol emulsifiers. *Journal of Controlled Release, 76*(3), 239–254.
103. Muntean, S. G., Nistor, M. A., Muntean, E., Todea, A., Ianoş, R., & Păcurariu, C., (2018). Removal of colored organic pollutants from wastewaters by magnetite/carbon nanocomposites: Single and binary systems. *Journal of Chemistry, 2018*. 1–16.
104. Murray, R. G. E., Steed, P., & Elson, H. E., (1965). The location of the mucopeptide in sections of the cell wall of *Escherichia coli* and other gram-negative bacteria. *Canadian Journal of Microbiology, 11*(3), 547–560.
105. Murray, T. M., (2014). PBT (persistent, bioaccumulative, and toxic) chemicals. In: Philip, W., (ed.), *Encyclopedia of Toxicology* (3rd edn., pp. 762–764). Cambridge – USA: Academic Press, Elsevier Inc.
106. Naeem, H., Ajmal, M., Qureshi, R. B., Muntha, S. T., Farooq, M., & Siddiq, M., (2019). Facile synthesis of graphene oxide-silver nanocomposite for decontamination of water from multiple pollutants by adsorption, catalysis and antibacterial activity. *Journal of Environmental Management, 230*, 199–211.
107. Nair, A. S., & Pradeep, T., (2007). Extraction of chlorpyrifos and malathion from water by metal nanoparticles. *Journal of Nanoscience and Nanotechnology, 7*(6), 1871–1877.
108. Nair, A. S., & Pradeep, T., (2003). Halocarbon mineralization and catalytic destruction by metal nanoparticles. *Current Science-Bangalore, 84*(12), 1560–1563.
109. Nair, A. S., Tom, R. T., & Pradeep, T., (2003). Detection and extraction of endosulfan by metal nanoparticles. *Journal of Environmental Monitoring, 5*(2), 363–365.
110. Narayanan, P. M., Wilson, W. S., Abraham, A. T., & Sevanan, M., (2012). Synthesis, characterization, and antimicrobial activity of zinc oxide nanoparticles against human pathogens. *Bionanoscience, 2*(4), 329–335.
111. Nemati, Y., Zahedi, P., Baghdadi, M., & Ramezani, S., (2019). Microfluidics combined with ionic gelation method for production of nanoparticles based on thiol-functionalized chitosan to adsorb Hg (II) from aqueous solutions. *Journal of Environmental Management, 238*, 166–177.
112. Neouze, M. A., & Schubert, U., (2008). Surface modification and functionalization of metal and metal oxide nanoparticles by organic ligands. *Monatshefte Für Chemie-Chemical Monthly, 139*(3), 183–195.
113. Ojha, A., (2020). Nanomaterials for removal of waterborne pathogens: Opportunities and challenges. Chapter 19. In: Prasad, M. N. V., & Grobelak, A., (eds.), *Waterborne Pathogens* (pp. 385–432). Oxford – UK: Butterworth-Heinemann Publishing.
114. Padmavathy, N., & Vijayaraghavan, R., (2008). Enhanced bioactivity of ZnO nanoparticles-an antimicrobial study. *Science and Technology of Advanced Materials, 9*(3), 035004.
115. Phillips, D. J., (1995). The chemistries and environmental fates of trace metals and organochlorines in aquatic ecosystems. *Marine Pollution Bulletin, 31*(4–12), 193–200.
116. Pintucci, J. P., Corno, S., & Garotta, M., (2010). Biofilms and infections of the upper respiratory tract. *European Review for Medical and Pharmacological Sciences, 14*(8), 683–690.

117. Pradeep, T., (2009). Noble metal nanoparticles for water purification: A critical review. *Thin Solid Films, 517*(24), 6441–6478.
118. Qu, F., Li, N. B., & Luo, H. Q., (2013). Highly sensitive fluorescent and colorimetric pH sensor based on polyethylenimine-capped silver nanoclusters. *Langmuir, 29*(4), 1199–1205.
119. Rajasulochana, P., & Preethy, V., (2016). Comparison on efficiency of various techniques in treatment of waste and sewage water-a comprehensive review. *Resource-Efficient Technologies, 2*(4), 175–184.
120. Rani, M., & Shanker, U., (2018). Degradation of traditional and new emerging pesticides in water by nanomaterials: Recent trends and future recommendations. *International Journal of Environmental Science and Technology, 15*(6), 1347–1380.
121. Rao, C. R., & Trivedi, D. C., (2006). Biphasic synthesis of fatty acids stabilized silver nanoparticles: Role of experimental conditions on particle size. *Materials Chemistry and Physics, 99*(2, 3), 354–360.
122. Ray, S. C., & Jana, N. R., (2017). Application of carbon-based nanomaterials for removal of biologically toxic materials. Chapter 2. In: Ray, S. C., & Jana, N. R., (eds.), *Micro and Nano Technologies, Carbon Nanomaterials for Biological and Medical Applications* (pp. 43–86). Amsterdam – Netherlands: Elsevier Inc.
123. Refat, M. S., (2013). Synthesis and characterization of ligational behavior of curcumin drug towards some transition metal ions: Chelation effect on their thermal stability and biological activity. *Spectrochimica Acta Part A: Molecular and Biomolecular Spectroscopy, 105*, 326–337.
124. Rogers, H. R., (1996). Sources, behavior and fate of organic contaminants during sewage treatment and in sewage sludges. *Science of the Total Environment, 185*(1–3), 3–26.
125. Ruan, W., Hu, J., Qi, J., Hou, Y., Zhou, C., & Wei, X., (2019). Removal of dyes from wastewater by nanomaterials: A review. *Advance Material Letter, 10*(1), 09–20.
126. Ruckenstein, E., & Li, Z. F., (2005). Surface modification and functionalization through the self-assembled monolayer and graft polymerization. *Advances in Colloid and Interface Science, 113*(1), 43–63.
127. Sankar, M. U., Aigal, S., Maliyekkal, S. M., Chaudhary, A., Kumar, A. A., Chaudhari, K., & Pradeep, T., (2013). Biopolymer-reinforced synthetic granular nanocomposites for affordable point-of-use water purification. *Proceedings of the National Academy of Sciences, 110*(21), 8459–8464.
128. Santhosh, C., Velmurugan, V., Jacob, G., Jeong, S. K., Grace, A. N., & Bhatnagar, A., (2016). Role of nanomaterials in water treatment applications: A review. *Chemical Engineering Journal, 306*, 1116–1137.
129. Santo, C. E., Taudte, N., Nies, D. H., & Grass, G., (2007). Contribution of copper ion resistance for survival of *Escherichia coli* on metallic copper surfaces. *Applied and Environmental Microbiology, 74*, 977–986.
130. Sarkar, S., & Das, R., (2017). PVP capped silver nanocubes assisted removal of glyphosate from water-a photoluminescence study. *Journal of Hazardous Materials, 339*, 54–62.
131. Schulz-Dobrick, M., Sarathy, K. V., & Jansen, M., (2005). Surfactant-free synthesis and functionalization of gold nanoparticles. *Journal of the American Chemical Society, 127*(37), 12816–12817.

132. Seisenbaeva, G. A., Geoffrey, D., Kessler, V. G., & Jean-Marie, N., (2015). General facile approach to transition-metal oxides with highly uniform mesoporosity and their application as adsorbents for heavy-metal-ion sequestration. *Chemistry, 20*, 10732–10736.
133. Sen, M., (2020). Nanocomposite materials. Chapter 6. In: Sen, M., (ed.), *Nanotechnology and the Environment* (pp. 1–12). London – UK: IntechOpen Limited.
134. Shah, I., Adnan, R., Ngah, W. S. W., & Mohamed, N., (2015). Iron impregnated activated carbon as an efficient adsorbent for the removal of methylene blue: Regeneration and kinetics studies. *Plos One, 10*(4), 0122603.
135. Shanker, U., Jassal, V., & Rani, M., (2016). Catalytic removal of organic colorants from water using some transition metal oxide nanoparticles synthesized under sunlight. *RSC Advances, 6*(97), 94989–94999.
136. Sharma, G., Pathania, D., & Naushad, M., (2015). Preparation, characterization, and ion exchange behavior of nanocomposite polyaniline zirconium (IV) selenotungstophosphate for the separation of toxic metal ions. *Ionics, 21*(4), 1045–1055.
137. Shi, Z., Neoh, K. G., Kang, E. T., & Wang, W., (2006). Antibacterial and mechanical properties of bone cement impregnated with chitosan nanoparticles. *Biomaterials, 27*(11), 2440–2449.
138. Shieh, Y. T., Yu, T. Y., Wang, T. L., & Yang, C. H., (2012). Effects of pH on electrocatalytic activity of carbon nanotubes in polyethylenimine composites. *Journal of Electroanalytical Chemistry, 664*, 139–145.
139. Shin, W. H., Jeong, H. M., Kim, B. G., Kang, J. K., & Choi, J. W., (2012). Nitrogen-doped multiwall carbon nanotubes for lithium storage with extremely high capacity. *Nano Letters, 12*(5), 2283–2288.
140. Shkodenko, L., Kassirov, I., & Koshel, E., (2020). Metal oxide nanoparticles against bacterial biofilms: Perspectives and limitations. *Microorganisms, 8*(10), 1545–1566.
141. Shrivas, K., Ghosale, A., Nirmalkar, N., Srivastava, A., Singh, S. K., & Shinde, S. S., (2017). Removal of endrin and dieldrin isomeric pesticides through stereoselective adsorption behavior on the graphene oxide-magnetic nanoparticles. *Environmental Science and Pollution Research, 24*(32), 24980–24988.
142. Singhal, A., & Lind, M. L., (2018). Removal of pesticide toxicity by cysteine-capped Ag nanoparticles and study of their adsorption kinetics. *International Journal of Nanomedicine, 13*, 25–29.
143. Somu, P., & Paul, S., (2018). Casein based biogenic-synthesized zinc oxide nanoparticles simultaneously decontaminate heavy metals, dyes, and pathogenic microbes: A rational strategy for wastewater treatment. *Journal of Chemical Technology & Biotechnology, 93*(10), 2962–2976.
144. Sondi, I., & Salopek-Sondi, B., (2004). Silver nanoparticles as antimicrobial agent: A case study on *E. coli* as a Model for gram-negative bacteria. *Journal of Colloid and Interface Science, 275*(1), 177–182.
145. Song, S. T., Saman, N., Johari, K., & Mat, H., (2014). Surface chemistry modifications of rice husk toward enhancement of Hg (II) adsorption from aqueous solution. *Clean Technologies and Environmental Policy, 16*(8), 1747–1755.
146. Srivastava, N. K., & Majumder, C. B., (2008). Novel biofiltration methods for the treatment of heavy metals from industrial wastewater. *Journal of Hazardous Materials, 151*(1), 1–8.

147. Stoimenov, P. K., Klinger, R. L., Marchin, G. L., & Klabunde, K. J., (2002). Metal oxide nanoparticles as bactericidal agents. *Langmuir, 18*(17), 6679–6686.
148. Tadic, M., Panjan, M., Damnjanovic, V., & Milosevic, I., (2014). Magnetic properties of hematite (A-$Fe_2O_3$) nanoparticles prepared by hydrothermal synthesis method. *Applied Surface Science, 320*, 183–187.
149. Turkevich, J., Stevenson, P. C., & Hillier, J., (1951). A study of the nucleation and growth processes in the synthesis of colloidal gold. *Discussion Faraday Society, 11*, 55–75.
150. Tuutijärvi, T., Lu, J., Sillanpää, M., & Chen, G., (2009). As (V) adsorption on maghemite nanoparticles. *Journal of Hazardous Materials, 166*(2, 3), 1415–1420.
151. Ulman, A., (1996). Formation and structure of self-assembled monolayers. *Chemical Reviews, 96*(4), 1533–1554.
152. Vasimalai, N., Sheeba, G., & John, S. A., (2012). Ultrasensitive fluorescence-quenched chemosensor for Hg (II) in aqueous solution based on mercaptothiadiazole capped silver nanoparticles. *Journal of Hazardous Materials, 213*, 193–199.
153. Venkateswarlu, S., & Yoon, M., (2015). Rapid removal of cadmium ions using green-synthesized $Fe_3O_4$ nanoparticles capped with diethyl-4-(4 amino-5-mercapto-4 H-1, 2, 4-triazol-3-Yl) phenyl phosphonate. *RSC Advances, 5*(80), 65444–65453.
154. Yang, J., Hou, B., Wang, J., Tian, B., Bi, J., Wang, N., Li, X., & Huang, X., (2019). Nanomaterials for the removal of heavy metals from wastewater. *Nanomaterials, 9*(3), 424. doi: 10.3390/nano9030424.
155. Yu, X., Tong, S., Ge, M., Wu, L., Zuo, J., Cao, C., & Song, W., (2013). Adsorption of heavy metal ions from aqueous solution by carboxylated cellulose nanocrystals. *Journal of Environmental Sciences, 25*(5), 933–943.
156. Zamiri, R., Zakaria, A., Abbastabar, H., Darroudi, M., Husin, M. S., & Mahdi, M. A., (2011). Laser-fabricated castor oil-capped silver nanoparticles. *International Journal of Nanomedicine, 6*, 565–568.
157. Zare, E. N., & Lakouraj, M. M., (2014). Biodegradable polyaniline/dextrin conductive nanocomposites: Synthesis, characterization, and study of antioxidant activity and sorption of heavy metal ions. *Iranian Polymer Journal, 23*(4), 257–266.
158. Zare, E. N., Motahari, A., & Sillanpää, M., (2018). Nanoadsorbents based on conducting polymer nanocomposites with main focus on polyaniline and its derivatives for removal of heavy metal ions/dyes: A review. *Environmental Research, 162*, 173–195.
159. Zhai, Y., Zhang, S., & Pang, H., (2007). Preparation, characterization, and photocatalytic activity of $CeO_2$ nanocrystalline using ammonium bicarbonate as precipitant. *Materials Letters, 61*(8, 9), 1863–1866.
160. Zhang, W. X., (2003). Nanoscale iron particles for environmental remediation: An overview. *Journal of Nanoparticle Research, 5*(3, 4), 323–332.

# INDEX

## B

## C

## O

## S

## T

## U

## V



## W



## X



## Z